# SIGMA DELTA MODULATORS
## *Nonlinear Decoding Algorithms and Stability Analysis*

## THE KLUWER INTERNATIONAL SERIES IN ENGINEERING AND COMPUTER SCIENCE

### VLSI, COMPUTER ARCHITECTURE AND DIGITAL SIGNAL PROCESSING

*Consulting Editor*
**Jonathan Allen**

**Latest Titles**

*Sequential Logic Testing and Verification,* A. Ghosh, S. Devadas, A. R. Newton
ISBN: 0-7923-9188-8
*Introduction to the Design of Transconductor-Capacitor Filters,* J. E. Kardontchik
ISBN: 0-7923-9195-0
*The Synthesis Approach to Digital System Design,* P. Michel, U. Lauther, P. Duzy
ISBN: 0-7923-9199-3
*Fault Covering Problems in Reconfigurable VLSI Systems,* R.Libeskind-Hadas, N. Hassan, J. Cong, P. McKinley, C. L. Liu
ISBN: 0-7923-9231-0
*High Level Synthesis of ASICs Under Timing and Synchronization Constraints* D.C. Ku, G. De Micheli
ISBN: 0-7923-9244-2
*The SECD Microprocessor, A Verification Case Study,* B.T. Graham
ISBN: 0-7923-9245-0
*Field-Programmable Gate Arrays,* S.D. Brown, R. J. Francis, J. Rose, Z.G. Vranesic
ISBN: 0-7923-9248-5
*Anatomy of A Silicon Compiler,* R.W. Brodersen
ISBN: 0-7923-9249-3
*Electronic CAD Frameworks,* T.J. Barnes, D. Harrison, A.R. Newton, R.L. Spickelmier
ISBN: 0-7923-9252-3
*VHDL for Simulation, Synthesis and Formal Proofs of Hardware,* J. Mermet
ISBN: 0-7923-9253-1
*Wavelet Theory and its Applications,* R. K. Young
ISBN: 0-7923-9271-X
*Digital BiCMOS Integrated Circuit Design,* S.H.K. Embabi, A. Bellaouar, M.I Elmasry
ISBN: 0-7923-9276-0
*Design Automation for Timing-Driven Layout Synthesis,* S. S. Sapatnekar, S. Kang
ISBN: 0-7923-9281-7
*Acoustical and Environmental Robustness in Automatic Speech Recognition,* A. Acero
ISBN: 0-7923-9284-1
*Logic Synthesis and Optimization,* T. Sasao
ISBN: 0-7923-9308-2

# SIGMA DELTA MODULATORS
*Nonlinear Decoding Algorithms and Stability Analysis*

*by*

**Søren Hein**
*Siemens AG*

**Avideh Zakhor**
*University of California*

KLUWER ACADEMIC PUBLISHERS
**Boston/Dordrecht/London**

**Distributors for North America:**
Kluwer Academic Publishers
101 Philip Drive
Assinippi Park
Norwell, Massachusetts 02061 USA

**Distributors for all other countries:**
Kluwer Academic Publishers Group
Distribution Centre
Post Office Box 322
3300 AH Dordrecht, THE NETHERLANDS

**Library of Congress Cataloging-in-Publication Data**
Hein, Søren, 1968-
Sigma Delta modulators : nonlinear decoding algorithms and stability analysis / by Søren Hein, Avideh Zakhor.
p. cm. -- (The Kluwer international series in engineering and computer science ; SECS 0213. VLSI, computer architecture and digital signal processing)
Includes bibliographical references and index.
ISBN 0-7923-9309-0
1. Analog-to-digital converters. 2. Coding theory. I. Zakhor, Avideh, 1963- . II. Title. III. Series : Kluwer international series in engineering and computer science ; SECS 0123. IV. Series: Kluwer international series in engineering and computer science. VLSI, computer architecture, and digital signal processing.
TK7887 . 6 . H45 1993
621 . 3815 ' 36--dc20 92-39218
CIP



*Printed on acid-free paper.*

Printed in the United States of America

# Contents

# Preface

Analog-to-Digital (A/D) converters are key components in Digital Signal Processing (DSP) systems, and are therefore receiving much attention as DSP becomes increasingly prevalent in telephony, audio, video, consumer products *etc.* The varying demands on conversion rate, resolution and other characteristics have inspired a large number of competing A/D conversion techniques. This book is concerned with the particular class of A/D techniques called Oversampled Noise-Shaping (ONS) or Sigma Delta ($\Sigma\Delta$) modulation that has recently come into prominence for a number of applications. The popularity of ONS converters is due to their compatibility with present-day CMOS VLSI technology and their robustness towards circuit imperfections.

As explained in chapter 1, an ONS converter consists of an encoder which generates a high-rate low-resolution digital signal, and a decoder which produces a low-rate higher-resolution digital approximation to the analog encoder input. The dual purpose of this book is to provide a unified treatment of new nonlinear decoding algorithms as well as to address the stability problem for ONS encoders. The stability problem is already recognized as a difficult and nonlinear one, so our interest in the problem requires little justification. Decoding, on the other hand, is done almost exclusively by linear filtering today, and in this book we develop a class of nonlinear decoding algorithms whose performance is shown to be superior to that of linear filtering.

The idea behind our algorithms is to extract all the information present in the binary output sequence of an ONS encoder, and to ensure that a decoded signal generates the observed binary output sequence when applied to the encoder. There are a number of motivations behind our nonlinear decoding approach. To begin with, there is no principle of signal processing which states that linear decoding is the best way to extract information from the binary output sequence of an ONS encoder. In fact, since the class of nonlinear decoders includes that of linear decoders, the best nonlinear decoder cannot be worse than the best linear one. In addition, linear decoding can only be argued to be optimal when linearized models of the ONS encoders are applied. However, it is well known that linearized models are inadequate in many respects, as described in more detail in chapter 1. Indeed, we will show in this book that there exist nonlinear decoding algorithms which perform significantly better than linear decoders. Another motivation behind our nonlinear decoding algorithms is that nonlinear signal reconstruction techniques are already prevalent in the signal processing community, and therefore their application to ONS decoding is quite natural. The computational complexity of nonlinear decoders compared to linear ones is a potential problem. However, since ONS converters are gaining much popularity in DSP systems, it is conceivable to use the embedded DSP system for the computations needed for nonlinear decoding. This becomes increasingly true as the advances of VLSI technology make hardware implementation of signal processing algorithms more feasible.

The book is organized as follows. In chapter 1 we present an introduction to ONS conversion, and emphasize the particular trade-offs involved in the design and analysis of ONS converters. We describe the aspects in which linearized analysis is known to be inadequate. We also discuss the signal processing viewpoint on ONS conversion, particularly the trade-off between oversampling ratio and amplitude resolution.

In chapter 2 we derive results to describe and bound the performance of ONS converters with constant inputs. The results are based on an examination of the time-domain outputs or codewords that ONS converters are capable of producing for different constant inputs. We show that the results can conceptually be used in a nonlinear decoding algorithm where each codeword is looked up in a large table and is mapped to an optimal estimate of the corresponding constant input. This approach is compared to the conventional decoding approach which is based on linear filtering.

In chapter 3 we build on the results of chapter 2 and derive a simple optimal nonlinear decoder which achieves the exact same performance as the table look-up decoder. We find that nonlinear decoding can achieve significant gains in Signal-to-Noise Ratio (SNR) over linear decoding, depending on the encoder architecture. We also determine numerically the sensitivity of the nonlinear algorithm towards a number of circuit imperfections.

In chapter 4 we momentarily digress and describe a novel method for band-limiting noisy finite-extent sequences. This method is later used in chapter 5. The method is based on Singular Value Decomposition (SVD) of a certain signal-independent matrix. The SVD method is intended to improve upon the performance of the Discrete Fourier Transform when the signal frequency components are not known to be periodic over the observation interval.

In chapter 5 we generalize the decoding algorithm from chapter 3 to band-limited inputs. The general algorithm is more computationally intensive than the algorithm for constant inputs; it is based on the theory of projections onto convex sets, and makes use of the fact that the encoder input generated the particular binary signal observed at the encoder output. We find significant SNR gains over linear decoding for all encoder architectures. We also present numerical results on the sensitivity towards circuit imperfections.

In chapter 6 we exploit the known connection between ONS converters and the error diffusion algorithm for halftoning of continuous-tone images. We extend the decoding algorithm of chapter 5 to solve the problem of reconstructing an image from its halftoned version. We find that our algorithm provides subjectively superior reconstruction to conventional low-pass filtering.

Finally, in chapter 7 we address the instability problem that plagues higher-order ONS encoders. We introduce a new stability concept which is well-suited to ONS encoders. The idea is not only to determine whether or not the signal levels in an encoder are bounded, but also to estimate their maximum levels. This way, an implementation of the encoder can be scaled to avoid integrator clipping and performance degradation. We apply the stability concept to the double-loop encoder as well as to the class of interpolative encoders, and show that there exists a trade-off between stability and SNR performance. Based on the results, we present explicit design examples.

Large parts of the material in this book have appeared or will appear in journals and conference proceedings. Our hope is that this monograph will make the results more easily accessible to a larger range of readers. The book can profitably be supplemented with the recent collection of papers entitled *Oversampling Delta-Sigma Data Converters: Theory and Simulation*, J. C. Candy and G. C. Temes (Eds.), IEEE Press, 1992. This collection contains many of the seminal papers on ONS converters. The analysis in a number of these papers is frequency-domain based and approximate; subject to the dubious white-noise assumption, these papers provide valuable insight into the function and performance of ONS converters. In contrast, the analysis in this book is mostly time-domain based and rigorous, although we mostly restrict ourselves to simple inputs such as constant or band-limited ones. Our emphasis on the nonlinear behavior of ONS converters was inspired by the papers by Gray and his colleagues which are represented in the above collection. While we hope that this book will advance the understanding of the nonlinear characteristics of ONS converters, we emphasize that the final word has not yet been said, and that much work remains to be done. Most notably, the interplay between stability, tones and performance deserves more attention.

It is a pleasure to acknowledge the assistance and kindness of many of our colleagues. At the risk of inadvertently leaving out someone, we would like to thank the following for useful discussions, advice, preprints *etc.*: Berkeley Professors P. R. Gray, C. A. Desoer, B. N. Parlett, R. W. Brodersen, B. E. Boser, J. Demmel, J. Rabaey, S. R. Sanders and C. Stone; Professors L. R. Carley (Carnegie-Mellon University), O. Feely (University College Dublin), R. M. Gray (Stanford University), B. H. Leung (University of Waterloo), C. G. Sodini (Massachusetts Institute of Technology), R. Schreier and C. G. Temes (both Oregon State University); Drs. W. Chou, P. C. Hansen (Technical University of Denmark), M. Hauser, D. B. Ribner (General Electric), A. Ganesan and D. A. Kerth (both Crystal Semiconductor); P. F. Ferguson, Jr. and Wai Lee (both Analog Devices); and finally M. Analoui (Purdue University) and Shujaat Nadeem (M. I. T.). For the Lanczos code, we thank Prof. B. N. Parlett, and for the QP code, we thank C. Geyer (University of Minnesota) and A. Tits (University of Maryland). For being helpful and making things go smoothly, we thank Heather Brown, Lisa Lloyd-Maffei and Geneviève Thiebaut. For being a responsive system administrator and help in final stages of manuscript preparation, we thank Max Holm.

The present book grew out of SH's doctoral dissertation. We would like to express our gratitude to the Berkeley department of Electrical Engineering and Computer Sciences for providing a challenging and stimulating research environment. AZ would like to thank Professor P. R. Gray of U. C. Berkeley for his encouragement and support throughout the project. AZ also expresses her deepest appreciation to Analog Devices for the Junior Faculty Development Award in 1989 which has supported much of this work during the past few years. In addition, we gratefully acknowledge the financial support provided by a National Science Foundation (NSF) Presidential Young Investigator award, MIP-9057466, and NSF grant MIP-8911017.

SH is happy to acknowledge his family for their immense moral and practical support across the Atlantic, Hélène Hein for putting up with him, and Liselotte Elsborg-Theisen and Griet Schils for remaining close friends in spite of the distance.

AZ would like to express her deepest thanks to Seth Sanders for his love, support and understanding, and to Malihe and Hooshang Zakhor for tolerating their daughter's extended periods of absence from home during the past fourteen years.

We thank IEEE for permission to base this book on a number of papers and conference papers, as indicated on the first pages of the appropriate chapters.

Soren Hein and Avideh Zakhor
Berkeley, California

# Chapter 1

# Introduction

## 1.1 Analog-to-Digital Conversion

Analog-to-Digital (A/D) converters are needed in all Digital Signal Processing (DSP) applications which interface with the analog world. As DSP continues to gain ground over analog signal processing, the importance of these converters increases correspondingly. A partial list of A/D converter applications includes DC instrumentation, process control, automatic test equipment, weigh scales [1], links with thermo-couple sensors, strain gauges and transducers [2], [3], voice-band codecs for telephony and the ISDN U-interface [4], [5], [6], modems, audio recording for Compact Disc (CD) and Digital Audio Tape (DAT) [7], [8], [9], [10], audio interfaces in Personal Computers (PCs) and workstations, digital radio [5], video signal acquisition [11] and storage oscilloscopes [12].

These diverse applications present a wide variety of demands on their A/D converters, with signal rates ranging from DC to 10's or 100's of MHz and required resolutions from 6-8 bits to 20+ bits.[1] In addition, it is commonly required that the A/D converters be easily implementable in present-day technology such as CMOS VLSI. This makes it desirable that the converters exhibit low sensitivity to component accuracy and nonlinearities.

There exist a large number of competing approaches to the A/D conversion problem. This work is concerned with a particular approach known as Oversampled Noise-Shaping (ONS) conversion, $\Sigma\Delta$ or $\Delta\Sigma$ modulation [13], [14], in which the conversion proceeds in two steps: First, a digital discrete-time signal is generated which roughly resembles the analog signal to be digitized, but which on the other hand is sampled at a rate exceeding the Nyquist rate of the analog signal. Second, the high-rate digital signal is processed to produce a digital signal which more closely resembles the analog signal, and which is sampled at a rate close to the Nyquist rate of the analog signal.

In this introductory chapter we provide the setting for the problems that this

[1] Other important characteristics include power consumption, die size and supply voltage.

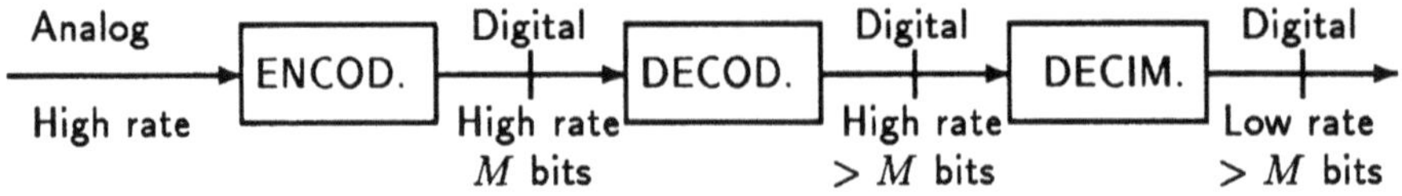


Figure 1.1: *Generic discrete-time model of an ONS converter.*

work addresses. Specifically, section 1.2 briefly describes two popular conventional A/D conversion techniques. Section 1.3 introduces ONS converters and discusses the modeling problems and performance trade-offs involved in ONS conversion. Section 1.4 describes the areas to which this work contributes, and outlines the organization of the work.

## 1.2 Conventional Conversion

In this section we briefly describe two conventional techniques for A/D conversion which may be viewed as opposites in terms of their requirements on clock rate and amount of circuitry. General introductions to A/D conversion are readily available in the literature [11], [15], [16], [17], [18], [19], [20], [21], [22], [23], [24], [25], [26].

A flash A/D converter [11], [22] is a uniform quantizer and achieves $N$-bit resolution by employing $2^N$ equal resistors, $2^N$ comparators and some control logic. In addition, a Sample-and-Hold (S/H) circuit may be used for practical reasons [11], although it is not required. A reference voltage $V_{\text{ref}}$ is applied across the series connection of the resistors, yielding $2^N$ equally spaced voltage levels between 0 volts and $V_{\text{ref}}$. Each of the $2^N$ comparators compares one voltage level to the analog voltage that is to be digitized, and the digital control logic subsequently determines the analog voltage with $N$-bit resolution. A flash converter can be made very fast because the conversion time is essentially independent of the required resolution $N$. On the other hand, the required circuitry grows exponentially with $N$.

A Successive Approximation (SA) A/D converter [22] is also a uniform quantizer and achieves $N$-bit resolution by employing only one comparator. A capacitor voltage is linearly charged up in steps of $V_{\text{ref}}/2^N$ from 0 volts towards a reference voltage $V_{\text{ref}}$, until the capacitor voltage exceeds the analog input voltage. The number of charge-up steps can be converted into an $N$-bit digital measure of the analog input. An SA converter requires little analog circuitry, but its conversion time grows exponentially with $N$ for a fixed signal rate and time step. In addition, an S/H circuit is required to keep the analog input signal constant for the duration of the conversion.

## 1.3 Oversampled Noise-Shaping Conversion

ONS A/D converters operate on a completely different principle than flash and SA converters, and are not uniform quantizers. An ONS converter, also known as a $\Sigma\Delta$ or $\Delta\Sigma$ modulator, contains an encoder and a decoder as illustrated in figure 1.1. The encoder accepts as its input an analog low-pass discrete-time signal sampled at a rate exceeding the Nyquist rate [24]. The encoder, which is described in more detail below, produces as its output a digital signal at the same rate as that of the input. Each digital output sample is an $M$-bit number, where $M$ is usually a small integer such as 1. The digital encoder output thus represents a coarse-amplitude high-rate approximation to the analog input. The decoder, which is also described in more detail below, accepts as its input the digital encoder output, and produces as its output a digital approximation to the analog input with a higher amplitude resolution. The decoder output is typically decimated to the Nyquist rate corresponding to the highest possible input signal frequency.

In an ONS converter there exists a trade-off between the encoder resolution $M$ and the OverSampling Ratio (OSR), where the OSR is defined as the ratio between the sampling frequency and the Nyquist rate of largest signal frequency for which the converter is designed to operate. If $M = \infty$, an OSR of 1 is sufficient to allow error-free acquisition of the analog input. If on the other hand $M = 1$, a large OSR is required to achieve a large Signal-to-Noise Ratio (SNR) in the A/D conversion. ONS converters thus introduce a trade-off between resolution in amplitude and time, and provide a conceptual means of bridging the gap between flash and SA conversion. This type of trade-off may be considered fundamental to signal processing, and is described in a similar context in [27], [28], [29].

In this section we provide an introduction to ONS converters with an emphasis on the problem areas that the work addresses. In section 1.3.1 we describe the connection between ONS converters and the well-known Delta modulators, thus explaining the often-used names $\Sigma\Delta$ and $\Delta\Sigma$ modulators for ONS converters. In section 1.3.2 we describe some important analysis and implementation issues, and in section 1.3.3 we discuss the architectural variations and performance trade-offs involved in ONS conversion. In section 1.3.4 we consider ONS conversion from a signal processing point of view.

### 1.3.1 Background

In this section we first describe Delta modulators, and then show heuristically how $\Sigma\Delta$ modulators represent a natural development. A more detailed introduction to Delta modulation may be found in [23].

Figure 1.2 shows three equivalent discrete-time models of the well-known simplest Delta modulator [23] to be explained in more detail below. Consistent with the terminology introduced above, we refer to the circuits as Delta encoders rather than modulators, because subsequent digital decoding is required to obtain high-resolution digital approximations to the analog input signals. This decoding usually takes the form of linear low-pass filtering. In the figure, $z^{-1}$

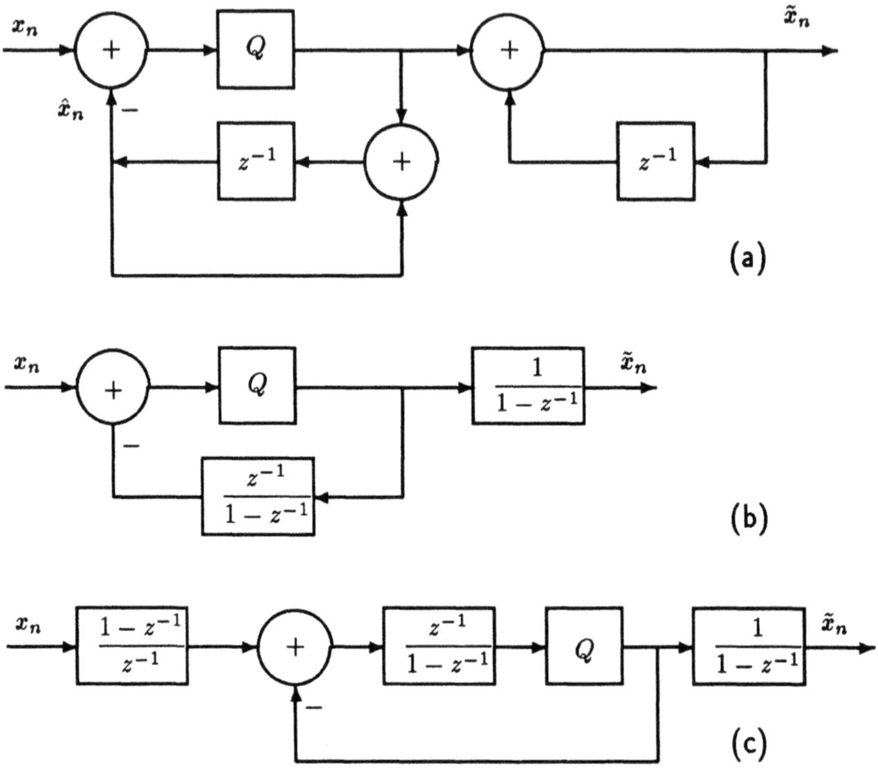


Figure 1.2: *Discrete-time models of the simplest Delta encoder.*

represents a delay element and $Q$ represents a quantizer or an A/D converter; typically, the resolution of the Delta modulator exceeds the resolution of the quantizer.[2]

Figure 1.2(a) shows the standard diagram of the simplest Delta encoder. The first stage produces a digital signal that tracks the derivative of the analog input signal $x_n$, and the second stage integrates the digital output of the first stage. The quantizer input is the difference ($\Delta$) between the analog input $x_n$ and a coarse prediction $\hat{x}_n$ of $x_n$ based on previous quantizer outputs. If the input sequence $\{x_n\}$ is highly correlated and the quantizer has a large number of levels, the predictions tend to be close to the true values. Therefore the quantizer only needs to convert the small prediction errors, and can do so accurately with a smaller number of bits than would be needed to quantize $x_n$ directly. The predictor that produces $\hat{x}_n$ accepts as its input a sequence which is the sum of the predictions and the quantized prediction errors, that

[2]Delta modulation is referred to as Differential Pulse Code Modulation (DPCM) when $Q$ is a multi-bit quantizer. This distinction is ignored here for convenience.

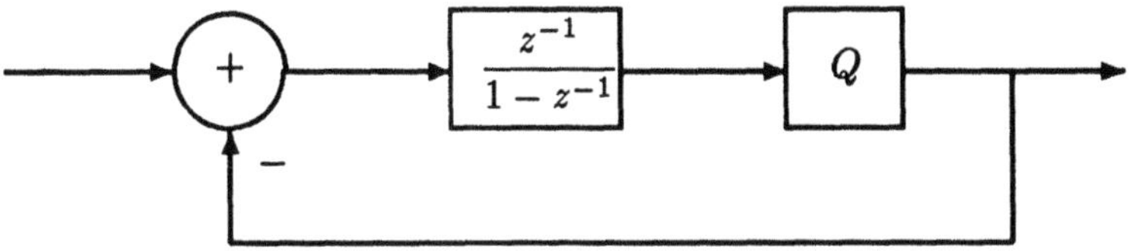


Figure 1.3: *Discrete-time model of the single-loop ΣΔ encoder.*

is, a sequence which approximates the input sequence $\{x_n\}$. In figure 1.2(a) the predictor is a simple one-sample delay, but much more elaborate predictors are possible [23], [30]. The quantizer output sequence enters the second stage which effectively mirrors the first one, and which outputs an approximation $\{\tilde{x}_n\}$ to the analog input sequence $\{x_n\}$. Note that the delay element in the second stage is also a first-order predictor; its input signal equals $\{\tilde{x}_n\}$ which is the sum of the previous predictions of the second stage and the quantized prediction errors of the first stage.

Figure 1.2(b) shows a redrawn version of figure 1.2(a) which emphasizes that the quantizer input is the difference between the analog input $x_n$ and the integrated quantizer output, and that the quantizer is embedded in a negative feedback loop. Therefore the integrated quantizer output tracks the analog input.

Finally, figure 1.2(c) shows a redrawn version of figure 1.2(b) which may be impractical for implementation, but which is useful in explaining the connection between Delta and ΣΔ modulation. Specifically, the central part of the Delta encoder in figure 1.2(c) is in fact the so-called single-loop ΣΔ encoder shown separately in figure 1.3. The output of the single-loop ΣΔ encoder tracks its input because the quantizer input equals the integrated difference between the encoder input and output, and the quantizer is embedded in a negative feedback loop. In figure 1.2(c) the single-loop ΣΔ encoder is preceded by an analog differentiator and followed by a digital integrator, so that the output of the Delta encoder also tracks its input.

As argued above, the justification for Delta modulation is that it exploits correlations such as those arising from oversampling the input sequence. It is natural to ask whether additional benefit can be derived by introducing more correlation in the input sequence. A simple way to increase inter-sample correlation is by integration. Preceding the Delta encoder in figure 1.2(c) by an analog integrator and following it by a compensating digital differentiator cancels out the corresponding differentiator and integrator in figure 1.2(c), and thus gives rise to the single-loop ΣΔ encoder shown in figure 1.3 except for an overall delay. We might therefore expect that the single-loop ΣΔ encoder is a better encoder than the simplest Delta encoder. This turns out to be true for a number of reasons. Among these, we mention that unlike the ΣΔ encoder, the Delta encoder suffers from error propagation, that is, a single bit error will

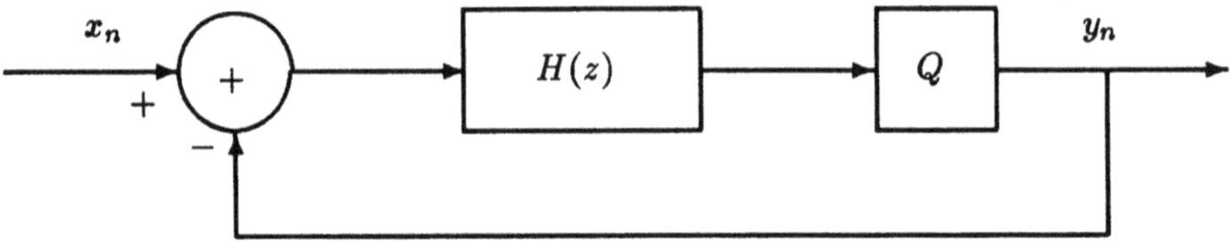


Figure 1.4: *Discrete-time model of the interpolative ΣΔ encoder.*

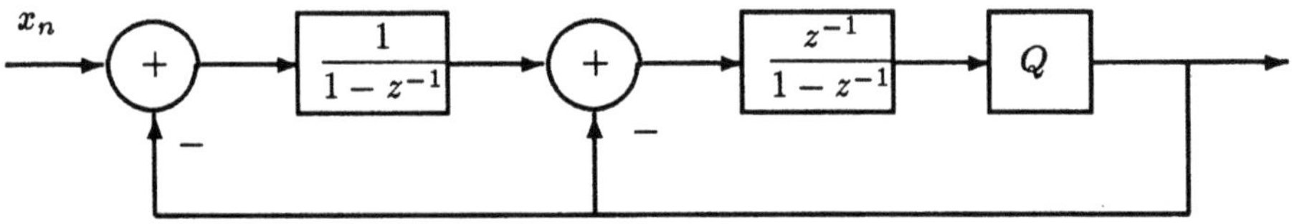


Figure 1.5: *Discrete-time model of the double-loop ΣΔ encoder.*

get the first and second stages out of synchronization, and consistent errors between $x_n$ and $\tilde{x}_n$ will occur. In addition, choosing the quantizer step size and resolution in a Delta encoder implies a complicated trade-off between granular noise and slope overload. Granular noise occurs because the quantizer has a finite number of levels. Slope overload occurs when the derivative of the analog input exceeds the largest possible quantizer output, and results in large differences between $x_n$ and $\tilde{x}_n$; slope overload depends on the frequency of the input signal, because high-frequency signals have larger gradients than low-frequency ones. Complicated schemes such as adaptive quantization and prediction may be necessary for high-quality Delta encoding [23]. In contrast, the ΣΔ encoder does not suffer from slope overload and strong signal frequency dependence.

There exist natural generalizations of both Delta and ΣΔ encoders to orders above one. For instance, a double-integration Delta encoder is mentioned in [23], and a double-loop ΣΔ encoder is described in [14], [31]. In general, Delta and ΣΔ encoders belong to the classes of predictive and interpolative coders, respectively [23]. Figure 1.4 shows a general interpolative ΣΔ encoder [32]. Figure 1.5 shows the double-loop ΣΔ encoder from [31] which is special in that it is a hand-crafted variation on the second-order interpolative encoder including an internal feedback path; it is thus neither predictive nor interpolative.

In concluding this section, we comment on some confusing terminology for ONS converters. In the original paper [14] on the circuit shown in figure 1.3, the circuit is called a ΔΣ modulator; the "Δ" refers to the difference operation at the input summing node, and the "Σ" presumably refers to the pre-integration that converts a simple Delta modulator into a single-loop ΔΣ mod-

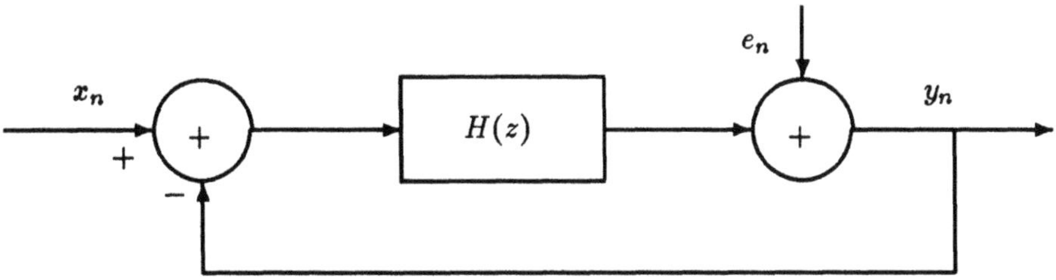


Figure 1.6: *Linearized model of the interpolative ΣΔ encoder.*

ulator. The word "modulator" refers to the pulse-density modulating function of the encoder.[3] Following the paper [14], a series of papers by Candy and his co-workers in the 1970's and 1980's [4], [31], [34], [35], [36], [37], [38], [39] introduced and used the term "ΣΔ modulator", and this term came to be dominating. Recently Hauser has led a campaign [40] in favor of the original term, and in favor of abolishing both and instead using "ONS converter" or, most recently, "OSADC" (OverSampled Analog-to-Digital Converter) [21]. We find that although "ΔΣ" is the original term, "ΣΔ" is more descriptive because the integration precedes the Delta modulator. The term "ONS converter" however has the advantage of sounding less arcane. We therefore use "ΣΔ modulator" and "ONS converter" interchangeably in this work. We also use both terms to refer to converters with multi-bit quantizers, even though the original ΔΣ modulator has a 1-bit quantizer.

## 1.3.2 Analysis and Implementation

This section discusses ONS converters in more detail, and illustrates how they differ from other A/D converters. Figure 1.4 shows a discrete-time model of the general interpolative ΣΔ encoder. The analog low-pass input sequence $\{x_n\}$ is oversampled; $H(z)$ represents an analog discrete-time filter, and the quantizer $Q$ is an $M$-bit A/D converter that produces a digital output sequence $\{y_n\}$. The filter and the quantizer are embedded in a negative feedback loop. In an implementation, it would be necessary to have an $M$-bit D/A converter in the feedback path; this D/A converter is omitted for clarity.

The discrete-time system shown in figure 1.4 is very difficult to analyze in its full generality due to the nonlinear quantizer, and in spite of the work of many researchers, there exists no general and exact analysis which can predict important performance parameters with certainty. It only appears possible to arrive at useful results when assumptions are made about the input sequence,

[3] As stated in [33], modulation implies trading off bandwidth for increased SNR.

or when the quantizer is modeled as a combination of a linear gain and a white noise source. Historically, both these approaches have been taken. For analytic purposes as well as for testing and characterizing $\Sigma\Delta$ modulators, DC and sinusoidal inputs are almost universally used; this is also the case in this work. In addition, it is common practice to model the quantizer as shown in figure 1.6, which we will refer to as the linearized model for short. Comparing with figure 1.4, the quantizer has been replaced by a noise sequence $\{e_n\}$; it is assumed that the quantization noise is uniformly distributed, white and statistically uncorrelated with the quantizer input, even though the noise is in fact a deterministic function of the input. Similar assumptions on quantization noise have been successfully applied in many digital filtering and communication applications [23], [30], and it has been shown that under certain conditions, the assumptions are true in a mathematical sense [26]. On the other hand, it has been recognized since the inception of $\Sigma\Delta$ modulators [14] that the linearized model is inadequate in many respects, particularly when $Q$ is a 1-bit quantizer as is usually the case.

This work is based on the viewpoint that although linearized analysis can be useful within its bounds, rigorous conclusions should be drawn based on a nonlinear circuit model such as figure 1.4 when at all possible. The credit for conveying this idea to the ONS converter community goes to Gray and his students for a series of rigorous papers [18], [19], [41], [42], [43], [44], [45], [46], [47], [48], [49], although the earlier paper [37] by Candy and Benjamin also avoids the linearized model. However, the methods of those papers appear to have been exploited fully, and to be limited in their applicability.

The remainder of this section is organized as follows. In section 1.3.2.1 we use the linearized model in figure 1.6 to gain intuition about the interpolative encoder in figure 1.4, and in section 1.3.2.2 we list a number of shortcomings of the linearized model. In section 1.3.2.3 we describe some implementation characteristics of ONS converters.

### 1.3.2.1 Linearized Model

We will assume in this section that the quantization noise sequence $\{e_n\}$ in figure 1.6 is white and uncorrelated with the quantizer input, and that the input signal $\{x_n\}$ is low-pass and oversampled. Denoting the $z$-transforms of the input, noise and output sequences by $X(z)$, $E(z)$ and $Y(z)$, respectively, we find the linearized signal and noise transfer functions

$$
\begin{aligned}
H_X(z) &\triangleq \frac{Y(z)}{X(z)}\Big|_{E(z)=0} = \frac{H(z)}{1+H(z)} \\
H_E(z) &\triangleq \frac{Y(z)}{E(z)}\Big|_{X(z)=0} = \frac{1}{1+H(z)}
\end{aligned}
\tag{1.1}
$$

If the open-loop filter $H(z)$ is chosen to have a large gain over signal baseband and a gain much less than 1 outside of baseband, the gain of the signal transfer function is close to 1 over baseband and close to 0 outside of baseband. On the other hand, the gain of the noise transfer function is close to 0 over baseband and close to 1 outside of baseband. Therefore the spectrum of the encoder output is approximately equal to the signal spectrum over baseband, and contains noise outside of baseband. This redistribution of the white quantization noise towards high frequencies is referred to as noise-shaping, which explains the term "ONS conversion". Noise-shaping is advantageous because the signal spectrum can be recovered quite accurately by low-pass filtering the encoder output. The operation of removing quantization noise from the encoder output is referred to as decoding.

The single-loop $\Sigma\Delta$ encoder in figure 1.3 is a special case of the interpolative encoder in figure 1.4 where the open-loop filter is an integrator, $H(z) = z^{-1}/(1-z^{-1})$. This filter has infinite DC gain and large gain close to DC. The signal and noise transfer functions for the single loop encoder are $H_X(z) = z^{-1}$ and $H_E(z) = 1 - z^{-1}$, that is, a delay and a differentiator, respectively. As a differentiator is a high-pass operator, noise-shaping is achieved. It has been shown that assuming the validity of the linearized model, the single-loop modulator can achieve an extra 9 dB of SNR with each doubling of the OSR [14], [34]. This relation has been verified experimentally [14], [34].

The double-loop $\Sigma\Delta$ encoder shown in figure 1.5 is not an interpolative encoder, but it can be shown that the signal and noise transfer functions are $H_X(z) = z^{-1}$ and $H_E(z) = (1-z^{-1})^2$, respectively. Thus the quantization noise is differentiated twice. For the double-loop modulator with proper decoding, the slope of the SNR vs. OSR characteristic is 15 dB/octave [31].

More general interpolative $\Sigma\Delta$ encoders have been proposed and fabricated. For instance, a fourth order encoder in which the poles of $H(z)$ are not at DC has been investigated [32], and third and fifth order encoders are described in [50] and [51].

#### 1.3.2.2 Shortcomings of Linearized Model

The assumption of white and uncorrelated quantization noise used in the previous section is convenient, but often misleading. This is particularly so when $Q$ is a 1-bit quantizer, that is,

$$Q(u) = \begin{cases} -b & \text{for } u \leq 0 \\ +b & \text{for } u > 0 \end{cases} \tag{1.2}$$

where $b$ is referred to as the quantizer step size. As pointed out by Gray [18], [19], there is no analytic reason to expect the linearized model to be valid. The main result in the area is due to Bennett [52], who shows that the linearized quantizer model is a good approximation in a precise sense when a number of conditions are satisfied. Specifically, the quantizer must have a large number of closely spaced output levels, the quantizer must not overload, and the quantizer inputs must have smooth joint probability density functions [52]. As none of these conditions are typically satisfied in a 1-bit $\Sigma\Delta$ encoder [18], it is perhaps surprising that the linearized model can provide accurate results at all.

In this section we present some mostly well-known examples of incorrect results predicted by the linearized quantizer model.

- Consider the effect of multiplying the open-loop filter $H(z)$ in figure 1.4 by a positive constant $K$. The function of the circuit remains unchanged because $Q(Ku) = Q(u)$, whereas the linearized transfer functions (1.1) are affected.

- The linearized model does not predict the obvious fact that the encoder output is binary. This characteristic is important for linear decoders, as multiplications can be avoided; another consequence is that the power of the encoder output signal is independent of the input signal.

- The interpolative $\Sigma\Delta$ encoder cannot accurately encode arbitrarily large input signals. The input signal levels are usually limited to $(-b, +b)$ or less. The linearized model predicts no such limitation.

- The quantization noise spectra of the single-loop encoder have been investigated extensively [42], [44]. It has been shown that under certain mild assumptions, the noise spectra are discrete and the noise is correlated with the input. These results confirm knowledge obtained by practical implementation. In contrast, the linearized model predicts a smooth noise spectrum.

- The linearized model erroneously predicts that linear decoding is optimal. This work shows that nonlinear decoding can improve substantially upon linear decoding in terms of SNR.

- The linearized model predicts that the stability of the interpolative encoder depends on the stability of the linearized transfer functions (1.1), that is, on the zeros of $1 + H(z)$. In contrast, it is well known that encoders with stable linear transfer functions can be unstable. Conversely, we conjecture that encoders with unstable linear transfer functions can be stable. This conjecture has not been tested in the literature because there do not exist design methods for such encoders. This work shows that the stability of the interpolative encoder is closely linked to the stability of $H(z)$ itself.

#### 1.3.2.3 Implementation Characteristics

In this section we describe some implementation characteristics of ONS converters. The main advantages of ONS converters are as follows [21], [34]: They are relatively robust towards circuit imperfections such as component inaccuracies, and to a lesser extent nonlinearities. That is, the components need not be specified with the accuracy of the targeted A/D converter resolution. ONS converters typically do not require an S/H front end. In addition, the converters do not need stringent anti-alias filtering; the anti-alias filter may be a simple first-order RC circuit, or may be omitted altogether. This advantage is a result of oversampling: Frequencies that alias into signal baseband must exceed half the sampling rate, as opposed to half the maximum signal Nyquist rate. Finally, a 1-bit ONS encoder avoids the need for internal multi-bit A/D and D/A converters. This is advantageous because practical multi-bit converters have non-idealities that create harmonic distortion; in contrast, offsets in a two-level converter only have linear effects.

These advantages of ONS converters make the converters well suited for on-chip CMOS VLSI implementation. With the exception of digital radio, video signal acquisition and storage oscilloscopes, the applications mentioned in the introduction to this chapter are either actual applications or within the reach of present ONS A/D converter technology. Several hundred papers have been published on the subject of ONS conversion, and several commercial products are available [1], [9], [51], [53], [54]. The main practical applications today are in the areas of low-frequency conversion [1], voiceband codecs [4] and audio [9].

The main limitation of ONS converters is that they require oversampling which restricts their applicability to low and moderate signal frequencies. For instance, oversampling a cellular-mobile radio signal of about 600 MHz by even a factor of 8 would require a sampling rate of almost 10 GHz. Another problem is that as the targeted converter resolution increases, the design and implementation of the decoder become correspondingly more involved. For instance, the use of a 4096-tap Finite Impulse Response (FIR) filter is reported in [55] for an audio-quality A/D converter. For multi-channel applications, it is difficult to multiplex an ONS converter because of the transients involved in the encoder as well as in the decoder filter.

### 1.3.3 Variations and Trade-Offs

Section 1.3.2 above presents the principles that distinguish ONS converters from other A/D converters. This section describes the considerable variation within the class of ONS converters, and the performance trade-offs involved.

The OSR of a $\Sigma\Delta$ modulator directly influences its SNR performance, with a larger OSR corresponding to a larger SNR. On the other hand, the largest signal rate for which a modulator can operate is inversely proportional to the OSR. In addition, modulators such as the fourth order interpolative one in [32] are designed for a specific OSR, and will not experience notably increased performance with increasing OSR. Another consideration is that the difficulties of designing and implementing the decoder and decimator increase with the OSR, other things being equal.

The sampling rate of a modulator is proportional to the largest allowable input signal frequency for a fixed OSR, and thus a large sampling rate is desirable. On the other hand, the difficulty of designing components such as operational amplifiers with a given gain increases with the bandwidth over which the components must function.

The resolution $M$ in bits of the in-loop quantizer $Q$ affects the SNR per-

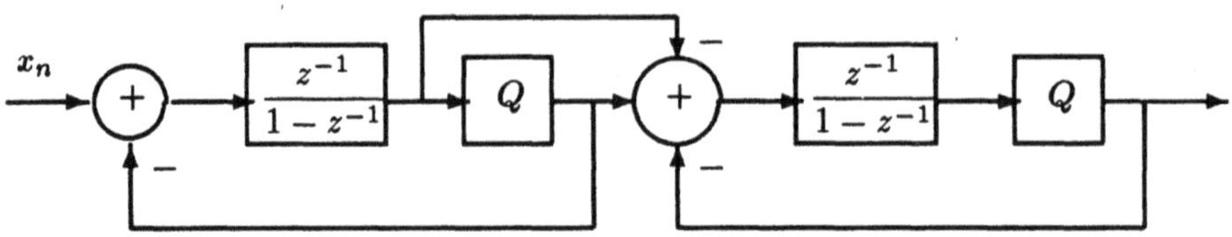


Figure 1.7: *Discrete-time model of the two-stage ΣΔ encoder.*

formance, with larger $M$ corresponding to larger SNR. On the other hand, the quantizer and the $M$-bit D/A converter in the feedback path must both be very linear[4] in order not to create harmonic distortion. For this reason, the inherently linear 1-bit A/D and D/A converters are most often employed. Another advantage of 1-bit quantizers is that FIR decoding can be performed without multiplications. Nonetheless, examples can be found in the literature where the quantizer $Q$ is tri-level [56], multi-bit [57], [58], [59], [60], [61], [62], [63] or adaptive [64].

The SNR of a modulator is in general an increasing function of the order of its encoder. For instance, the double-loop encoder is preferred over the single-loop one for this as well as other reasons. On the other hand, higher-order encoders require more analog circuitry than lower-order ones. In addition, higher-order encoders experience instability problems that are insufficiently understood at present. Instability problems are described in more detail in section 1.4 and in chapter 7.

The specific architecture of a ΣΔ encoder influences its SNR performance, sensitivity towards imperfections, stability *etc.* There exist several alternatives to and variants of the interpolative encoder shown in figure 1.4. For instance, a second analog filter $G(z)$ may be included in the feedback path [50]. The chain of integrators around which the analog filter $H(z)$ is usually implemented may have intentional nonlinear limiters inserted in order to improve stability [65], [66]. Several ΣΔ encoders may be cascaded, with their individual digital outputs combined into one digital output sequence; for example, figure 1.7 shows the so-called two-stage or second-order MASH encoder in which the input to the second stage is the quantization noise of the first stage. The comparison of competing architectures is impeded by the lack of reliable high-level mechanisms for assessing the important performance parameters of a given modulator.

The open-loop filter of an interpolative encoder may be implemented in continuous rather than discrete time, in which case the quantizer is clocked at the sampling rate. A continuous-time encoder can be shown to be equivalent to a discrete-time encoder with an analog pre-filter and a sampler [67]. However, continuous-time encoders suffer from timing jitter in the quantizer and feedback

[4] Of course, any quantizer performs a nonlinear operation on its input. The linearity of a quantizer refers to the extent to which the steps of the quantizer characteristic are linearly aligned. This is described in more detail in [11].

path, whereas discrete-time encoders suffer from sampling jitter on the input. Discrete-time converters are currently favored in implementations because they are easily implemented in switched-capacitor technology.

The open-loop filter $H(z)$ may be band-pass rather than low-pass [68], [69], [70], [71], [72]. For band-pass encoders to be easily implementable, the sampling rate must be a fixed simple multiple of the band-pass center frequency such as 4 or 8. A linear decoder should be a band-pass filter over the same frequency band as the open-loop filter. Band-pass encoders enjoy the same trade-offs between OSR and SNR as their low-pass counterparts, for example 9 dB/octave and 15 dB/octave for first and second order encoders, respectively.[5] It appears that a compelling case for band-pass $\Sigma\Delta$ modulation rather than analog quadrature demodulation has yet to be made.

Finally, a $\Sigma\Delta$ modulator can also be used as a Digital-to-Analog (D/A) converter by interchanging the roles of analog and digital circuitry [6], [39], [73], [74], [75], [76], [77]. Referring to figure 1.4, the input signal is then a high-rate interpolation of a digital signal sampled at a rate close to the Nyquist rate, and the encoder outputs a digital bit stream which can subsequently be decoded and decimated by an analog low-pass filter. The 1-bit quantizer $Q$ can be implemented by stripping off the most significant bit of the digital quantizer input.

### 1.3.4 Signal Processing Perspective

As described in section 1.3.3, the SNR performance of an ONS converter is affected by the OSR and the quantizer resolution $M$ in bits. In fact, a trade-off between the two can be said to exist in the sense that a given SNR performance can be achieved for a number of combinations of OSR and $M$. In practice, $M$ is usually 1 and only occasionally some other small integer, so the trade-off is only exploited over a small part of the possible range of combinations.

Nevertheless, the existence of the trade-off is interesting from a signal processing point of view. In fact, this work was motivated by the demonstration that for signals in two or more dimensions, there exists a spectrum of sampling and reconstruction schemes [27]; for one-dimensional signals, the analogous result only holds for band-pass signals [78]. At one end of the sampling spectrum lies Nyquist sampling with its requirement of infinite amplitude resolution, but minimal position resolution. At the other end lies reconstruction from zero-crossings in which signal values are sampled with only a single bit of amplitude resolution, but the locations of the zero-crossings must be specified with infinite precision.

There is a close analogy between the described trade-off and the trade-off between OSR and $M$ in ONS converters, even though these converters operate on one-dimensional signals. Specifically, in ONS converters, a signal is sampled

[5] For band-pass modulators, the OSR is defined as the ratio between the sampling rate and twice the (one-sided) bandwidth of the modulator. Note that a first-order band-pass $H(z)$ has one complex-conjugate pole pair, that is, two poles.

with a fixed time resolution given by the OSR, and an amplitude resolution given by $M$. Moreover, the sampling is implicit in the sense that the quantizer input equals the difference between a function of the encoder input and a function of the signal-dependent encoder output; thus the level-crossings that the quantizer provides are time-varying and signal-dependent. The same comments apply in principle to error diffusion coding for halftoning of images, that is, two-dimensional signals [79]. However, the quantizer resolution is necessarily fixed at $M = 1$ in such coders. Error diffusion coding is described in more detail in chapter 6.

Another example of a trade-off between oversampling and amplitude resolution can be found in time-frequency analysis [80]. The Discrete Short-Time Fourier Transform (DSTFT) of a continuous-time function $f$ is given by

$$c_{mn}(f) = < f, g_{mn} > \triangleq \int_{-\infty}^{+\infty} f(t) g_{mn}(t)\, dt$$

where $m, n \in \{\cdots, -1, 0, +1, \cdots\}$, and

$$g_{mn}(t) = \exp(-jmp_0 t) g(t - nq_0)$$

The function $g$ is a so-called window function, and $p_0$ and $q_0$ are frequency and time shifts; $|c_{mn}(f)|^2$ is a measure of the energy of $f$ in the vicinity of time $nq_0$ and frequency $mp_0$. The number $1/(p_0 q_0)$ equals the sampling density of phase space. To permit reconstruction of $f$ from its DSTFT coefficients $c_{mn}(f)$, the sampling density must exceed the Nyquist density $1/(2\pi)$ [80]. As $1/(p_0 q_0)$ decreases towards $1/(2\pi)$, the reconstruction of $f$ from its DSTFT coefficients requires that the coefficients be known with ever increasing accuracy [80]. In fact, it is shown in [81] that if the DSTFT coefficients are subject to stationary noise, the noise in the reconstruction of $f$ is inversely proportional to the square of the oversampling factor $2\pi/(p_0 q_0)$. Taking the noise on the DSTFT coefficients to be analogous to a limitation in amplitude resolution, we see the described trade-off with oversampling. A similar trade-off has been observed empirically in the context of signal reconstruction from wavelet coefficients, but no theoretical justification is known at present [80].

## 1.4 Contributions and Outline of Work

As pointed out by Hauser [21], the increasing popularity of ONS converters is to a large extent technology driven. However, it appears fair to say that the theoretical understanding of the converters has been wanting, and that only recently, the area has attracted the attention of rigorously inclined researchers. The theoretical papers on ONS converters that have appeared may not all have affected the design and implementation process directly, but the theoretical momentum that these papers has generated has had an impact also on the practical development of the field. The purpose of this work is to make contributions on a rigorous theoretical basis which either have direct practical consequences or which add to the theoretical understanding that is one of the driving forces of the field. The contributions to the areas of optimal decoding and stability are described in section 1.4.1 and 1.4.2 below, and the organization of the main body of the work is indicated.

### 1.4.1 Optimal Decoding

As described in section 1.3, the conventional decoding technique for $\Sigma\Delta$ encoders is linear low-pass filtering. However, there is no reason to expect linear decoding to be optimal. In fact, since the encoder is a highly nonlinear circuit, it is natural to expect nonlinear decoding to be superior in SNR performance. Another argument in favor of this expectation is the following: A linear decoder only makes use of the fact that the encoder input signal is assumed low-pass with a known bandwidth. A linear decoder considers the high-frequency components of the output spectrum to be noise, and does not use the information that the input signal gave rise to the particular observed binary encoder output. Since a linear decoder discards information, it is expected that improvements are possible.

This work demonstrates that nonlinear decoding does indeed achieve gains in SNR over linear decoding; the gains are typically 10-30 dB, depending on the encoder structure, OSR and other parameters. As a preliminary to this demonstration, chapter 2 presents some theoretical concepts and results relating to $\Sigma\Delta$ encoders with constant inputs. The concepts are applicable to all encoder structures, and we obtain analytical results for the single-loop and double-loop encoders. We also present a cascade structure designed for constant inputs, and describe a decoding method. In chapter 3 we then present an optimal and conceptually simple nonlinear decoding algorithm, dubbed the zoomer, for the special case of constant encoder inputs. The algorithm is applicable to all current $\Sigma\Delta$ encoders. Chapter 3 also addresses the sensitivity of the algorithm with respect to a number of circuit imperfections.

Chapter 5 describes the generalization of the optimal decoding idea to band-limited input signals. As a preliminary to this, chapter 4 presents a novel method for band-limiting finite extent sequences which is used in chapter 5. The optimal nonlinear algorithm of chapter 5 is based on the theory of projections onto convex sets [82], [83], and determines a signal which is both band-limited and which gives rise to the observed binary encoder output when used as the encoder input. Chapter 5 also addresses sensitivity and computational issues. Similar to the DC decoding algorithm, the band-limited decoding method is applicable to all current $\Sigma\Delta$ encoders, including multi-bit ones. Both algorithms are also applicable to predictive coders, although this is not pursued. In fact, the concept of optimal nonlinear decoding is general enough that it can be applied to the seemingly unrelated problem of recovering a continuous-tone image from an image that has been half-toned using error diffusion. Results of exploiting this connection are presented in chapter 6.

### 1.4.2 Stability

Chapter 7 addresses the instability problems encountered in $\Sigma\Delta$ encoders. In mathematical models of dynamic systems, instability typically refers to the situation where one or more state variables are unbounded [84], [85]. This concept is not well suited for the analysis of $\Sigma\Delta$ encoders, because practical circuits have physical limitations on their internal voltages, and so unbounded state variables cannot occur. Instability of a $\Sigma\Delta$ encoder therefore loosely refers to the situation where the quantizer input increases or decreases to its physical limit, and the quantizer output saturates at one of its two extreme levels. This instability behavior prevents the encoder from functioning normally, and is found to be excitable in some encoders even though a linearized model would predict stability.

In this work we argue that stability considerations must take into account the nonlinearity of the circuit, and that as a starting point, the full complexity of the circuit must be incorporated in the analysis. To address the nonlinear stability problem, we introduce a new stability concept which takes into account that integrator outputs cannot in practice exceed circuit-defined bounds. We call a system "very stable" if the largest occurring integrator outputs in the mathematical model of the circuit are small compared to the integrator saturation levels of the actual circuit, and conversely, we call a system "very unstable" if the largest integrator outputs significantly exceed the saturation levels. We thus do not ask *Is a given system stable?*, but rather *How stable is it?*

To simplify the analysis, we consider the stability problem for the special case of constant inputs. This is justified in chapter 7. Using the new stability concept, we show that there exists a trade-off between stability and SNR performance, and thus stability must be considered an integral part of the design process. As an example, we analyze a general version of the double-loop encoder which includes four scaling factors that affect SNR performance as well as stability. This encoder is not usually subject to stability analysis as its integrator outputs are bounded. We address the design problem of choosing the scaling factors to mediate between two objectives: On one hand, it is desirable to choose the scaling factors such that integrator saturation does not occur for the projected range of inputs. On the other hand, it is desirable to design an encoder that utilizes the full dynamic ranges of its integrators and maximizes the noise immunity.

As another example of stability analysis, we consider a specific fourth order interpolative encoder. We derive a numerical method for detecting specific periodic behavior in the encoder, and for finding the associated largest integrator outputs. We also develop a describing function method which results in an approximate relation between the open-loop filter and the largest occurring integrator outputs. We use this relation to explicitly analyze the trade-off between SNR performance and stability for the fourth order encoder, and we present an approximately optimal method for sacrificing SNR to increase stability.

# Chapter 2

# Properties of ONS Converters with Constant Inputs

In this chapter[1] we derive new properties of the single-loop and double-loop $\Sigma\Delta$ encoders with constant inputs by exploiting the inherent structure of the encoder output sequences or codewords. Specifically, we derive upper bounds of $O(N^2)$ and $O(N^3)$ on the number of $N$-bit codewords for the single-loop and double-loop encoders, respectively. We then derive analytical lower bounds on the Mean Squared Error (MSE) obtainable with any decoder, linear or non-linear, in approximating the constant input; based on $N$-bit codewords, the bounds are $O(N^{-3})$ and $O(N^{-6})$ for the single-loop and double-loop encoders, respectively. Optimal nonlinear decoders for constant inputs can be based on a table look-up approach which operates directly on the non-uniform quantization intervals. Numerical results show that if the constant input is uniformly distributed, the MSE of such nonlinear decoders are $O(N^{-3})$ and $O(N^{-5})$ for the single-loop and double-loop encoders, respectively. Simulations show that the optimal nonlinear decoders perform better than linear decoders by about 3 dB and 20 dB for the single-loop and double-loop encoders, respectively. We also introduce a cascade structure specifically for constant inputs, and derive its corresponding decoding algorithm. The idea behind the cascade structure is to requantize the residue of each stage in order to fully utilize the dynamic range of the next stage. We show that when comparing our two-stage cascade structure to the conventional two-stage MASH encoder for a fixed latency, our structure has 12 dB superior MSE performance and twice the throughput.

---

[1] Based on [86], ©1990 IEEE, and [87], ©1992 IEEE.

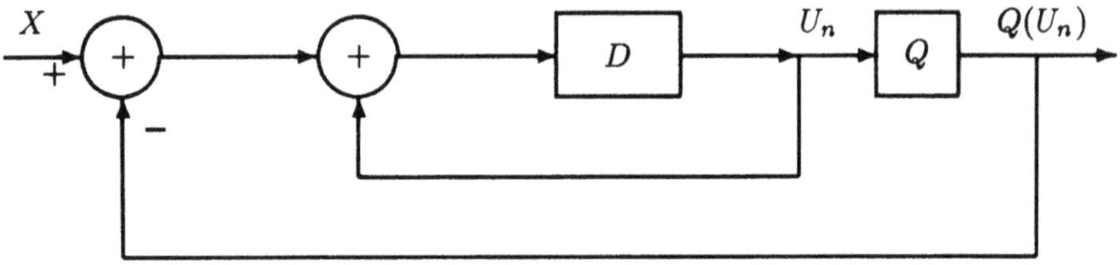


Figure 2.1: *Discrete-time model of the single-loop ΣΔ encoder.*

## 2.1 Introduction

In this chapter we consider the application of the single-loop and double-loop ΣΔ modulators to data acquisition applications in which the analog input is approximately constant. The results also provide insight into the more general case of time-varying inputs, since such inputs are typically heavily oversampled. To introduce the problems that we address, let us consider the discrete-time model of the single-loop ΣΔ encoder shown in figure 2.1; the figure is equivalent to figure 1.3 in chapter 1. In figure 2.1, $D$ represents a delay element and $Q$ is a 1-bit quantizer specified by (1.1) which is repeated here for convenience:

$$Q(U_n) = \begin{cases} -b & U_n \leq 0 \\ +b & U_n > 0 \end{cases}$$

The constant input $X$ is assumed to be a stochastic variable in the dynamic range $B = (-b, +b)$, and the encoder output is a binary sequence of $\pm b$'s. The discrete integrator in the feed-forward path accumulates the error between the input and the quantizer output, and the negative feedback serves to make this error small.

The task of a decoder is to produce an estimate $\hat{X}$ of the constant input $X$, given a number $N$ of output bits, $\{Q(U_0), \cdots, Q(U_{N-1})\}$. We refer to $N$ as the Sample Size (SS).[2] The collection of estimates $\hat{X}$ that a given decoder is capable of producing for a given SS and all constant inputs $X \in B$ is referred to as the decoder's reproduction alphabet for that SS [41]. The Mean Squared Error (MSE) of a decoder is defined as

$$\text{MSE} \triangleq E\left[(X - \hat{X})^2\right]$$

[2] Although there are similarities between the sample size and the oversampling ratio, the two should not be confused. The oversampling ratio is only defined for input signal classes with non-zero bandwidth.

The MSE is related to the Signal-to-Noise Ratio (SNR) via

$$\text{SNR} \triangleq \frac{E[X^2]}{\text{MSE}}$$

One of the simplest possible decoders is the averaging decoder given by

$$\hat{X} = \frac{1}{N}\sum_{n=0}^{N-1} Q(U_n) \tag{2.1}$$

The size of its reproduction alphabet is $N+1$ [41], and if the constant input has a smooth probability distribution function, the MSE of the averaging decoder is upper bounded by [41]

$$\text{MSE} \leq \frac{4b^2}{3N^2}$$

For the averaging decoder, Gray [42] has shown that if the input is uniformly distributed on $B$ and $N$ is large, the MSE is given by

$$\text{MSE} \approx \frac{2b^2}{3N^2}$$

The accuracy of the single-loop modulator as an A/D converter of constant inputs is intuitively linked to the number of output values, that is, the size of the reproduction alphabet at the disposal of its decoder. For instance, the reproduction alphabet size is $O(N^2)$ for a linear decoder consisting of an $N$-tap Finite Impulse Response (FIR) filter with a triangular impulse response, commonly referred to as a $sinc^2$ filter.[3] This decoder achieves an asymptotic MSE of $O(N^{-3})$, which may be contrasted with the $O(N^{-2})$ performance of the simpler averaging decoder quoted above. It is conceivable that there exist a variety of other decoders with larger reproduction alphabets and smaller MSEs than the $sinc^2$ filter. The question arises as to the fundamental limits on reproduction alphabet size and MSE performance of the single-loop modulator, as well as more general modulators, with constant inputs.

In this chapter we show that the answers to the stated performance questions can be found by decoupling the encoder and decoder parts of a modulator, and studying separately the structure of the output sequences that an encoder is capable of producing. Because the $N$-bit output sequences that an encoder can generate for constant inputs $X \in B$ and fixed initial integrator states play an important role in this investigation, we will refer to them as the codewords for the given initial states, or codewords for short. The size of a modulator's reproduction alphabet is clearly upper bounded by the number of different codewords of its encoder, and the number of codewords is bounded, since there are at most $2^N$ of them. We will show that for given initial integrator states,

[3] The impulse response of a $sinc^k$ filter ($k > 1$) can be obtained recursively by convolving the averaging or $sinc^1$ filter with the $sinc^{k-1}$ filter. The name "$sinc^k$" refers to the form of the filter transfer function.

many of the $2^N$ binary $N$-bit sequences are in fact not possible codewords. Using results on codewords, we can bound the resolution and hence the MSE performance of a modulator. This is because each codeword can be generated by a range of constant inputs, and no decoder can distinguish between constant inputs in such ranges. In principle, an optimal nonlinear decoder can be based on a table look-up approach in which each codeword is mapped to the mean value of the input over the range corresponding to the codeword.[4]

Although the above discussion is pertinent to all $\Sigma\Delta$ modulators, we restrict our analysis in the following to the single-loop and double-loop modulators, because these permit an analytical rather than a numerical approach. The chapter is organized as follows: In section 2.2, we derive an upper bound of $O(N^2)$ on the number of codewords for the single-loop encoder, and an $O(N^{-3})$ lower bound on the MSE performance of any single-loop decoder with constant inputs. We present simulation results to show that an MSE performance of $O(N^{-3})$ is achievable with the optimal nonlinear decoder, as well as with a number of linear filters. Similar analysis and simulation results for the double-loop encoder are presented in section 2.3. In section 2.4, we first derive the MSE performance of the averaging decoder for the single-loop encoder. Even though the expression for the MSE is known [42], our particular derivation enables us to develop a decoding scheme for a modified cascade structure with DC inputs; the structure may be viewed as a modification of the multistage noise-shaping (MASH) modulator [77]. In section 2.4, we discuss this modified cascade structure and its corresponding decoding algorithm. We also compare its performance to that of the MASH modulator with linear decoding. Finally, section 2.5 summarizes the chapter.

Our focus on $N$-bit codewords and fixed initial integrator states bears some similarity to the approach in [88]. In that paper, $N$-tap linear decoders are derived which are in some sense optimal. However, the work in [88] is based on an assumption of white, uncorrelated quantization noise—an assumption which is now known not to be valid for low-order encoders with one-bit quantizers [41].

## 2.2 Single-Loop Encoder

In section 2.2.1 we derive an upper bound on the number of codewords of the single-loop $\Sigma\Delta$ encoder with constant inputs and a fixed initial integrator state, and we compare the bound to simulation results. In section 2.2.2 we use the results of section 2.2.1 to derive a lower bound on the MSE performance of any single-loop decoder, and compare the bound to the simulated performance of an optimal nonlinear decoder and two linear decoders.

[4] In practice, the table would be prohibitively large. In chapter 3 we describe a more practical alternative.

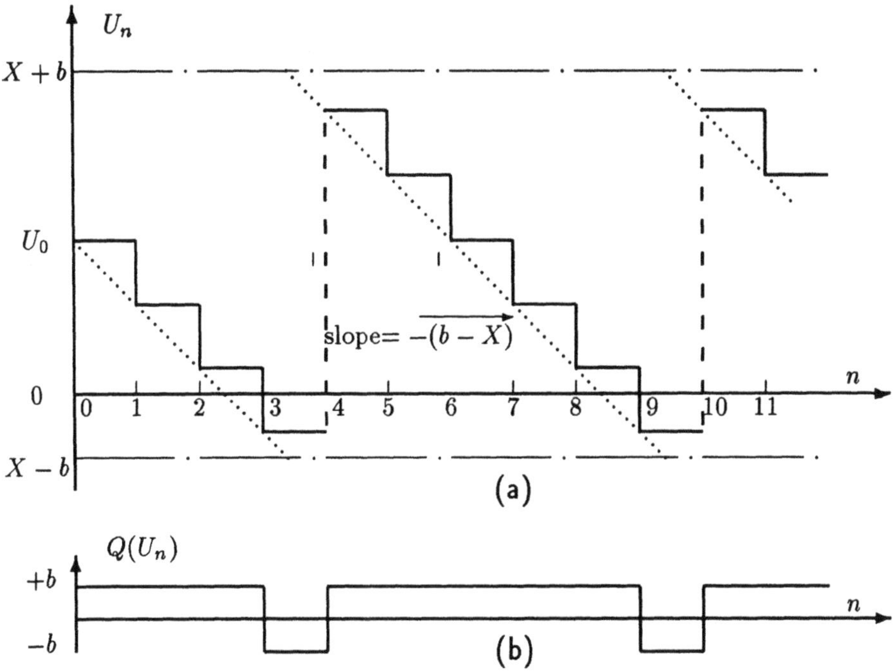


Figure 2.2: *Typical waveforms for: (a) the state variable $U_n$, (b) the quantizer output $Q(U_n)$.*

### 2.2.1 Codewords

In section 2.2.1.1 we derive an upper bound on the number of codewords, and in section 2.2.1.2 we show simulation results on the actual number of codewords.

#### 2.2.1.1 Upper Bound

Our approach is to examine the behavior of the encoder state variable $U_n$ in figure 2.1. From the figure, the nonlinear difference equation relating $U_n$ to $U_{n-1}$ is

$$U_n = U_{n-1} + X - Q(U_{n-1}) \tag{2.2}$$

where $X$ is the constant input. It has been shown [41] that if the initial integrator state $U_0$ is in the range $(X-b, X+b)$, the integrator state remains in that range at all future times, that is,

$$U_0 \in (X-b, X+b) \quad \Rightarrow \quad U_n \in (X-b, X+b) \text{ for all } n > 0 \tag{2.3}$$

To gain some intuition about the behavior of the state variable, figure 2.2 shows a typical time sequence for $U_n$. Equation (2.2) says that if $U_{n-1}$ is positive, the state variable is decremented by $b - X$ in the next time step, whereas if $U_{n-1}$ is negative, the state variable is incremented by $b + X = 2b - (b - X)$. Thus the state variable is always decremented by $b - X$, but if $U_{n-1} < 0$, it is additionally incremented by $2b$, and a negative output bit is generated. To satisfy (2.3), the number of negative bits $j$ among the first $n$ output bits must therefore satisfy

$$X - b \leq U_0 - n(b - X) + 2jb \leq X + b \tag{2.4}$$

In appendix A.1.1 we show that equation (2.4) leads to the following conclusions for fixed values of $N$ and $U_0$: The dynamic range $(-b, +b)$ is divided into quantization intervals whose width and position depend on $U_0$. Within each interval, all constant inputs generate the same $N$-bit codeword, but distinct intervals correspond to distinct codewords. The edges of the quantization intervals, referred to as transitions points, are given by

$$X = \frac{pb - (2jb + U_0)}{p} \tag{2.5}$$

where $1 \leq p \leq N - 1$, and $j$ is a positive integer in the appropriate range such that $X \in (-b, +b)$. As an example, the transition points for $N = 12$ and an initial state of $U_0 = 0$ are shown in table 2.1. By counting the number of parameter combinations $(p, j)$ in equation (2.5), we find that the number of quantization intervals, hence the number of codewords, is upper bounded by

$$C_{\max} = \begin{cases} \frac{1}{2}N(N-1) + 1 & U_0 \neq 0 \\ \frac{1}{2}(N-1)(N-2) + 1 & U_0 = 0 \end{cases} \tag{2.6}$$

This is an $O(N^2)$ upper bound. Clearly, the number of codewords is a diminishingly small fraction of $2^N$ for large sample sizes.

The actual number of codewords may be less than indicated by (2.6) because some parameter combinations may correspond to the same transition point. Such degeneration occurs only if $U_0/b$ is rational, which happens with probability zero if $U_0$ has a smooth probability distribution function on $(X - b, X + b)$. If $U_0/b$ is irrational, the exact number of codewords is $\frac{1}{2}N(N-1) + 1$. In the special case $U_0 = 0$, the transition points of equation (2.5) are known as the Farey series of number theory [89], and an asymptotic expansion of their number is available [89],

$$C \rightarrow \frac{3}{\pi^2}N^2 + O(N \log N) \approx 0.304N^2 \quad \text{as} \quad N \rightarrow \infty \tag{2.7}$$

The upper bound (2.6) is thus asymptotically 65% too high for the case $U_0 = 0$, but it has the correct dependence on $N$.

| p j | 1 | 2 | 3 | 4 | 5 | 6 | 7 | 8 | 9 | 10 |
|---|---|---|---|---|---|---|---|---|---|---|
| 2 | 0 | | | | | | | | | |
| 3 | 1/3 | −1/3 | | | | | | | | |
| 4 | 1/2 | 0 | −1/2 | | | | | | | |
| 5 | 3/5 | 1/5 | −1/5 | -3/5 | | | | | | |
| 6 | 2/3 | 1/3 | 0 | −1/3 | -2/3 | | | | | |
| 7 | 5/7 | 3/7 | 1/7 | −1/7 | -3/7 | -5/7 | | | | |
| 8 | 3/4 | 1/2 | 1/4 | 0 | −1/4 | −1/2 | -3/4 | | | |
| 9 | 7/9 | 5/9 | 1/3 | 1/9 | −1/9 | −1/3 | -5/9 | -7/9 | | |
| 10 | 4/5 | 3/5 | 2/5 | 1/5 | 0 | −1/5 | -2/5 | -3/5 | -4/5 | |
| 11 | 9/11 | 7/11 | 5/11 | 3/11 | 1/11 | -1/11 | -3/11 | -5/11 | -7/11 | -9/11 |

Table 2.1: *Single-loop encoder: Transition points for* $N = 12$ *and* $U_0 = 0$. *The points are normalized by* $b$.

#### 2.2.1.2 Numerical Results

For the special case of zero initial integrator state, $U_0 = 0$, we have used computer simulations to find the actual number of codewords for the single-loop encoder as a function of sample size. Figure 2.3 shows the results as well as the upper bound (2.6) derived in section 2.2.1.1. The upper bound is seen to be rather tight. The analytical approximation (2.7) is not plotted, since it is indistinguishable from the simulated results.

### 2.2.2 Mean Squared Error

In section 2.2.2.1 we use the results of section 2.2.1 to derive a number of lower bounds on the MSE obtainable with any decoder operating on a single-loop encoder. In section 2.2.2.2 we show numerical results on the actual MSE of a number of decoders.

#### 2.2.2.1 Lower Bounds

Our first lower bound on the MSE follows directly from the upper bound on the number of codewords derived in section 2.2.1.1. To state the bound, we quote a well-known result from quantization theory [90]: If the constant input $X$ is restricted to $(-b, +b)$ and has probability density function $p(x)$, the minimum MSE obtainable with a large number $C$ of quantization levels is

$$\mathrm{MSE}_{\min} = \frac{1}{12C^2}\left[\int_{-b}^{+b} p(x)^{1/3}\,dx\right]^3 \tag{2.8}$$

If in particular $X$ is uniformly distributed on $(-b, +b)$, the minimum MSE is $b^2/(3C^2)$. Since the single-loop $\Sigma\Delta$ encoder has a maximum of $O(N^2)$ codewords, equation (2.8) implies that a lower bound on its MSE for any given

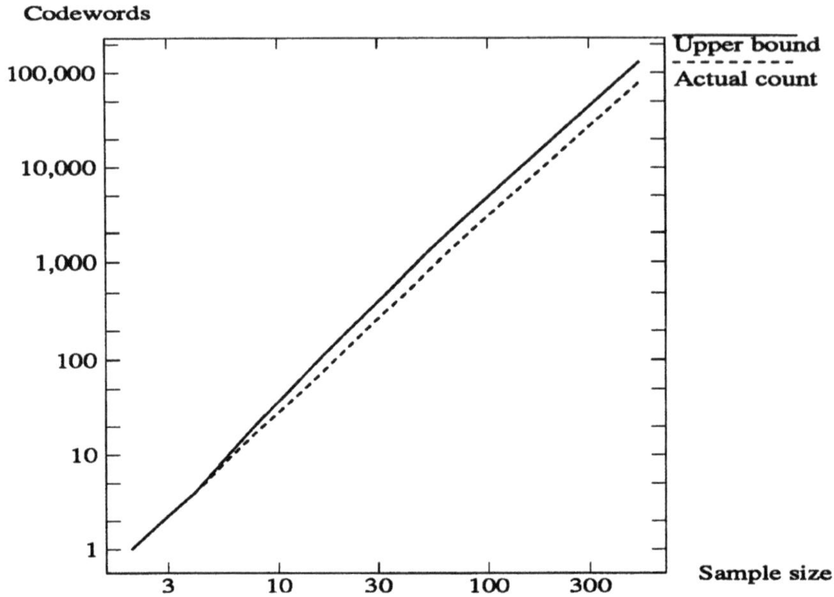


Figure 2.3: *Single-loop encoder: Actual number of codewords and the upper bound (2.6) as functions of sample size.*

$p(x)$ is $O(N^{-4})$. If the input is uniformly distributed on $(-b, +b)$, the following $O(N^{-4})$ lower bound holds on the MSE:

$$\text{MSE}_{\text{min}} = \frac{4b^2}{3N^4}$$

To strengthen this result, we now derive an $O(N^{-3})$ lower bound on the MSE by considering in more detail the quantization intervals of the single-loop encoder. If the number of codewords $C$ is large, and the density function $p(x)$ is smooth, the constant input $X$ is approximately uniformly distributed on each quantization interval $I_i$, $1 \leq i \leq C$. An optimal decoder will therefore decode any codeword into the midpoint of the corresponding quantization interval. Denoting the width of the $i$th interval by $d_i$, a lower bound on the MSE is thus

$$\text{MSE}_{\text{optimal}} = \sum_{i=1}^{C} P[X \in I_i] \cdot \frac{d_i^2}{12} = \sum_{i=1}^{C} \frac{d_i^3}{24b} \tag{2.9}$$

where the last equality holds if $X$ is uniformly distributed on $(-b, +b)$. Clearly, the optimal MSE is lower-bounded by an expression similar to (2.9) in which only some of the $C$ terms are included. Our strategy for lower-bounding the MSE is to find a few large interval widths $d_i$ of order $O(N^{-1})$. As long as there is non-zero probability of inputs in these intervals, we arrive at an $O(N^{-3})$ lower bound on the MSE, but for simplicity we state the MSE bounds for the case of uniformly distributed input. Appendix A.1.2 shows that if the initial state is zero, the two intervals closest to the extreme inputs $-b$ and $+b$ are the largest intervals, and they have widths

$$d_1 = d_C = \frac{2b}{N-1} \tag{2.10}$$

while the two intervals immediately above and below $X = 0$ have widths

$$d_{\frac{C}{2}-1} = d_{\frac{C}{2}} \geq \frac{b}{N-1} \tag{2.11}$$

Including only the four terms corresponding to (2.10) and (2.11) in (2.9), we thus find the $O(N^{-3})$ lower bound

$$\text{MSE}_{\text{optimal}} \geq \frac{d_1^3 + d_C^3 + d_{\frac{C}{2}}^3 + d_{\frac{C}{2}+1}^3}{24b} \geq \frac{3b^2}{4(N-1)^3} \approx \frac{0.75b^2}{N^3} \tag{2.12}$$

As an aside, it is also shown in appendix A.1.2 that the smallest quantization intervals are located next to the two largest ones, near $+b$ and $-b$, and have widths

$$d_2 = d_{C-1} = \frac{2b}{(N-1)(N-2)} \tag{2.13}$$

This shows that the ratio between the largest and smallest interval widths is $N-2$, and gives a bound on the non-uniformity of the quantization intervals of the single-loop encoder.

The $O(N^{-3})$ lower bound (2.12) on the MSE was based on the assumption that $U_0 = 0$. For $U_0 \neq 0$ a similar bound can be obtained, although the intervals of width $O(N^{-1})$ are harder to find. To demonstrate their existence, appendix A.1.2 shows that for $U_0 > 0$, the interval near $+b$ has width

$$d_1 = \frac{b}{N-1} U_0$$

A similar result holds for $d_C$ if $U_0 < 0$.

#### 2.2.2.2 Numerical Results

To illustrate the results in equations (2.10), (2.11), and (2.13), figure 2.4 shows a plot of quantization interval widths for a sample size of 16 under the assumption of zero initial state. The results seem to be in excellent agreement with the experimental results in [37]. The figure indicates that the single-loop encoder is a highly non-uniform quantizer.

Figure 2.5 shows several curves of MSE performance versus sample size, assuming the input is uniformly distributed on $(-b, +b)$. One curve shows the

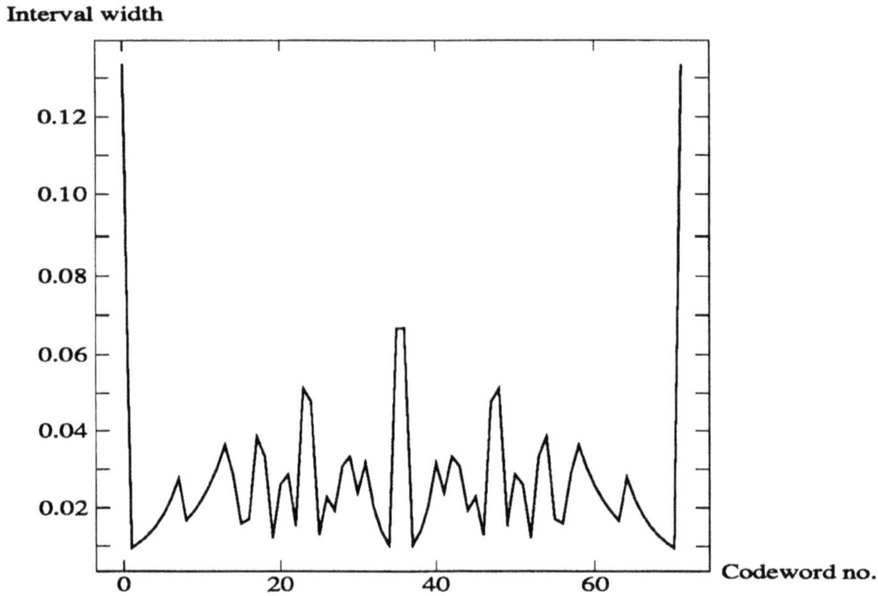


Figure 2.4: *Single-loop encoder: Width of quantization interval number i, normalized by b, vs. i. The sample size is $N = 16$.*

actual performance of the optimal nonlinear decoder whose MSE is given by (2.9); numerically, we find that

$$\mathrm{MSE}_{\mathrm{optimal}} \approx \frac{0.91b^2}{N^3} \tag{2.14}$$

Another curve shows the lower bound (2.12), which is rather tight. This indicates that the four terms used to derive (2.12) cont[illegible]ute most of the error to the MSE summation (2.9). A third curve shows the residual MSE obtained by ignoring the contributions from the edge intervals $d_1$ and $d_C$, corresponding to a slight decrease in dynamic range. The reduction in dynamic range improves the MSE by 5.8 dB, but the performance remains at $O(N^{-3})$, in part because of the central intervals of width $O(N^{-1})$ in (2.11).

The last curve in figure 5 shows the MSE performance achievable with a specific linear $N$-tap FIR filter as the decoder; the filter is derived by Gray [42] as the asymptotically optimal linear decoder for constant inputs. The MSE of the FIR filter is asymptotically $2b^2/N^3$ for large $N$. At a fixed sample size, the optimal nonlinear decoder is 3.4 dB superior to the optimal linear decoder.

The simulation results presented in this section assume that $U_0 = 0$. If we choose a different, but known initial state, the codewords and transition points change, but similar conclusions are reached. Specifically, the bounds of $O(N^2)$

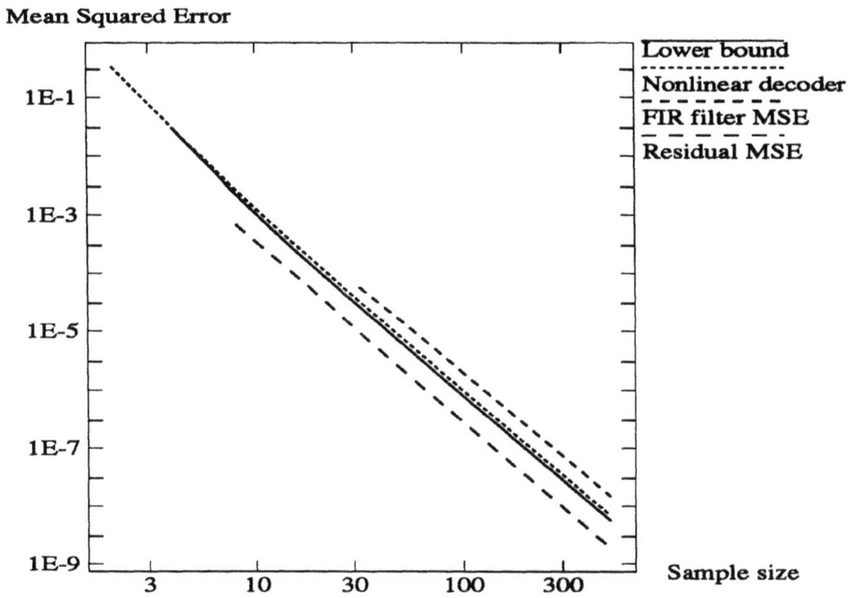


Figure 2.5: *Single-loop encoder: MSE performance (2.9) of the optimal nonlinear decoder as a function of sample size. Also shown is the lower bound (2.12) and the residual MSE obtained by discarding the two large edge intervals. The last curve shows the performance achievable with the asymptotically optimal linear decoder derived by Gray [42]. The MSE is normalized by* $b^2$.

on the number of codewords and $O(N^{-3})$ on the MSE of any decoder still hold. Choosing $U_0 = 0$ is merely a convenient way to ensure that $U_0 \in (X - b, X + b)$ for any input $X \in (-b, +b)$.

## 2.3 Double-Loop Encoder

In this section we consider the double-loop $\Sigma\Delta$ encoder whose discrete-time model is shown in figure 2.6; the figure is equivalent to figure 1.5 in chapter 1. In section 2.3.1 we derive an upper bound on the number of codewords of the double-loop encoder with constant inputs and fixed initial integrator states, and we compare the bound to simulation results. In section 2.3.2 we use the results of section 2.3.1 to derive lower bounds on the MSE performance of any decoder.

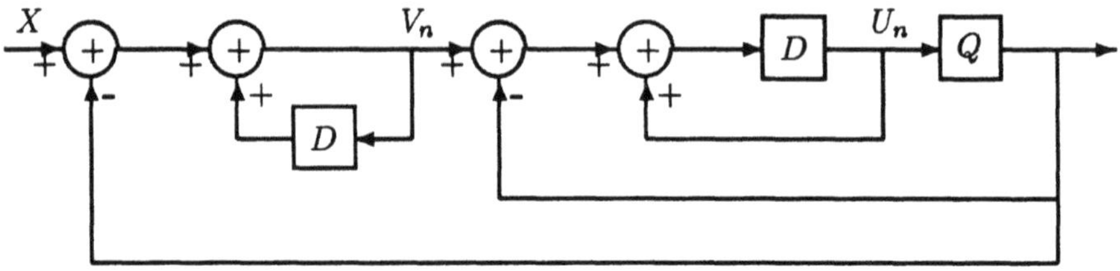


Figure 2.6: *Discrete-time model of the double-loop* $\Sigma\Delta$ *encoder.*

### 2.3.1 Codewords

In section 2.3.1.1 we derive an upper bound on the number of codewords, and in section 2.3.1.2 we show simulation results on the actual number of codewords.

#### 2.3.1.1 Upper Bound

Our approach is to examine the state variables $U_n$ and $V_n$ in figure 2.6. We obtain the state equations

$$\begin{aligned} U_n &= U_{n-1} + V_{n-1} - Q(U_{n-1}) \\ V_n &= X + V_{n-1} - Q(U_n) \end{aligned} \qquad (2.15)$$

Appendix A.2.1 contains a derivation of an expression for the transition points, analogous to (2.5) for the single-loop encoder. For all transition points $X$ there exist integers $(j, n)$ such that

$$X = \frac{\frac{1}{2}n(n+1)b - (2jb + U_0 + nV_0)}{\frac{1}{2}n(n-1)}, \qquad 2 \leq n \leq N-1 \qquad (2.16)$$

where $U_0$ and $V_0$ are initial integrator states. Unlike for the single-loop encoder, (2.16) only gives all the values of the constant input $X$ that *might be* transition points, but any transition point corresponds to some parameter set $(j, n)$. The absence of a better formula is due to the fact that no result analogous to (2.3) exists for the double-loop encoder.

From (2.16) we can find the range of integers $j$ such that $X \in B$. Counting the number of possible parameter sets $(p, j)$, we obtain the following upper bound on the number of codewords, derived in more detail in appendix A.2.1:

$$C_{max} = \frac{1}{6}N(N-1)(N-2) + 1, \qquad N \geq 3 \qquad (2.17)$$

This is an $O(N^3)$ upper bound, in contrast with the upper bound of $O(N^2)$ for the single-loop encoder. For large $N$, the upper bound is a diminishingly small fraction of $2^N$.

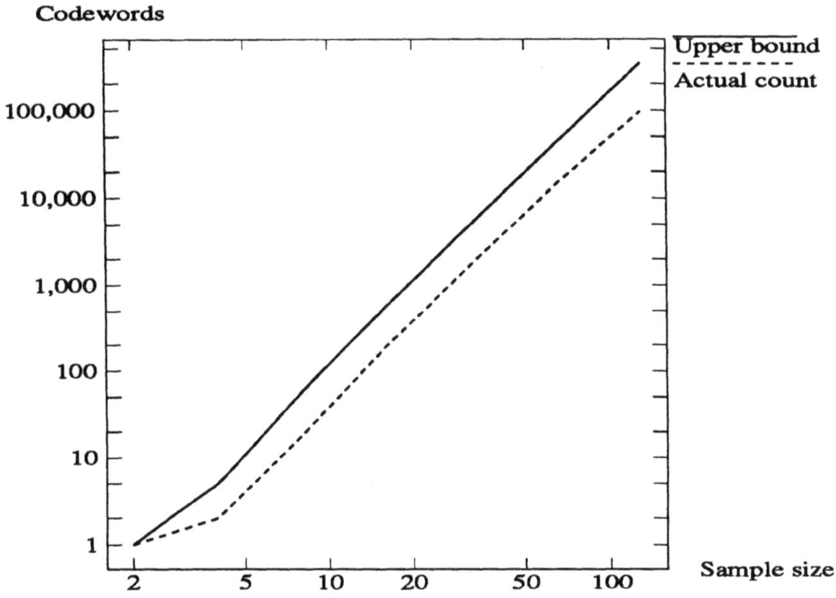


Figure 2.7: *Double-loop encoder: Actual number of codewords and the upper bound (2.17) as functions of sample size.*

#### 2.3.1.2 Numerical Results

We have used computer simulations to find the actual number of codewords for the case $U_0 = V_0 = 0$. Figure 2.7 shows the results, as well as the $O(N^3)$ upper bound derived in section 2.3.1.1. The actual number of codewords is seen to also be $O(N^3)$; the upper bound is about three times larger than the actual count.

### 2.3.2 Mean Squared Error

In section 2.3.2.1 we derive lower bounds on the MSE obtainable with any decoder operating on a double-loop encoder. In section 2.3.2.2 we show numerical results on the actual MSE of the optimal nonlinear decoder.

#### 2.3.2.1 Lower Bounds

Our first lower bound is based on (2.8) as for the single-loop encoder. Since the number of codewords is upper-bounded by $O(N^3)$, the minimum MSE is lower-bounded by $O(N^{-6})$. If in particular the constant input $X$ is uniformly distributed on $(-b, +b)$, then

$$\text{MSE}_{\text{min}} \geq \frac{b^2}{3\left[\frac{1}{6}N(N-1)(N-2)+1\right]^2} \approx \frac{12b^2}{N^6} \tag{2.18}$$

We next derive a tighter lower bound on the MSE by finding the widths of some actual quantization intervals. Appendix A.2.2 shows that for initial integrator states $U_0 = V_0 = 0$, the intervals closest to the extreme inputs are the largest intervals, and they have widths

$$d_1 = d_C = \frac{2b}{N-2}$$

We can use these interval widths in equation (2.9) to derive the following $O(N^{-3})$ lower bound on the MSE, assuming uniformly distributed input:

$$\text{MSE} \geq \frac{1}{24b}\left[2\left(\frac{2b}{N-2}\right)^3\right] = \frac{2b^2}{3(N-2)^3} \approx \frac{0.67b^2}{N^3}, \quad N \geq 3 \tag{2.19}$$

At first sight, this bound is disappointing since it does not improve on the $O(N^{-3})$ lower bound for the single-loop encoder, except for a proportionality constant. However, if we restrict the dynamic range of the input so that the two largest quantization intervals in the neighborhood of $+b$ and $-b$ are avoided, the performance can be improved drastically, as shown numerically in the next section.

#### 2.3.2.2 Numerical Results

To demonstrate the non-uniformity of the double-loop encoder, figure 2.8 shows a plot of the quantization interval widths for a sample size of 16 under the assumption of zero initial states. Comparing with figure 2.4 for the single-loop encoder, we see that the double-loop encoder is a more uniform quantizer, the only significant non-uniformities being located at the edges of the dynamic range.

Figure 2.9 shows a number of theoretical and computed MSE curves for the double-loop encoder, assuming uniformly distributed input. One curve shows the performance of an optimal nonlinear decoder with an MSE computed from (2.9) using the actual interval widths. Another curve represents the $O(N^{-3})$ lower bound of equation (2.19). The lower bound and the simulated curve are extremely close, indicating that almost all the MSE comes from the two edge intervals with widths $O(N^{-1})$. The MSE can be significantly reduced by excluding these two edge intervals. A third curve in figure 2.9 shows simulated values of the residual MSE obtained by limiting the dynamic range to $(-0.9b, +0.9b)$; the corresponding MSE performance is approximately $O(N^{-5})$. This improvement provides a quantitative reason for not fully utilizing the dynamic range, since avoiding the edge intervals is an advantageous sacrifice of dynamic range

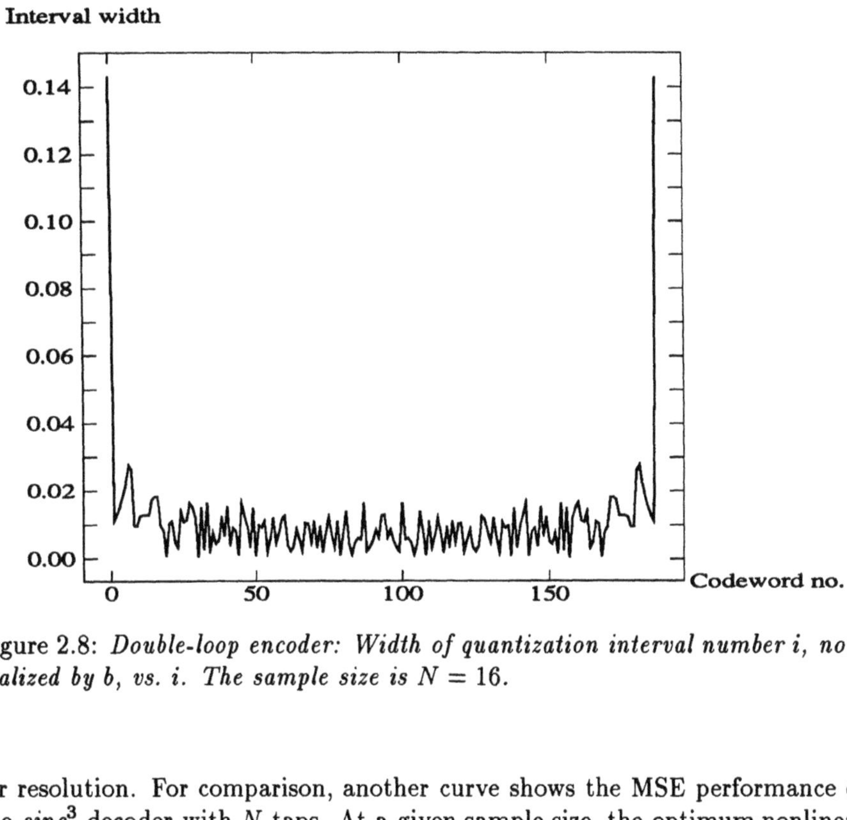


Figure 2.8: *Double-loop encoder: Width of quantization interval number i, normalized by b, vs. i. The sample size is $N = 16$.*

for resolution. For comparison, another curve shows the MSE performance of the $sinc^3$ decoder with $N$ taps. At a given sample size, the optimum nonlinear decoder is about 20 dB better than the $sinc^3$ decoder. Finally, the fifth curve in the figure corresponds to the $O(N^{-6})$ lower bound of equation (2.18).

## 2.4 A Modified Cascade Structure for Constant Inputs

In section 2.4.1 we present a derivation of the MSE performance of the single-loop encoder with an averaging decoder. Although the result is known for the general case of arbitrary FIR filtering [42], our derivation is necessary to understand the cascade structure introduced in the following sections. In sections 2.4.2 and 2.4.3 we describe a modified cascade encoder structure suitable for DC inputs, and a decoding algorithm for it. The encoder structure can have arbitrarily many stages, but for convenience we only consider two stages. Section 2.4.4 contains numerical results for the structure. We compare our structure to that of the MASH modulator [77] which is applicable to more general band-limited inputs, and which is not designed specifically for DC inputs. Similar to the MASH modulator, our structure does not suffer from instability problems

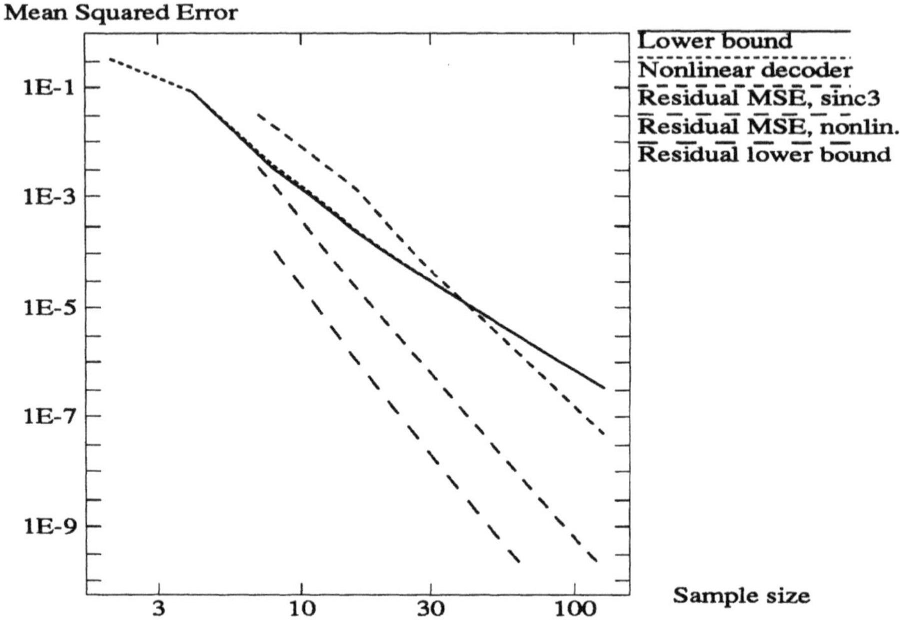


Figure 2.9: *Double-loop encoder: MSE performance of the optimal nonlinear decoder as a function of sample size, and the $O(N^{-3})$ lower bound (2.19). Also shown is the $O(N^{-5})$ residual MSE obtained by limiting the dynamic range by 10%. For comparison, a curve shows the residual MSE of a* $\text{sinc}^3$ *decoder. The last curve shows the $O(N^{-6})$ lower bound (2.18). The MSE is normalized by $b^2$.*

since it consists of stable single-loop encoders.

### 2.4.1 Analysis of Averaging Decoder

We begin by decomposing the smoothly distributed constant input $X \in (-b, +b)$ into an integer multiple $P$ of the step size

$$\Delta = \frac{2b}{N}$$

and a residue $R$ in the range $[0, \Delta)$. Specifically,

$$X = -b + P\Delta - R \tag{2.20}$$

From (2.20),

$$1 \leq P = \left\lceil \frac{N(X+b)}{2b} \right\rceil \leq N-1$$

For large values of $N$, the random variable $R$ is uniformly distributed on $[0, \Delta)$. The residue can be considered the quantization error in representing the input $X$ by the integer $P$.

Consider now the single-loop encoder with state variable $U_n$. We assume that the initial state $U_0$ is uniformly distributed on the range $(X - b, X + b)$. Equation (2.3) then shows that $U_N \in (X - b, X + b)$. From (2.4), the number of negative output bits $A$ in an $N$-bit codeword is specified by

$$U_N = U_0 - N(b - X) + 2Ab \tag{2.21}$$

Using the decomposition (2.20), equation (2.21) can be rewritten

$$RN = U_0 - U_N - 2b(N - P - A) \tag{2.22}$$

The product $RN$ is in the range $[0, 2b)$, and since both $U_0$ and $U_N$ are in the range $(X - b, X + b)$, their difference is in the range $(-2b, +2b)$. The last term on the right hand side of (2.22) is an integer multiple of $2b$. We therefore find that

$$RN = \begin{cases} U_0 - U_N & \text{if } U_0 \leq U_N \\ U_0 - U_N + 2b & \text{if } U_0 > U_N \end{cases}$$

or more concisely,

$$RN - b = (U_0 - U_N) - Q(U_0 - U_N) \tag{2.23}$$

Inserting (2.23) in (2.22),

$$N - P - A = \frac{b - Q(U_0 - U_N)}{2b} \triangleq \Theta \tag{2.24}$$

where the random variable $\Theta$ takes on the values 0 and 1. The following three theorems investigate the statistical properties of $\Theta$; their proofs can be found in appendix A.3.

**Theorem 1** *If $U_0$ is uniformly distributed on $(X - b, X + b)$, then for a given input $X$,*

$$P[\Theta = 1 \mid X] = \frac{R}{\Delta}$$

**Theorem 2** *If $U_0$ is uniformly distributed on $(X - b, X + b)$, then for large sample sizes, the MSE of the averaging decoder (2.1) for a given $X$ is*

$$E[(X - \hat{X})^2 \mid X] = \Delta^2 \cdot P[\Theta = 1 \mid X] \cdot P[\Theta = 0 \mid X] = R(\Delta - R)$$

**Theorem 3** *If the constant input $X$ has a smooth probability distribution on $(-b, +b)$, and the initial state $U_0$ is uniformly distributed on $(X - b, X + b)$, then for large values of $N$, the MSE of the averaging decoder is*

$$\text{MSE} = E[(X - \hat{X})^2] = \frac{\Delta^2}{6}$$

The result of theorem 3 agrees with that of [42]. However, in [42] the input is assumed to be uniformly distributed on $(-b, +b)$, whereas we only assume smoothly distributed input; furthermore the result in [42] is independent of the distribution of $U_0$, while we assume $U_0$ to be uniformly distributed. There are also fundamental differences between our MSE derivation and those in [34], [37], [38]. The latter derivations assume that the error sequence $\{U_n - Q(U_n)\}$ is white. Gray showed in [42] that in fact this assumption does not hold in situations where $Q$ quantizes to as few bits as one.

### 2.4.2 Cascade Encoder

In this section we describe a cascade structure specifically for constant inputs, based on the derivations of section 2.4.1. The idea is to feed a constant input $X_1$ into a single-loop encoder with a sample size of $N_1$, and subsequently use the residual error of this first stage as the constant input $X_2$ to another single-loop encoder with sample size $N_2$. Using the output sequences of both encoders, the input $X_1$ can be estimated with greater accuracy than if only the output of the first stage is used. The idea generalizes easily to more than two stages, but we restrict attention to two stages for simplicity. The cascade structure may be viewed as a variation on the MASH encoder proposed by Uchimura *et al.* [77].

We recall that by (2.20), any constant input $X_1$ to a single-loop encoder with sample size $N_1$ can be decomposed as

$$X_1 = -b + P_1\Delta_1 - R_1, \quad \text{where} \quad \Delta_1 \triangleq \frac{2b}{N_1} \tag{2.25}$$

For large sample sizes and a smoothly distributed constant input $X_1$, the residue $R_1$ is uniformly distributed on $[0, \Delta_1)$, so the random variable

$$X_2 \triangleq R_1 N_1 - b \tag{2.26}$$

is uniformly distributed on $(-b, +b)$. It can therefore be used as the constant input to a second single-loop encoder. We refer to $X_2$ as the requantized residue. Denoting the state variable of the single-loop encoder by $U_n$, the requantized residue can be written

$$X_2 = (U_0 - U_{N_1}) - Q(U_0 - U_{N_1}) \tag{2.27}$$

To calculate $X_2$ we therefore need access to $U_{N_1}$, which is immediately available, and $U_0$, which can be stored in a sample-and-hold circuit. Alternatively, $U_0$ can be initialized to zero at the beginning at each conversion cycle; in that case, the residue $X_2$ is simply given by $Q(U_{N_1}) - U_{N_1}$. In either case, the requantized residue is used as the constant input to the second stage over $N_2$ cycles. The set-up can be contrasted with that of the two-stage MASH encoder, in which the input to the second stage is the quantization error of the first stage at *each* time step, that is

$$X_{2,n} = Q(U_n) - U_n$$

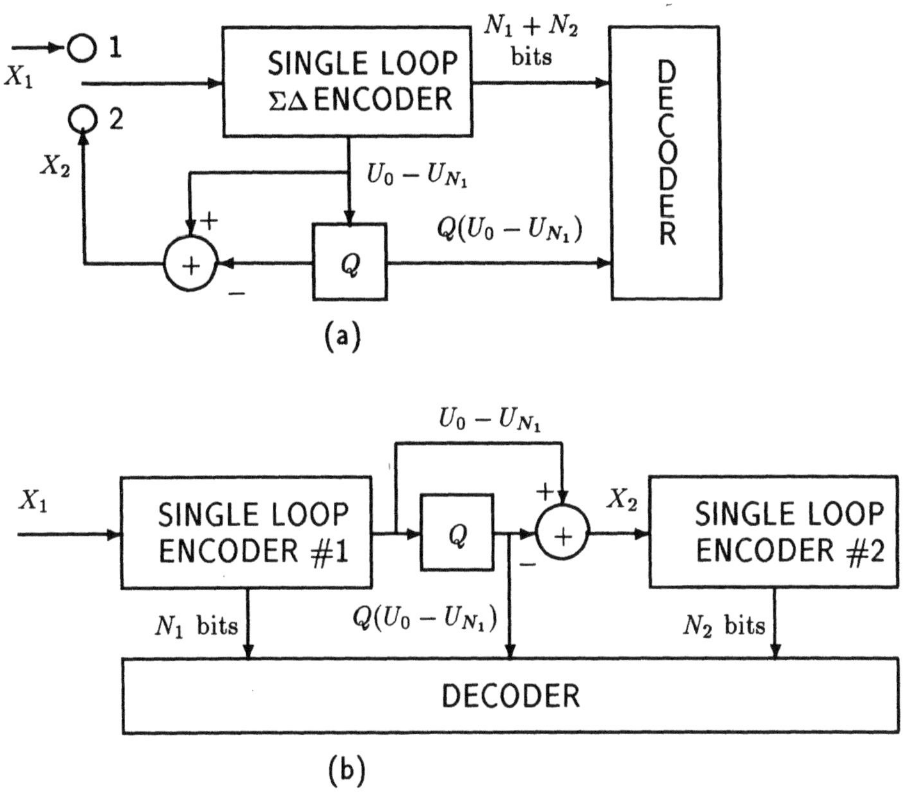


Figure 2.10: *Cascade structure: (a) One single-loop encoder performs the functions of both stages, controlled by a switch; (b) Two single-loop encoders are cascaded, and the first one passes a requantized residue as the constant input to the second one.*

To summarize, the operation of the modified cascade encoder structure is as follows:

1. The constant input $X_1$ to the first stage gives rise to an $N_1$-bit codeword.
2. The requantized residue $X_2$ is found using (2.27) and used as the constant input to the second stage over $N_2$ samples.

Decoding for this encoder is described in the following section. Our proposed cascade structure is shown in figure 2.10. In figure 2.10(a) the same single-loop encoder is used to perform the tasks of both stages, controlled by the position of the switch. In figure 2.10(b) two separate single-loop encoders are used, and the first encoder can start encoding the next input while the second encoder is encoding the requantized version of the first input.

### 2.4.3 Cascade Decoder

In this section we derive a decoding scheme for the cascade structure described in section 2.4.2. By (2.25) we need to determine the integer number of steps $P_1$ and to estimate the residue $R_1$ in order to achieve our goal of estimating the constant input $X_1$. Using equation (2.24), $P_1$ is given by

$$P_1 = N_1 - A_1 + \frac{Q(U_0 - U_{N_1}) - b}{2b} \tag{2.28}$$

where the number of negative bits $A_1$ is easily counted, and $Q(U_0 - U_{N_1})$ is already known because it is used to find the requantized residue $X_2$ in (2.27). Since

$$A_1 = \frac{1}{2}\left(N_1 - \frac{1}{b}\sum_{i=0}^{N_1-1} Q(U_i)\right)$$

equation (2.28) can also be written

$$P_1 = \frac{N_1+1}{2} + \frac{1}{2b}\left(Q(U_0 - U_{N_1}) + \sum_{i=0}^{N_1-1} Q(U_i)\right)$$

This shows that the algorithm for finding $P_1$ is linear in the output bits

$$\{Q(U_0), \cdots, Q(U_{N_1-1})\}$$

and also linear in $Q(U_{N_1})$ if $U_0$ is initialized to zero.

An estimate $\hat{X}_2$ of $X_2$ can be found from the $N_2$-bit codeword of the second stage using any decoder, linear or nonlinear. From (2.25) and (2.26) we then obtain the following estimate $\hat{X}_1$ of the input $X_1$:

$$\hat{X}_1 = P_1\Delta_1 - \frac{\hat{X}_2 + b}{N_1} = \Delta_1\left(P_1 - \frac{N_1+1}{2} - \frac{\hat{X}_2}{2b}\right)$$

### 2.4.4 Performance

In this section we consider the MSE performance of our proposed modified cascade structure shown in figure 2.10(a). From (2.26), the MSE in estimating the constant input $X_1$ is $1/N_1^2$ times the MSE in estimating the residue $R_1$. Any decoder can be used for the second stage, but for specificity we consider the optimal nonlinear decoder with MSE given by (2.14). The overall MSE is then

$$\text{MSE}_{\text{2stage}} \approx \frac{0.91b^2}{N_1^2 N_2^3} = \frac{0.91b^2}{N^5} \tag{2.29}$$

where the last equality follows if $N_1 = N_2 = N$. In this case, the cascade decoder can potentially process a different constant input every $N$ cycles, but it takes $2N$ cycles to obtain an estimate of any given input. For a given number of total samples $N_1 + N_2$, the optimal ratio of $N_1$ to $N_2$ is $2:3$.

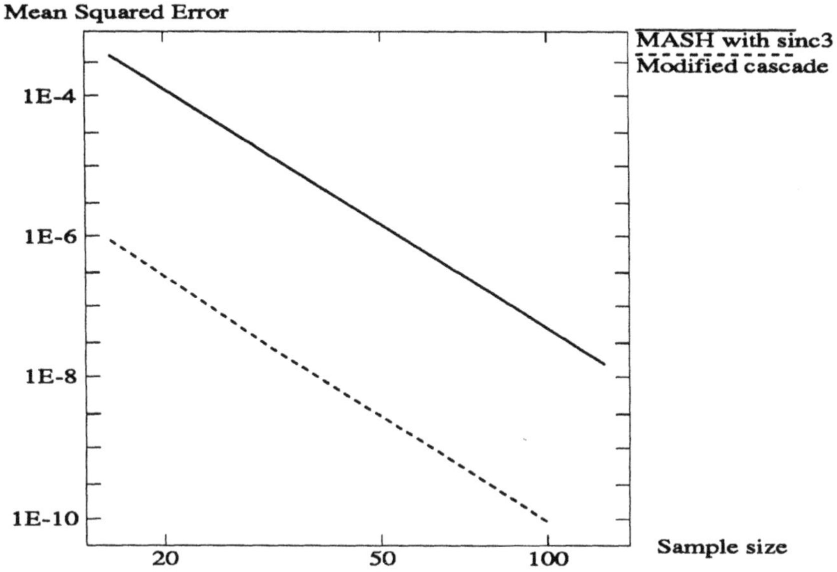


Figure 2.11: *MSE curves for the proposed cascade structure with optimal decoding of second stage, and for the MASH two stage modulator using a* sinc$^3$ *filter as the decoder. The MSE is normalized by* $b^2$.

The decoder for the first stage of our cascade encoder is fixed and described in the previous section; however, we can use any decoder for the second stage. If we use the optimal $N_2$-tap linear filter derived by Gray [42] instead of the optimal nonlinear decoder, we lose about 3 dB compared to (2.29), as shown in section 2.2.2.2.

The MSE performance (2.29) can be compared to that of the two-stage MASH modulator. We consider the case where a two-stage MASH encoder is operated for $N$ cycles, and a decoder must base its estimate on a total of $2N$ output bits, namely $N$ bits from each stage. Such a modulator produces an input estimate every $N$ cycles, and takes $N$ cycles to obtain an input estimate. It thus has the same throughput as our cascade structure, but only half the latency. Figure 2.11 shows the MSE performance of the two cascade modulators[5]: For the MASH modulator we use a standard $N$-tap $sinc^3$ filter operating on a linearly filtered combination of the output sequences of the two stages [47].

[5]The dynamic range of the MASH modulator has been reduced to $(-0.9b, +0.9b)$. The corresponding reduction is not necessary for our cascade structure, since the requantized residue (2.27) used as input to the second stage would still be in the larger range $(-b, +b)$, and only the decoder of the second stage would benefit from a reduction in dynamic range. This is because from (2.26), the MSE in estimating the constant input $X_1$ is independent of the integer $P_1$ defined in (2.25).

Our modified cascade structure performs about 27 dB better than the MASH modulator for DC inputs. If we reduce the total number of samples used in our cascade structure such that it achieves the same latency as the MASH modulator, the modified cascade structure still performs 12 dB better than the MASH modulator, and has twice its throughput.

## 2.5 Summary

In this chapter we analyzed the single-loop and double-loop $\Sigma\Delta$ modulators as quantizers of constant inputs in a set-up where $N$ encoder output bits are available at a time to a decoder. By considering in detail the structure of the encoders, we derived upper bounds of $O(N^2)$ and $O(N^3)$ on the number of codewords for the single-loop and double-loop encoders, respectively. Numerical results indicate that the actual numbers of codewords are also $O(N^2)$ and $O(N^3)$. These results bound the sizes of reproduction alphabets of all decoders operating on single-loop and double-loop encoder outputs under the defined set-up. Other simulations showed that the single-loop encoder is a highly nonuniform quantizer. In fact, the ratio between the largest and smallest quantization interval widths is approximately equal to the sample size. We found that the double-loop encoder is a more uniform quantizer than the single-loop one.

We used the codeword results to derive analytical lower bounds on the MSE of all single-loop and double-loop decoders, linear or nonlinear. The bounds are $O(N^{-3})$ for both the single-loop and double-loop modulators, but if the dynamic range of the double-loop modulator is decreased slightly, our best lower MSE bound is $O(N^{-6})$. We simulated the performance of an optimal nonlinear decoder, based on a table look-up approach in which each possible codeword is mapped to the midpoint of the corresponding quantization interval. Such a decoder, although impractical, represents a bound on the DC performance obtainable with any other decoder. Simulations indicate that the actual MSE performance of the optimal nonlinear decoder is $O(N^{-3})$ for the single-loop modulator, and $O(N^{-5})$ for the double-loop one when the dynamic range is slightly decreased. The same dependence on sample size can be obtained with linear decoders, but there is a constant gain in dB associated with the optimal decoder. The gain is independent of the sample size and is about 3 dB for the single-loop modulator and 20 dB for the double-loop encoder.

We described a cascade structure designed specifically for constant inputs, consisting of two single-loop encoders. The idea is to coarsely estimate the constant input using essentially an averaging decoder, and subsequently to estimate the error of the first stage using another modulator. For a fixed latency, the MSE performance of the cascade structure for constant inputs is about 12 dB better than that of the two-stage MASH modulator, and it achieves twice the throughput rate of the two-stage modulator. For the same throughput, our modified cascade structure shows a 27 dB improvement over the two-stage MASH modulator, but twice the latency.

# Chapter 3

# Optimal Decoding for ONS Encoders with Constant Inputs

In this chapter[1] we propose a class of optimal nonlinear decoding algorithms for data acquisition applications of ONS converters, that is, applications in which the analog input is approximately constant. The decoding algorithms are conceptually simple and are applicable to all current ONS encoders including single-loop and double-loop, cascade and interpolative encoders. The performance of the algorithms is identical to that of the table look-up technique mentioned in chapter 2, but the proposed algorithms are considerably simpler to implement. We present numerical results to compare the performance of the proposed nonlinear decoding algorithms to that of linear decoding. In the absence of circuit non-idealities, the proposed algorithms achieve an increase in Signal-to-Noise Ratio (SNR) over linear decoding of 8 dB for the single-loop encoder, 20-30 dB for the double-loop encoder and 20-30 dB for the two-stage MASH encoder. We examine the effects of circuit imperfections on performance.

## 3.1 Introduction

In this chapter we investigate the application of ONS converters or $\Sigma\Delta$ modulators to data acquisition. The particular set-up we consider for a conversion cycle of the $\Sigma\Delta$ modulator is the same as that of chapter 2, that is: All initial encoder states are set to zero, and the encoder is run for $N$ cycles with constant input. The resulting $N$-bit output sequence is fed to a decoder whose task is to estimate the input. $N$ is referred to as the sample size (SS). We emphasize that the sample size is different from the OverSampling Ratio (OSR) which is

[1] Based on [91], ©1993 IEEE, and [92], ©1990 IEEE.

only defined for dynamic inputs. Therefore results in papers on dynamic inputs [31], [34], [36], [38] are not directly comparable to the results of this chapter.

In chapter 2 we decoupled the $\Sigma\Delta$ modulator into its encoder and decoder parts and investigated the encoder separately. The idea was to view the encoder as a source coder or non-uniform quantizer that divides its dynamic range into quantization intervals separated by transition points, with each interval corresponding to a distinct $N$-bit output sequence. We showed that for fixed initial encoder states, only a small fraction of the $2^N$ possible $N$-bit sequences or codewords can appear as the constant input is swept over the dynamic range. The optimal performance in terms of minimizing the Mean Squared Error (MSE) is achieved by a decoder that takes a codeword as its input and outputs the midpoint of the corresponding interval.[2] Such a decoder is nonlinear, and exploits the specific bit patterns, rather than a frequency domain representation of them, to arrive at optimal estimates of the input.

The optimal decoder could in principle be implemented using a table in the form of a Programmable Logic Array (PLA). In practice this is not feasible as the table would be prohibitively large. In this chapter we present a general and conceptually simple algorithm, dubbed zooming, for optimal decoding [93]. We compare the performance of the zoomer to linear decoding under ideal circumstances in section 3.2, and in the presence of various circuit imperfections in section 3.3. The encoder structures we consider include the single-loop, double-loop, two-stage (MASH) and interpolative encoders.

## 3.2 Optimal Decoding under Ideal Conditions

This section presents detailed algorithms for optimal nonlinear decoding of a number of $\Sigma\Delta$ encoders including the single-loop and double-loop, two-stage MASH and interpolative encoders. All elements of the encoders are assumed to function ideally; non-idealities are considered in section 3.3. The performance measures used to compare decoders are the MSE and the worst-case error, or equivalently, the Signal-to-Noise Ratio (SNR) and the Worst-Case Resolution (WCR) in bits. These measures are defined in appendix B.1.1. A summary of the performance results is shown in table 3.1. Throughout the chapter we make the following assumptions:

- The quantizer $Q$ is 1-bit as given by (1.1). We point out, however, that the presented techniques generalize easily to multi-bit quantization.
- Known initial encoder states. In fact, we assume for convenience that these are all initialized to zero before the encoder is started.
- Constant input. The input $X$ is assumed to be a random variable that takes on constant values in the dynamic range $D \subset B = (-b, +b)$ where

[2] This is only strictly true if the random variable $X$ representing the constant input is uniformly distributed on the dynamic range $D$. It holds in the limit as $N \to \infty$ if $X$ has a smooth probability distribution function.

| | SNR slope (dB/oct) | | WCR slope (bits/oct) | |
|---|---|---|---|---|
| | Zoomer | Linear | Zoomer | Linear |
| Single-loop | 9.0 | 9.0 | 1.0 | 1.0 |
| Double-loop | 17.0 | 14.7 | 2.3 | 1.7 |
| Two-stage | 18.0 | 14.7 | 2.2 | 2.2 |

| | SNR (dB), $N = 128$ | | WCR (bits), $N = 128$ | |
|---|---|---|---|---|
| | Zoomer | Linear | Zoomer | Linear |
| Single-loop | 62.5 | 54.5 | 7.8 | 6.8 |
| Double-loop | 92.0 | 67.0 | 13.3 | 8.8 |
| Two-stage | 101.0 | 72.5 | 13.4 | 10.7 |

Table 3.1: *Comparison of the zoomer and linear decoding for the ideal single-loop, double-loop and two-stage modulators.*

$b$ is the quantizer step size. The assumption of constant input is made because we are focusing on data acquisition applications in which inputs can be assumed more or less constant. In practice, the full dynamic range $B$ is seldom used. One reason is to avoid the possibility of exceeding the dynamic range; another reason is that the largest estimation errors are generally made when the input is close to $\pm b$ [86] as shown also in chapter 2. We therefore restrict the dynamic range to $D = (-Kb, +Kb)$, where $K$ is chosen throughout to be 0.9.[3]

### 3.2.1 Single-Loop Modulator

The single-loop encoder is the simplest $\Sigma\Delta$ encoder. For convenience, its discrete-time model is repeated in figure 3.1. Section 3.2.1.1 presents an optimal decoding algorithm under the assumptions stated above, and section 3.2.1.2 presents numerical results.

#### 3.2.1.1 Zoomer Algorithm

In this section we present an optimal decoding algorithm under the assumptions of constant input $X$ and known initial integrator state $U_0$. For reasons detailed below, we refer to this as the zoomer algorithm. The basic idea is to derive a series of bounds on the constant input. The system shown in figure 3.1 satisfies

[3] Setting $K$ to 1 will decrease the average performance of both linear and nonlinear decoders, but the optimal nonlinear decoder will still be superior to the optimal linear decoder.

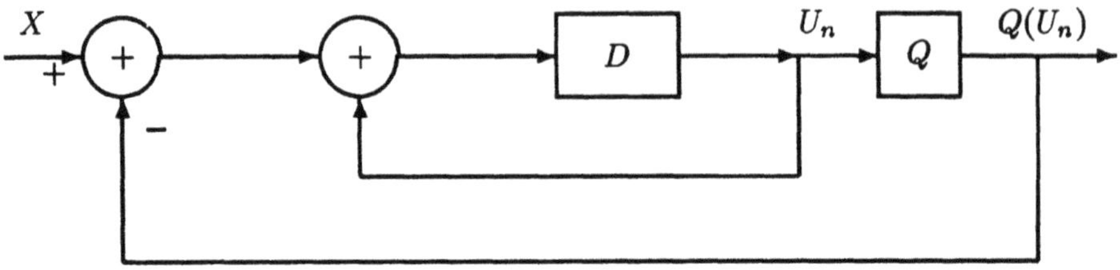


Figure 3.1: *Discrete-time model of the single-loop ΣΔ encoder.*

the following difference equation in terms of the state variable $U_n$:

$$U_n = U_{n-1} + X_{n-1} - Q(U_{n-1}), \quad n \geq 1 \tag{3.1}$$

Assuming that the initial state is $U_0 = 0$, the state at time $n$ is given by

$$U_n = \sum_{i=0}^{n-1} [X_i - Q(U_i)] = \left(\sum_{i=0}^{n-1} X_i\right) - S_n, \quad n \geq 1 \tag{3.2}$$

where $S_n$ is the running sum of output bits given by

$$S_n = \sum_{i=0}^{n-1} Q(U_i), \quad n \geq 1 \tag{3.3}$$

Assuming constant input, $X_i = X$ for $i \geq 0$, the first sum in (3.2) equals $nX$, and for any given codeword, $S_n$ can be found by summation of the known output sequence. With the definition $S_0 \triangleq 0$, we have the recursive relationship

$$S_n = S_{n-1} + Q(U_{n-1}), \quad n \geq 1$$

The only information available to the decoder is the $N$-bit encoder output sequence, $\{Q(U_n),\ 0 \leq n \leq N-1\}$, or equivalently, the signs of the quantizer input sequence. Taking (3.2) into account, this information determines whether the sum of inputs is greater or less than the sum of outputs at each time $n$. For each $n$ we can thus derive a bound on the input:

$$X > \bar{X}_n \text{ if } Q(U_n) = +b; \qquad X \leq \bar{X}_n \text{ if } Q(U_n) = -b \tag{3.4}$$

where $\bar{X}_n$ is the running average given by

$$\bar{X}_n = \frac{1}{n} S_n \tag{3.5}$$

The first two bits of any codeword are uninformative, since they are always $Q(U_0) = -b$ and $Q(U_1) = +b$. The reason for this is that $U_0 = 0$ and from (3.2), $U_1 = X_0 - Q(U_0) = X + b > 0$. But for each $2 \leq n \leq N-1$, equation (3.4) gives a lower or upper bound on $X$ for the known quantity $Q(U_n) = +b$ or $-b$, respectively; thus $Q(U_n)$ determines the type of the bound.

Because there are only a finite number of codewords for inputs in the dynamic range, each codeword can be generated by a specific range of inputs. The zoomer is the decoder that uses the succession of lower and upper bounds from (3.4) to arrive at the sharpest possible lower and upper bounds on the input resulting in a given codeword. Figure 3.2 shows a flowchart of the zoomer algorithm [93], consisting of an initialization phase and a loop with an update of running sums and an update of either the lower or the upper bound. The algorithm uses lower and upper bound registers $L$ and $U$ initialized to the endpoints of the dynamic range. Sweeping $n$ from 2 to $N-1$, the zoomer maintains the greatest lower bound and the least upper bound in the registers. At each time step, the new bound is compared with previous bounds, and the appropriate bound register updated accordingly. The algorithm extracts all information from the codeword, and thus results in an optimal decoding procedure. After processing the $N$ encoder output bits, the decoder outputs $(L+U)/2$ as its estimate of the input. From the above, any codeword will result in compatible bounds, that is, $L \leq U$. Conversely, all non-codewords will result in incompatible bounds, as shown in appendix B.2. The zoomer is mostly linear, but the conditional register updating is nonlinear.

The zoomer approach is reminiscent of successive approximation, but unlike that type of conversion, a new codeword bit is far from certain to produce new information. In fact, the number of codewords with $n$ bits is close to $n^2/2$ as shown in chapter 2, and so at time $n$, there are only approximately $[(n^2-(n-1)^2]/2 \approx n$ more codewords than at time $n-1$. This means that for most of the $n^2/2$ codewords, the last bit is uniquely determined by the previous bits, and for these codewords the last bit carries no information at all.

The calculations involved in the algorithm are quite simple, the division in (3.5) being the most time-consuming. However, it is not necessary to actually do the divisions until the end; for example, to decide whether or not $X_n > X_m$, we only need to check whether

$$mS_n > nS_m \tag{3.6}$$

This reduces to integer multiplications rather than floating-point divisions. Instead of storing the best lower and upper $\bar{X}_n$ in the bound registers $L$ and $U$, we then need to store the best pairs $(n, S_n)$.

Along the same lines, consider the case where $p$ negative bits are interspersed between two positive bits,

$$\{Q(U_n), Q(U_{n+1}), \cdots, Q(U_{n+p+1})\} = \{+b, -b, -b, \cdots, -b, +b\} \tag{3.7}$$

We wish to compare the lower bounds on $X$ resulting at time $n$ and $n+p+1$. It can be shown that the difference between the running averages at these times

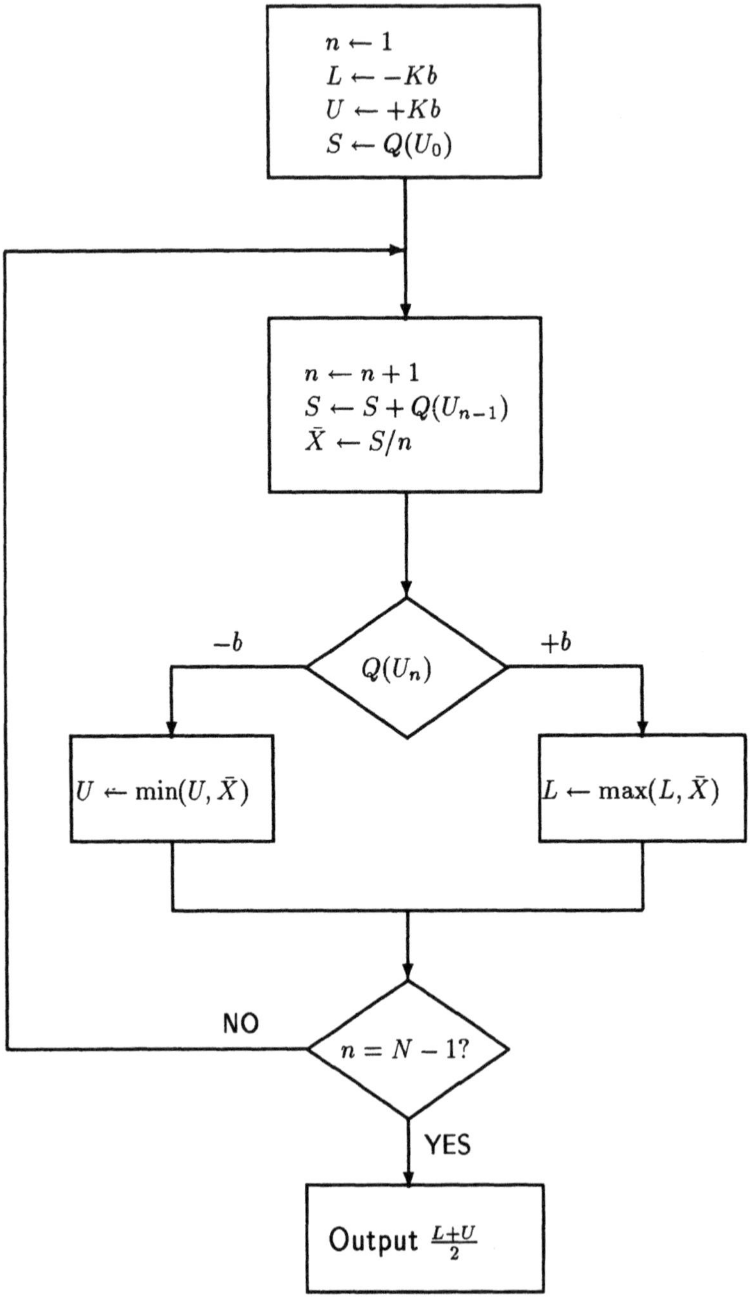


Figure 3.2: *Flowchart for the single-loop zoomer algorithm.*

is given by

$$\bar{X}_{n+p+1} - \bar{X}_n = \frac{b+\bar{X}_n}{n+p+1}\left[\frac{b-\bar{X}_n}{b+\bar{X}_n} - p\right]$$

Since the factor in front of the square brackets is positive, the new lower bound at time $n+p+1$ is at least as sharp as the old bound at time $n$ if and only if the term in the square bracket is also positive, that is

$$p \leq \frac{b-\bar{X}_n}{b+\bar{X}_n}$$

or equivalently

$$\bar{X}_n \leq -\frac{p-1}{p+1}\,b \tag{3.8}$$

or equivalently

$$(p+1)S_n \leq -n(p-1)b \tag{3.9}$$

These expressions can be used in two different ways. First, note that for $p=0$ the inequality for $\bar{X}_n$ in (3.8) is always satisfied. This means that if two adjacent bits are both positive, only the bound corresponding to the second bit needs to be calculated, since the bound due to the first bit is guaranteed to be inferior to that of the second one. The decoder can easily check for this by looking one bit ahead in the codeword.

We can also use (3.9) to compare two lower bounds for more general $p$. As a replacement for (3.6), this is only useful if the bound $\bar{X}_n$ at time $n$ is the best lower bound at time $n$. If so, the bound at time $n+p+1$ is at least as sharp at that at time $n$ if and only if (3.9) holds. Similar derivations can be used to compare two upper bounds for the case where $p$ positive bits are interspersed between two negative bits.

To assess the value of using (3.9) rather than (3.6), the work involved must be compared. The comparison in (3.9) calls for integer multiplications to replace (3.6) which also contains integer multiplications. However, the integer factors in (3.9) are simpler, especially for small $p$. For instance, all that needs to be checked for $p=1$ is the sign of $S_n$. For $p=2$, the factor $(p-1)$ reduces to 1, and for $p=3$, both $(p-1)$ and $(p+1)$ are powers of 2. Since small values of $p$ are more likely to occur, this could potentially save computation.

#### 3.2.1.2 Numerical Results

We now present numerical results on the performance of the zoomer and compare it to the asymptotically optimal linear $N$-tap Finite Impulse Response (FIR) decoder for constant inputs. This filter was derived by Gray [42] and has tap coefficients

$$h_n = 6\frac{(n+1)(N-n)}{N(N+1)(N+2)}, \qquad 0 \leq n \leq N-1 \tag{3.10}$$

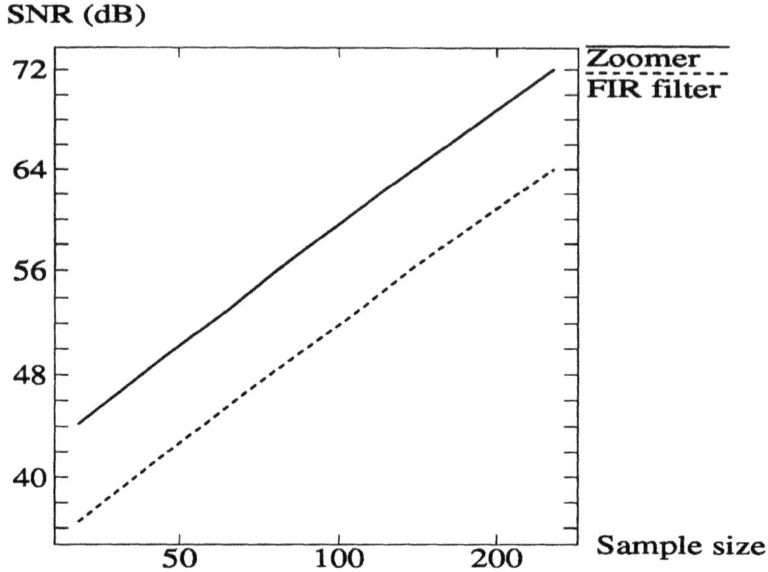


Figure 3.3: *Single-loop encoder: SNR as a function of sample size for the zoomer and the asymptotically optimal FIR filter.*

Figures 3.3 and 3.4 shows that for a given sample size, the zoomer gains about 8 dB or $1\frac{1}{2}$ bits of SNR over the FIR filter,[4] and increases the worst-case resolution by 1 bit. Alternatively, the zoomer requires half the sample size of the FIR filter to obtain essentially the same performance, resulting in shorter data acquisition times. Both SNR curves have a slope of 9 dB/octave, and both WCR curves have a slope of 1 bit/octave.

A direct comparison of these results to those obtainable with other types of A/D converters is difficult, but for purposes of illustration we consider the successive approximation (SA) converter described in section 1.2. We assume that the required number of clock cycles is comparable to the sample size of a $\Sigma\Delta$ modulator. An SA converter using $P$ clock cycles for full-scale inputs has the effect of dividing the dynamic range into $P$ intervals of width $2b/P$ each, leading to an MSE of $b^2/(3P^2)$ and an SNR of $20\log_{10}(P/2)$. To match the SNR of the single-loop zoomer at a sample size of 128, the SA converter thus requires approximately 1500 cycles.

### 3.2.2 Double-Loop Modulator

The double-loop encoder is a generalization of the single-loop encoder which has a more favorable trade-off between resolution and sample size [31]. Its

[4] This is not inconsistent with the gain of 3 dB reported in chapter 2, because the 3 dB gain is based on a dynamic range of $(-b, +b)$, not $(-Kb, +Kb)$ with $K = 0.9$.

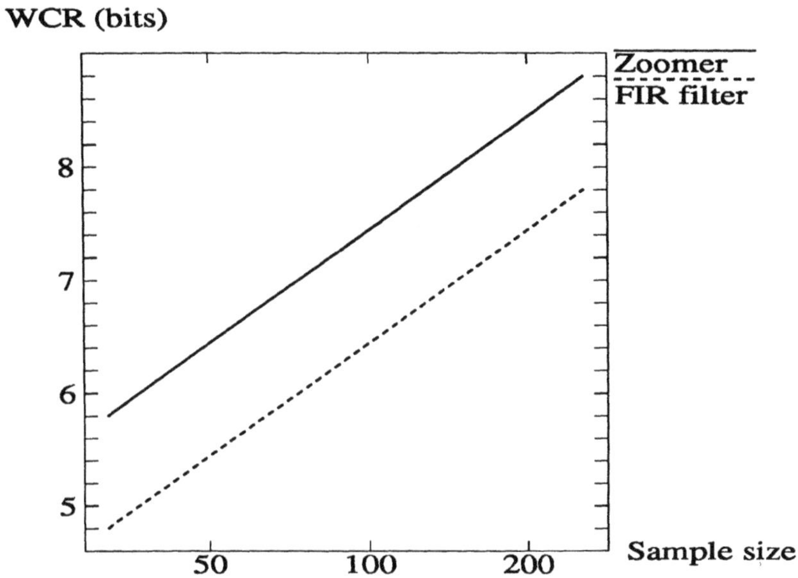


Figure 3.4: *Single-loop encoder: Worst-case resolution as a function of sample size for the zoomer and the asymptotically optimal FIR filter.*

discrete-time model is repeated in figure 3.5 for convenience. The encoder is important in its own right and also as a building block in cascaded modulators. Section 3.2.2.1 presents an optimal decoding scheme under the assumptions stated in the beginning of section 3.2, and section 3.2.2.2 presents simulation results.

#### 3.2.2.1 Zoomer Algorithm

The analysis of the double-loop encoder proceeds in a fashion similar to that of the single-loop encoder. The governing difference equations for the system shown in figure 3.5 are

$$\begin{aligned} U_n &= U_{n-1} + V_{n-1} - Q(U_{n-1}) \\ V_n &= V_{n-1} + X_n - Q(U_n), \qquad n \geq 1 \end{aligned}$$

where $U_n$ and $V_n$ are the two integrator state variables. Assuming zero initial states, $U_0 = V_0 = 0$, and constant input, $X_n = X$, the equation for $V_n$ implies

$$\sum_{i=0}^{n-1} V_i = \frac{1}{2}n(n-1)X - \sum_{i=1}^{n-1}(n-i)Q(U_i)$$

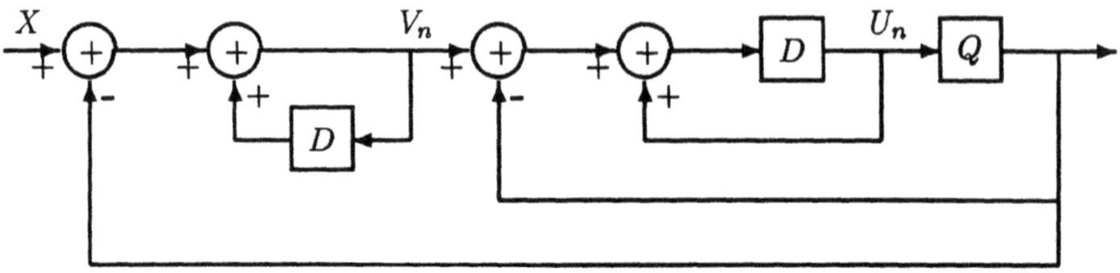


Figure 3.5: *Discrete-time model of the double-loop ΣΔ encoder.*

and the equation for $U_n$ implies

$$\begin{aligned} U_n &= \sum_{i=0}^{n-1} V_i - \sum_{i=0}^{n-1} Q(U_i) \\ &= \frac{1}{2}n(n-1)X - Q(U_0) - \sum_{i=1}^{n-1}(n-i+1)Q(U_i), \quad n \geq 2 \end{aligned} \tag{3.11}$$

To facilitate the calculation of the sum involving quantizer outputs, we define the running sums

$$S_n = \sum_{i=1}^{n-1} Q(U_i), \quad n \geq 2 \tag{3.12}$$

$$W_n = \sum_{i=1}^{n-1}(n-i+1)Q(U_i), \quad n \geq 2 \tag{3.13}$$

Defining $S_1 = W_1 \triangleq 0$, we then have the recursions

$$S_n = S_{n-1} + Q(U_{n-1}), \quad n \geq 2 \tag{3.14}$$

$$W_n = W_{n-1} + S_n + Q(U_{n-1}), \quad n \geq 2 \tag{3.15}$$

The quantity $W_n$ can thus be found as a weighted summation of the output sequence, as shown in (3.13).

The information available to the decoder is $\{Q(U_n), 0 \leq n \leq N-1\}$. From the difference equations we have $U_1 = +b > 0$, $V_1 = X - b$ and $U_2 = X - b < 0$, so the first three output bits are always $Q(U_0) = -b$, $Q(U_1) = +b$ and $Q(U_2) = -b$ regardless of the input, and the first informative bit is $Q(U_3)$. As in the single-loop case, we can use (3.11) to obtain a bound on the input at each time $3 \leq n \leq N-1$: We obtain a lower or an upper bound depending on whether $Q(U_n) = +b$ or $-b$. Specifically,

$$X > \bar{X}_n \ \text{ if } \ Q(U_n) = +b; \qquad X \leq \bar{X}_n \ \text{ if } \ Q(U_n) = -b$$

where $\bar{X}_n$ is given by

$$\bar{X}_n = \frac{Q(U_0) + W_n}{\frac{1}{2}n(n-1)}, \quad n \geq 3 \tag{3.16}$$

Recall that each codeword is generated by a specific range of input values. Analogous to the single-loop zoomer, the double-loop zoomer is the decoder that uses the output sequence to derive a succession of lower and upper bounds on the input. The sharpest of these bounds are the best possible bounds on the input resulting in the output sequence. These sharpest bounds are obtained by using the lower and upper bound registers $L$ and $U$, initialized to the endpoints of the dynamic range. Sweeping $n$ from 3 to $N-1$, the zoomer maintains the greatest lower bound and the least upper bound in the registers. To be specific, figure 3.6 show a flowchart of the double-loop zoomer algorithm. The variables $S$ and $W$ correspond to the quantities given by (3.14) and (3.15), respectively, and $p$ is the denominator in the bound fraction (3.16).

During the conversion cycle, the floating-point division in (3.16) can be replaced by integer multiplication, since comparisons of the form $\bar{X}_n > \bar{X}_m$ can be written in the simpler form

$$m(m-1)[Q(U_0) + W_n] \; > \; n(n-1)[Q(U_0) + W_m]$$

As in the single-loop case, there exist alternative formulas to decide whether or not a new bound is better than an old bound in the special case described by (3.7). However, the formulas are more involved for the double-loop than for the single-loop zoomer. Even when two successive bits are identical, there is no guarantee that the second bound is tighter than the first one.

#### 3.2.2.2 Numerical Results

This section compares the performance of the double-loop zoomer to that of linear decoding. There is no parallel in the literature to the asymptotically optimal FIR filter (3.10) for single-loop modulation. The linear decoder under consideration here is therefore chosen to be the $N$-tap filter with a $sinc^3$ transfer function which is believed to be close to optimal [38].

Figures 3.7 and 3.8 show that for a given sample size, the zoomer is superior to the $sinc^3$ filter in performance, in that the SNR gain is 20-30 dB and the WCR gain is 3-5 bits. The SNR and WCR achieved by the $sinc^3$ filter at a sample size of 256 are reached by the zoomer at sample sizes of approximately 100 and 64, respectively. This translates into shorter data acquisition times. For the zoomer, the slopes of the SNR and WCR curves are about 17 dB/octave and 2.3 bits/octave, respectively. For the linear filter, the slopes are 14.7 dB/octave and 1.7 bits/octave. It is thus seen that ideally, the zoomer achieves a better trade-off with sample size than the linear filter, and the gap between the curves widens as the sample size increases.

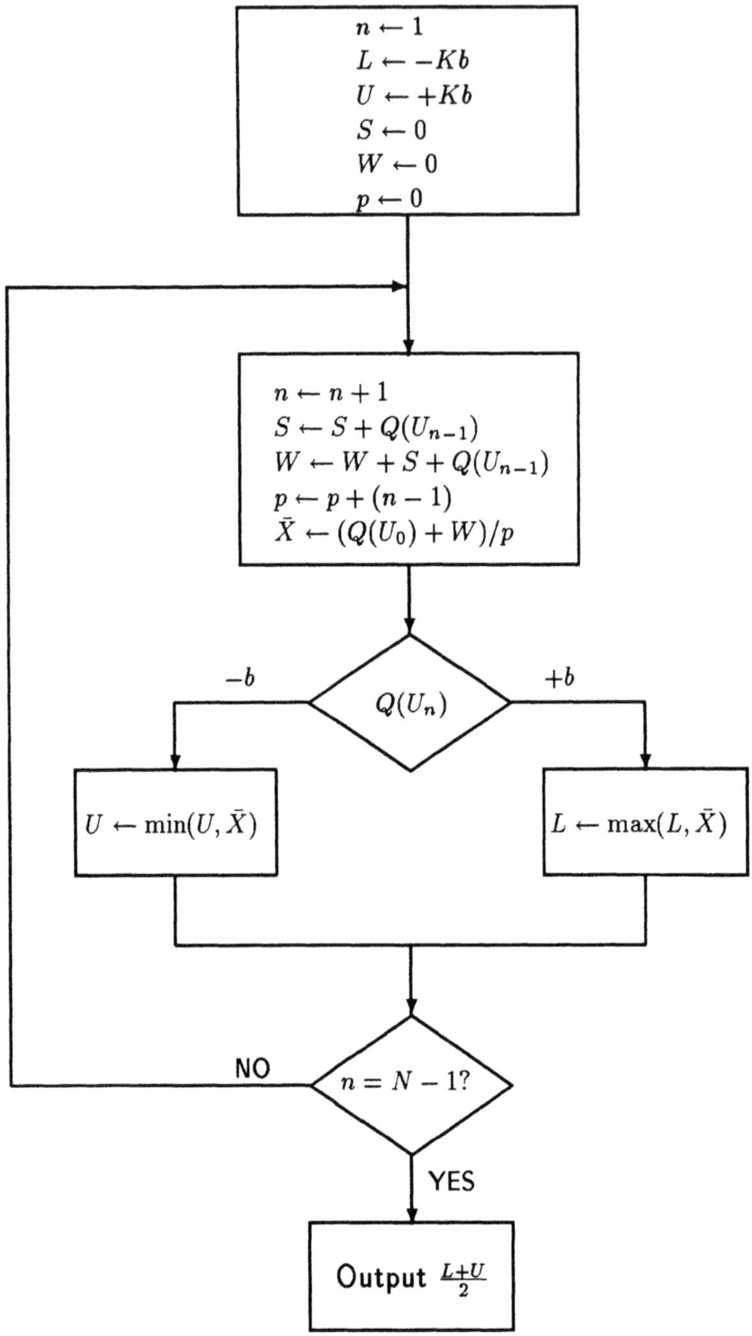


Figure 3.6: *Flowchart for the double-loop zoomer algorithm.*

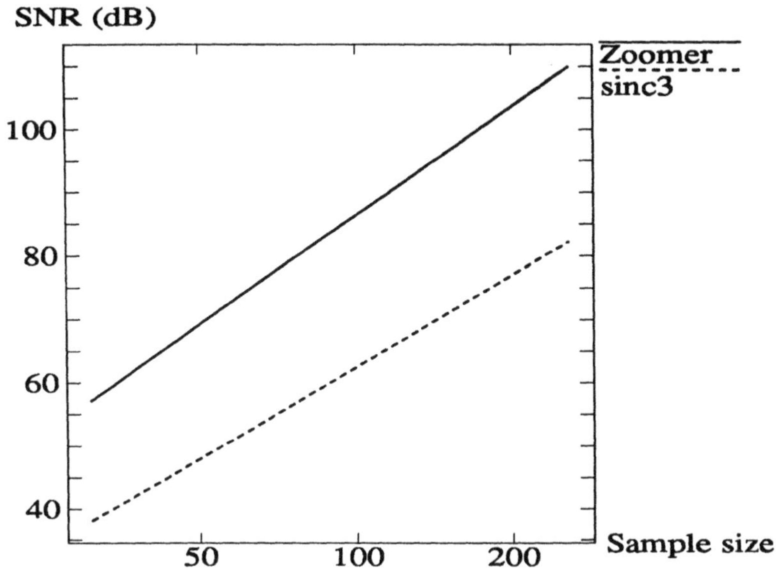


Figure 3.7: *Double-loop encoder: SNR as a function of sample size for the zoomer and the* $\text{sinc}^3$ *filter.*

### 3.2.3 Two-Stage Modulator

Figure 3.9 shows the discrete-time model of the two-stage MASH encoder. The MASH architecture was originally proposed by Uchimura *et al.* [77] and has been extensively analyzed by Wong, Chou and Gray in several papers, including [43] and [47]. The encoder consists of two single-loop stages of which the first is fed with the input and the second is fed with the quantization error sequence of the first stage.

The original papers on this cascade structure also include a noise cancelling circuit that performs noise shaping and combines the two binary output streams into one quaternary sequence [47], [77]. This circuit has the effect of eliminating the direct appearance of the first stage quantization error in the output sequence. It should be noted that although this elimination is a desirable characteristic, the circuit will in general be discarding information present in the separate stage outputs. In addition, for data acquisition applications, the noise shaping characteristic is of little importance since the input is more or less constant. Here we will adopt the viewpoint that the noise cancelling circuit is part of a decoder, and the decoder should not be limited to operating on the sequence obtained by irreversibly combining the two output sequences into one. We will therefore use $\{Q(U_i)\}$ and $\{Q(V_i)\}$ directly for decoding. Section 3.2.3.1 presents an optimal decoding scheme under the assumptions stated in

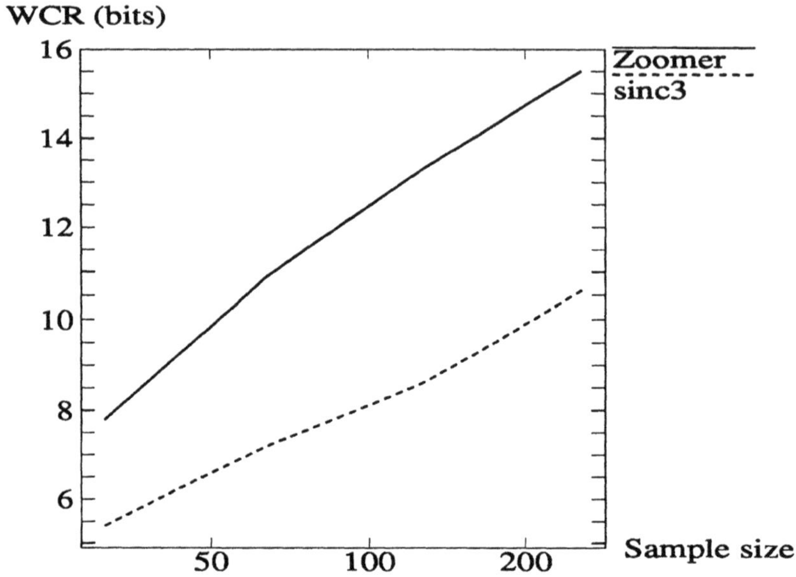


Figure 3.8: *Double-loop encoder: Worst-case resolution as a function of sample size for the zoomer and the* sinc$^3$ *filter.*

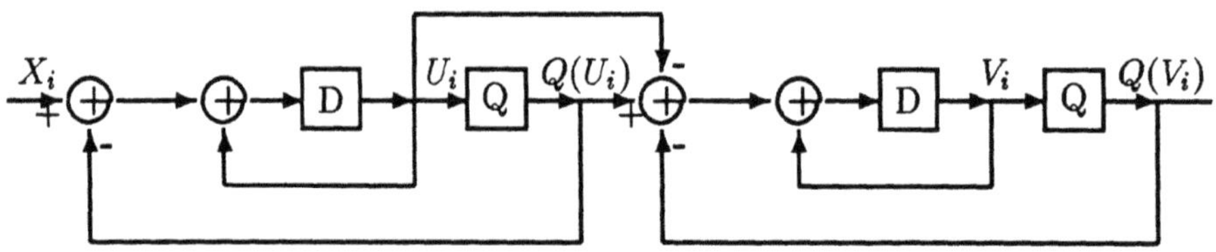


Figure 3.9: *Discrete-time model of the two-stage* $\Sigma\Delta$ *encoder.*

section 3.1, and section 3.2.3.2 presents simulation results.

#### 3.2.3.1 Zoomer Algorithm

The difference equations governing the state variables $U_n$ and $V_n$ of the two-stage encoder shown in figure 3.9 are

$$\begin{aligned} U_n &= U_{n-1} + X_{n-1} - Q(U_{n-1}) \\ V_n &= V_{n-1} - U_{n-1} + Q(U_{n-1}) - Q(V_{n-1}), \quad n \geq 1 \end{aligned}$$

Assuming that the initial states are $U_0 = V_0 = 0$, these can be solved to yield

$$U_n = \sum_{i=0}^{n-1} X_i - \sum_{i=0}^{n-1} Q(U_i), \quad n \geq 1 \tag{3.17}$$

$$V_n = -\sum_{i=0}^{n-2}(n-1-i)X_i + \sum_{i=0}^{n-1}(n-i)Q(U_i) - \sum_{i=0}^{n-1} Q(V_i), \quad n \geq 2 \qquad (3.18)$$

Let us define the running sums

$$S_n = \sum_{i=0}^{n-1} Q(U_i), \quad n \geq 1 \qquad (3.19)$$

$$T_n = \sum_{i=0}^{n-1} Q(V_i), \quad n \geq 1 \qquad (3.20)$$

$$W_n = \sum_{i=0}^{n-1}(n-i)Q(U_i), \quad n \geq 1 \qquad (3.21)$$

Defining $S_0 = T_0 = W_0 \triangleq 0$ we then have the recursions

$$\begin{aligned} S_n &= S_{n-1} + Q(U_{n-1}), \quad n \geq 1 \\ T_n &= T_{n-1} + Q(V_{n-1}), \quad n \geq 1 \\ W_n &= W_{n-1} + S_n, \quad n \geq 1 \end{aligned}$$

As before we assume that the input is constant, $X_i = X$ for $i \geq 0$. At time $n$, (3.17) and (3.18) each provide potential new bounds on this input: The new bound is an upper or a lower bound depending on whether $Q(U_n) = -b$ or $+b$. Specifically, (3.17) results in

$$X > \bar{X}_n^{(1)} \text{ if } Q(U_n) = +b; \qquad X \leq \bar{X}_n^{(1)} \text{ if } Q(U_n) = -b; \quad n \geq 1 \qquad (3.22)$$

where $\bar{X}_n^{(1)}$ is the running average

$$\bar{X}_n^{(1)} = \frac{1}{n} S_n \qquad (3.23)$$

Equation (3.18) results in an upper or lower bound on the input, depending on whether $Q(V_n)$ is positive or negative. Specifically,

$$X < \bar{X}_n^{(2)} \text{ if } Q(V_n) = +b; \qquad X \geq \bar{X}_n^{(2)} \text{ if } Q(V_n) = -b; \quad n \geq 2 \qquad (3.24)$$

where

$$\bar{X}_n^{(2)} = \frac{W_n - T_n}{\frac{1}{2}n(n-1)} \qquad (3.25)$$

Figure 3.10 shows a flowchart for the two-stage zoomer algorithm; the two-stage zoomer uses the succession of lower and upper bounds from both (3.22) and (3.24) to arrive at overall lower and upper bounds on the input resulting in a specific codeword. This is achieved by using lower and upper bound registers $L$ and $U$ initialized to the endpoints of the dynamic range. Sweeping $n$ from 3

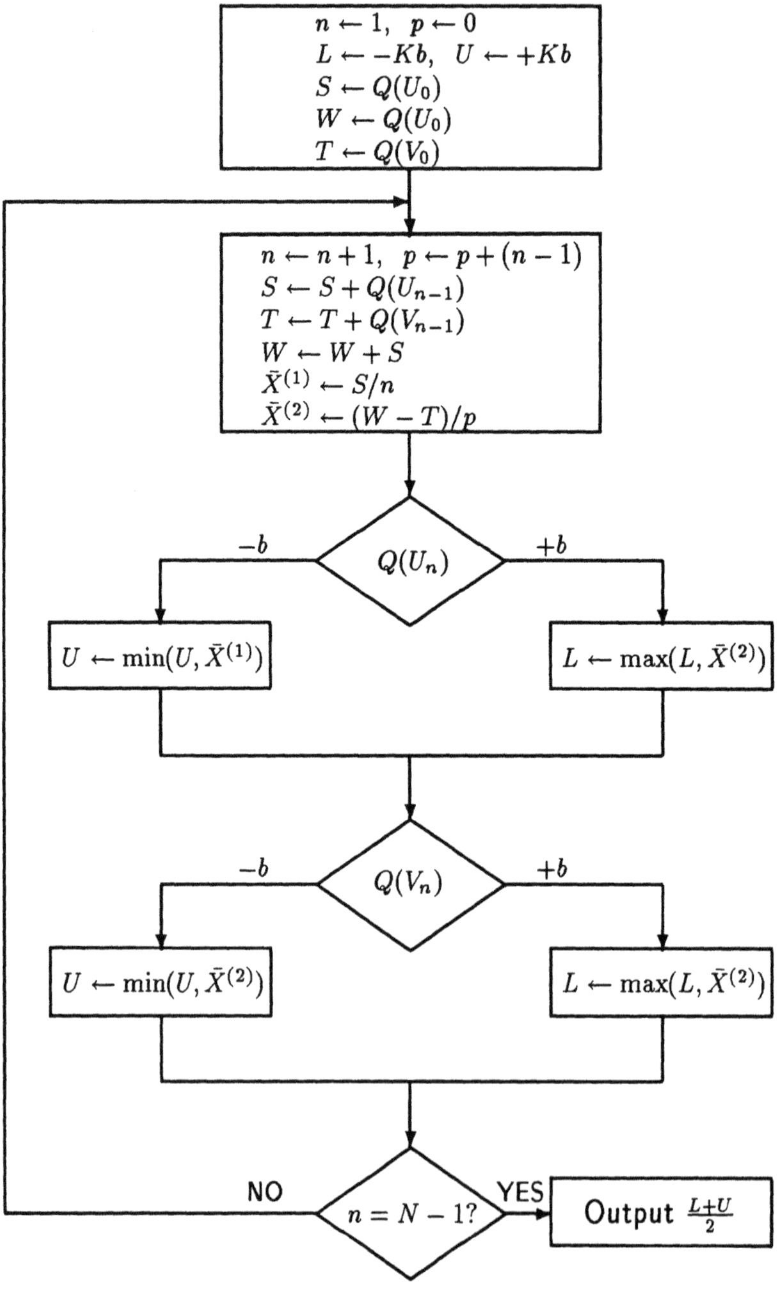


Figure 3.10: *Flowchart for the two-stage zoomer algorithm.*

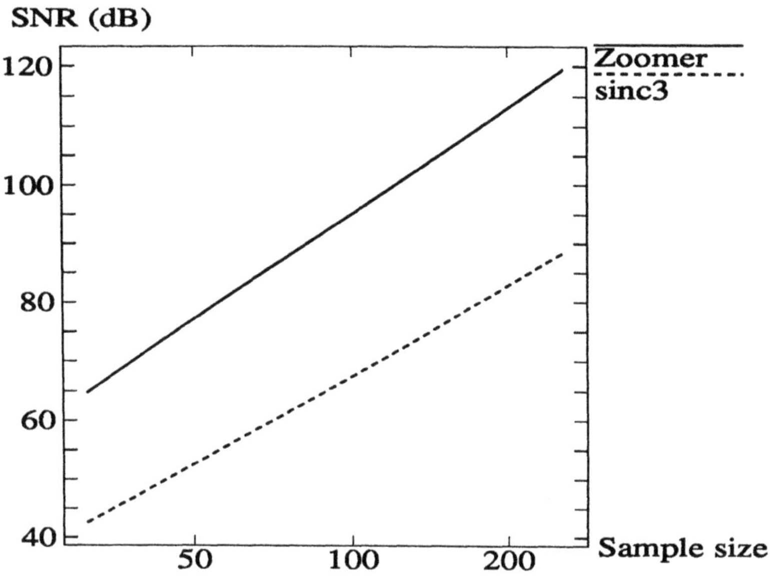


Figure 3.11: *Two-stage encoder: SNR as a function of sample size for the zoomer and the* sinc$^3$ *filter.*

to $N-1$, the zoomer maintains the greatest lower bound and the least upper bound in the registers. Variables $S$, $T$ and $W$ in figure 3.10 hold the sums in (3.19), (3.20) and (3.21), respectively, and $p$ is the denominator of (3.25), while $\bar{X}^{(1)}$ and $\bar{X}^{(2)}$ correspond to the quantities (3.23) and (3.25). At each time step, the flowchart contains two conditional updates of the bound registers, corresponding to equations (3.23) and (3.25).

#### 3.2.3.2 Numerical Results

This section compares the performance of the two-stage zoomer to an $N$-tap filter with a *sinc*$^3$ transfer function. It is shown in [47] that this filter achieves an MSE of $O(N^{-5})$, and that no *sinc*$^k$ filter can achieve a better asymptotic dependence on $N$.

Figures 3.11 and 3.12 show SNR and WCR curves as functions of the sample size. We see that the zoomer outperforms the *sinc*$^3$ filter by 20-30 dB of SNR and 2.5 bits of WCR. For the depicted range of sample sizes, these gains translate into a reduction by a factor of 2–3 in data acquisition times to achieve a given performance. For the zoomer, the slopes of the SNR and WCR curves are about 18 dB/octave and 2.2 bits/octave, respectively. For the linear filter, the slopes are 14.7 dB/octave and 2.2 bits/octave. We thus see that ideally, the zoomer achieves a more favorable trade-off with sample size than the linear filter, and the gaps between the curves widen as the sample size increases: The

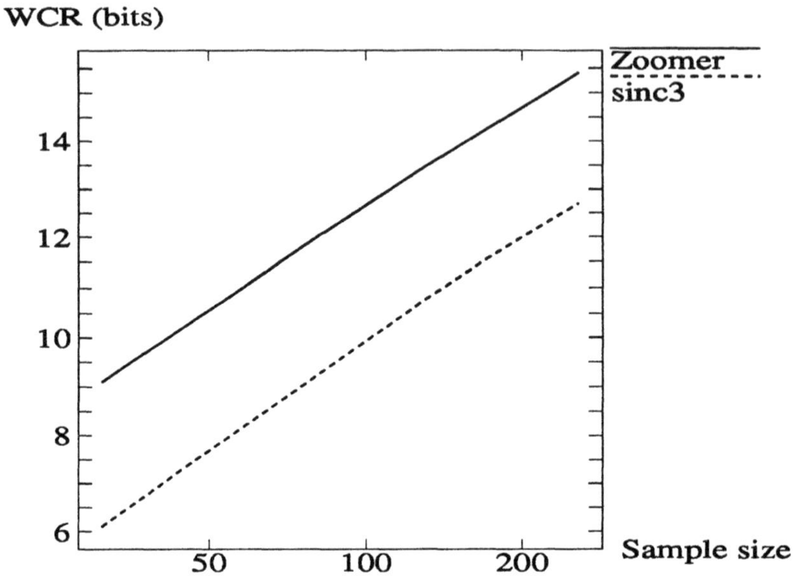


Figure 3.12: *Two-stage encoder: Worst-case resolution as a function of sample size for the zoomer and the* sinc$^3$ *filter.*

MSE goes as $O(N^{-6})$ and $O(N^{-5})$, respectively. Compared to the double-loop results in figure 3.7, the slope difference for the two-stage encoder is about 1 dB/octave greater for the SNR curves, while the slope difference for the WCR curves is about 0.6 bit/octave inferior.

### 3.2.4 Interpolative Modulators

The general interpolative encoder structure is shown in figure 3.13 [32]. It is characterized by the low-pass transfer function $H(z)$. This section demonstrates the applicability of the zoomer concept to interpolative encoders.

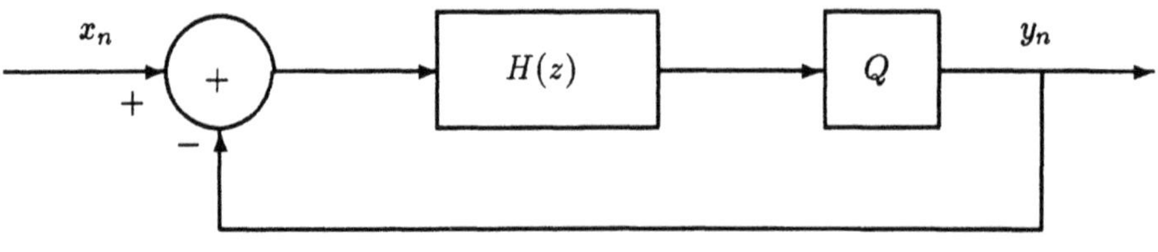


Figure 3.13: *Discrete-time model of the interpolative* $\Sigma\Delta$ *encoder.*

Let us denote the impulse response of $H(z)$ by $\{h_0, h_1, h_2, \cdots\}$. To avoid race-around we must have $h_0 = 0$. Under the constant input assumption, the state variable $U$ at time $n$ is given by

$$U_n = \sum_{i=0}^{n} h_{n-i}\left[X_i - Q(U_i)\right] = X\left(\sum_{i=0}^{n-1} h_{n-i}\right) - \sum_{i=0}^{n-1} h_{n-i}Q(U_i), \quad n \geq 0$$

The zoomer algorithm uses the output bits to derive a succession of upper and lower bounds on the input. The bound at time $n$ is a lower or an upper bound for $Q(U_n) = +b$ and $-b$, respectively. Specifically,

$$X > \bar{X}_n \text{ if } Q(U_n) = +b; \qquad X \leq \bar{X}_n \text{ if } Q(U_n) = -b$$

where $\bar{X}_n$ is given by

$$\bar{X}_n = \frac{\sum_{i=0}^{n-1} h_{n-i}Q(U_i)}{\sum_{i=1}^{n} h_n}$$

As an example, the single-loop encoder can be viewed as an interpolative encoder with $H(z) = z^{-1}/(1 - z^{-1})$. Therefore $\{h_0, h_1, h_2, \cdots\} = \{0, 1, 1, \cdots\}$ and

$$\bar{X}_n = \frac{1}{n}\sum_{i=0}^{n-1} Q(U_i)$$

in agreement with (3.5). Note that neither the double-loop nor the two-stage encoder are interpolative encoders.

The above technique readily generalizes to the case where the transfer function from input to quantizer input is different from the transfer function from quantizer output to quantizer input. This would be the case if there is also a filter in the feedback path from quantizer input to input summing node; see for example [50].

## 3.3 Decoding in the Presence of Non-Idealities

In this section we derive modifications to the zoomer algorithms that compensate for a number of circuit non-idealities, provided the numerical values of these non-idealities can be considered known in advance. For the case of unknown non-idealities, we also present simulation results that illustrate the effects of non-idealities on the performance of the zoomer algorithms. To realistically assess the performance of the zoomers, we consider a variety of circuit imperfections. As our results are based on simulation, they must be considered tentative; practical circuit implementations are needed to verify them experimentally. Sections 3.3.1, 3.3.2 and 3.3.3 address the single-loop, double-loop and two-stage modulators, respectively.

| | Non-ideality to cause 2 dB SNR loss at $N = 128$ | | |
|---|---|---|---|
| | Single loop | Double loop | Two stage |
| Gain | N/A | 0.1% | 0.02% |
| Leak | 2,500 | 50,000 | 100,000 |
| Initial state | 1.2% | 0.05% | 0.01% |
| DC offset | 1.2% | 0.5% | 0.01% |
| Noise | 0.5% | 0.02% | 0.005% |
| Input var. | 0.2% | 0.01% | 0.005% |

Table 3.2: *Sensitivity of the zoomer to various non-idealities. Quantities given in percent are measured relative to the quantizer step size $b$. Quantities given for the leak are the minimum acceptable op-amp gains.*

For each encoder, the following types of non-idealities have been investigated: Non-unity integrator gain, integrator leak, non-zero initial integrator state, DC quantizer offset, random noise and linearly time-varying input. Appendix B.3 quantifies the form the non-idealities are assumed to take. While it is recognized that practical encoders suffer to some degree from all of these non-idealities, their effects are considered separately here to facilitate assessments. For each combination of encoder structure and non-ideality, we derive a modification to the zoomer algorithm that can be used if the numerical value of the non-ideality is known in advance. To distinguish between the modified zoomers and the ones described in section 3.2, the latter ones are referred to as original zoomers when necessary. For brevity, we only include plots of simulation results for SNR and not WCR performance. The simulation results are summarized in table 3.2.

With the modifications of the original zoomers that we derive, the concepts of transition points and quantization intervals are still well defined; each non-ideal encoder can be specified by its quantization intervals, transition points and codewords. However, using an original zoomer to decode an output sequence of a non-ideal encoder will not in general result in an estimate which is the midpoint of the actual quantization interval. This is because the quantization intervals of a non-ideal encoder change as functions of the numerical values of the non-idealities. In fact, an original zoomer estimate based on an output sequence from a non-ideal encoder may be outside the actual quantization interval, and the original zoomer bounds $L$ and $U$ may be incompatible.[5] Using an original zoomer thus results in performance degradation in the presence of non-idealities. In appendix B.1.2 we modify the definitions of SNR and WCR to apply also to non-ideal encoders.

[5] In the case of incompatible bounds ($L \geq U$), the original zoomer estimate is still defined as the average $(L + U)/2$.

### 3.3.1 Single-Loop Modulator

This section considers the effects of non-idealities on the single loop zoomer algorithm. Throughout we compare the zoomer to the linear decoder given by (3.10).

#### 3.3.1.1 Non-Unity Gain

A non-unity integrator gain $g$ manifests itself in the difference equation for the single-loop encoder as

$$U_i = U_{i-1} + g(X_{i-1} - Q(U_{i-1})), \quad n \geq 1 \tag{3.26}$$

Using (3.26) instead of (3.1), equation (3.2) becomes

$$U_n = g \sum_{i=0}^{n-1} [X_i - Q(U_i)], \quad n \geq 1$$

Since the gain $g$ is always positive, it does not affect $Q(U_n)$. Hence non-unity integrator gain does not affect the performance of any single loop decoder. As shown in sections 3.3.2.1 and 3.3.3.1 below, this is not the case for the double-loop and two-stage encoders.

#### 3.3.1.2 Leaky Integrator

With an integrator leak factor $\alpha$, the difference equation governing the single-loop encoder becomes

$$U_i = \alpha U_{i-1} + X_{i-1} - Q(U_{i-1}), \quad i \geq 1 \tag{3.27}$$

Using (3.27) instead of (3.1), equation (3.2) becomes

$$U_n = \sum_{i=0}^{n-1} \alpha^{n-i-1} [X_i - Q(U_i)], \quad n \geq 1$$

Therefore integrator leak leaves the zoomer algorithm unchanged, except that (3.5) becomes

$$\bar{X}_n = \frac{1-\alpha}{1-\alpha^n} S_n$$

where

$$S_n = \sum_{i=0}^{n-1} \alpha^{n-i-1} Q(U_i)$$

Note that $S_n$ can also be computed recursively; setting $S_0 \triangleq 0$,

$$S_n = \alpha S_{n-1} + Q(U_{n-1}), \quad n \geq 1$$

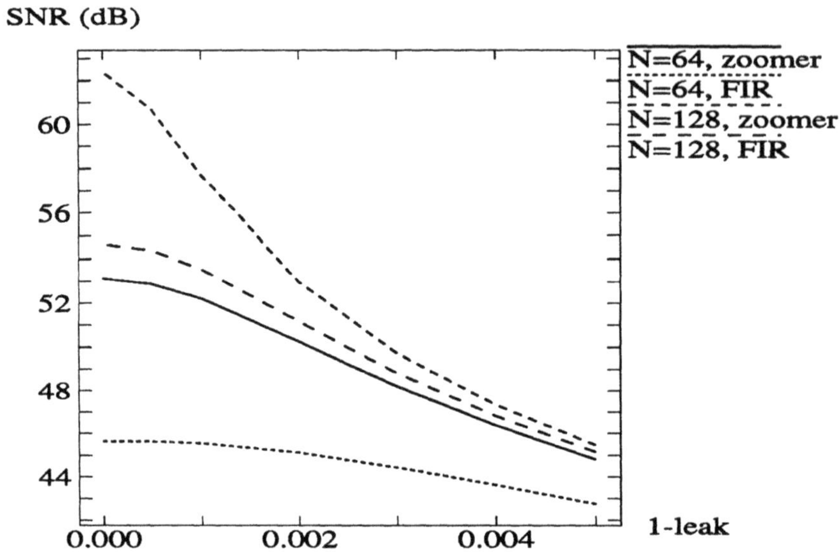


Figure 3.14: *Single-loop encoder: SNR for sample sizes 64 and 128 as a function of integrator leak* $\alpha$.

In fact, $\{S_n\}$ is the result of filtering $\{Q(U_0), \cdots, Q(U_{n-1})\}$ with the leaky integrator. From the above derivation we see that if the integrator leak is known, it can be compensated for by a slight modification of the zoomer algorithm. Under these conditions, simulation results show that the modified zoomer performs almost identically to the original zoomer operating on an ideal encoder.

We now consider the performance degradation if the leak factor is unknown. Figure 3.14 shows the SNR performance of the original zoomer versus the integrator leak for fixed sample sizes $N = 64$ and $N = 128$. We see that the leak sensitivity increases with the sample size, and that the losses at $\alpha = 0.999$ are 0.9 dB and 4.7 dB for $N = 64$ and $N = 128$, respectively. As the leak increases, the distance between curves for different sample sizes decreases; in other words, the benefit of larger sample sizes diminishes. The curves indicate that operational amplifiers (op-amps) with gains on the order of a few thousand should be used in conjunction with the original single loop zoomer.

#### 3.3.1.3 Non-Zero Initial State

If the initial state of the integrator is non-zero, that is, $U_0 \neq 0$, equation (3.2) becomes

$$U_n = U_0 + \sum_{i=0}^{n-1} [X_i - Q(U_i)], \qquad n \geq 1$$

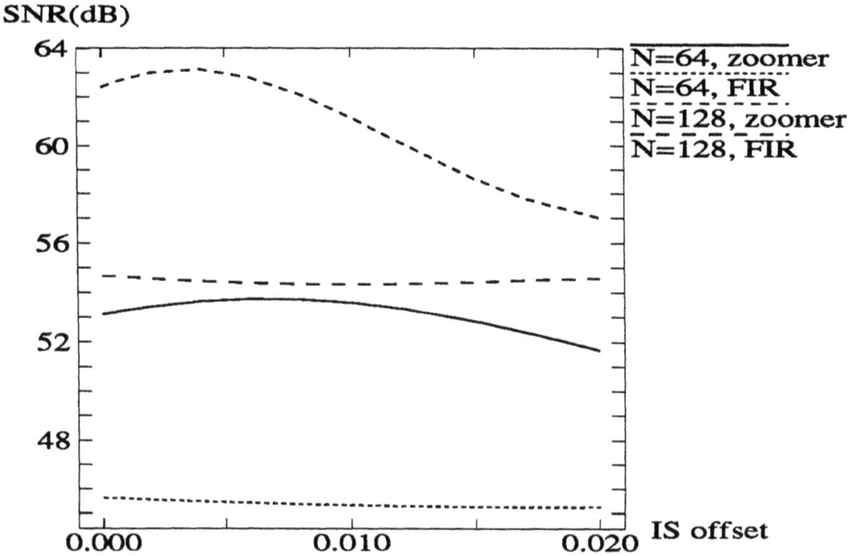


Figure 3.15: *Single-loop encoder: SNR for sample sizes 64 and 128 as a function of initial state offset* $U_0$*, or equivalently, of quantizer DC offset* $q_0 = -U_0$*. The offset is normalized by b.*

and

$$\bar{X}_n = \frac{-U_0 + S_n}{n} \tag{3.28}$$

with $S_n$ given by (3.3). Compensation for a known non-zero $U_0$ is thus straightforward. Under these conditions, simulation results show that the modified zoomer performs almost identically to the original zoomer operating on an ideal encoder, as described in section 3.2.1.1.

We now consider the performance degradation if the non-zero initial state is unknown; figure 3.15 shows the performance of the original zoomer as a function of the initial state $U_0$. We see that the SNR actually increases slightly as the initial state offset deviates from 0, and at $U_0 = 0.008b$, the performance is essentially the same as when $U_0 = 0$. The sensitivity increases with the sample size. The initial state offset corresponding to an SNR loss of 2 dB is about 1.2% of the dynamic range.

#### 3.3.1.4 Quantizer Offset

In the presence of a quantizer offset $q_0$, equation (3.5) becomes

$$\bar{X}_n = \frac{q_0 + S_n}{n}$$

with $S_n$ given by (3.3). By comparison with (3.28), a DC offset of $q_0$ has the same effect as an initial state offset of $-q_0$ for the single loop zoomer. If the offset is known, it can thus be compensated for; if not, figure 3.15 also demonstrates the effect of quantizer offset on performance, and the same comments apply as in section 3.3.1.3.

#### 3.3.1.5 Random Noise

The optimal procedure for taking into account noise on the constant input is difficult to derive. However, a simple way to take noise into account is as follows: Substituting (B.10) in (3.2), we obtain

$$U_n = \sum_{i=0}^{n-1} [X_i - Q(U_i)] = nX + \left( \sum_{i=0}^{n-1} N_i \right) - S_n, \quad n \geq 1$$

If the noise sequence were known, the modified zoomer algorithm would involve replacing (3.5) with

$$\bar{X}_n = \frac{1}{n} S_n - \frac{1}{n} \sum_{i=0}^{n-1} N_i \tag{3.29}$$

The variance of the right hand side of equation (3.29) is

$$\frac{1}{n^2} \sum_{i=0}^{n-1} \text{var}(N_i) = \frac{(Mb)^2}{3n}$$

Since the noise sequence is unknown in practice, one possible countermeasure against it is to loosen the bound in (3.5) by two noise standard deviations.[6] This changes (3.5) into

$$\bar{X}_n = \frac{1}{n} S_n - 2 \frac{MQ(U_n)}{\sqrt{3n}} \tag{3.30}$$

Note that the sign of the noise correction is chosen opposite that of $Q(U_n)$, so that the bound is loosened rather than tightened. Using (3.30) requires knowledge of the maximum noise level $Mb$.

In itself, this compensation method does not result in much gain for the single loop encoder; the gain is somewhat larger for the double-loop encoder. Figure 3.16 shows the SNR performance of the original zoomer in the presence of white uniform noise. In contrast with results given in previous sections, these results are generated by averaging the MSE over a large number of randomly generated constant inputs, adding random noise to each input sequence; this approach is necessary since equations such as (B.2) were derived under the assumption of deterministic non-idealities, or equivalently, fixed quantization intervals. At each sample size, two zoomer curves are shown: One for the original zoomer that does not utilize knowledge of the maximum noise level,

[6] Simulations indicate that it is better to use two standard deviations than one, but no claim is made that two is optimal.

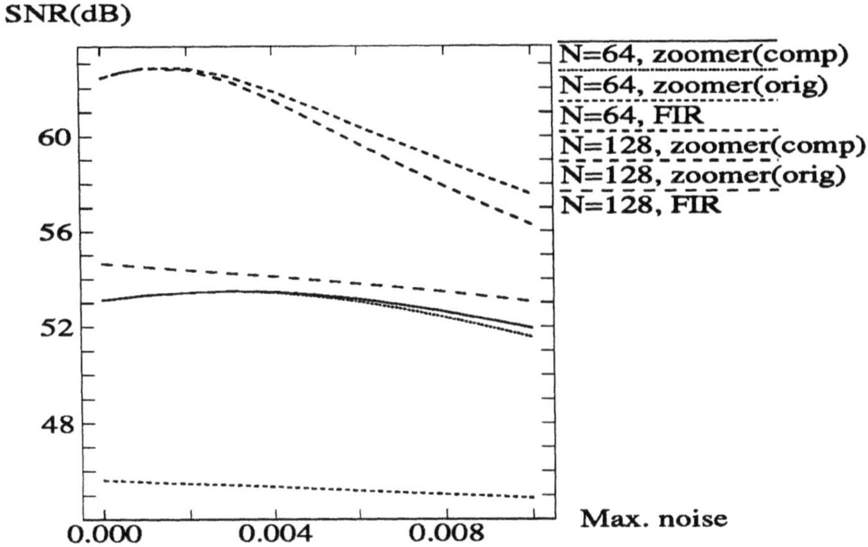


Figure 3.16: *Single-loop encoder: SNR for sample sizes 64 and 128 as a function of the maximum normalized random noise level M. Curves are shown for the original zoomer and the compensated zoomer which uses (3.30).*

and one for the compensated zoomer that uses (3.30) to partially overcome noise. We see that the compensated zoomer is marginally better than the original zoomer. At $N = 128$, the maximum noise level corresponding to an SNR loss of 2 dB is $M = 0.005$, that is, 0.5% of the dynamic range.

#### 3.3.1.6 Time-Varying Input

If the input sequence to the single-loop encoder has a slope $s$ and an average value $X_{\text{mid}}$ as defined in appendix B.3, we find that

$$\sum_{i=0}^{n-1} X_i = n\left(X_{\text{mid}} + \frac{n-N}{2}sb\right), \quad 1 \le n \le N \tag{3.31}$$

If the slope is known and we wish to estimate $X_{\text{mid}}$, equation (3.5) must thus be modified to

$$\bar{X}_n = \frac{1}{n}S_n + \frac{N-n}{2}sb$$

Simulation results show that under the assumption of known slope, the performance of the modified zoomer is essentially identical to that of the original zoomer described in section 3.2.1.1.

We now assume that the slope is unknown; figure 3.17 shows the performance of the original zoomer as a function of the total linear input variation

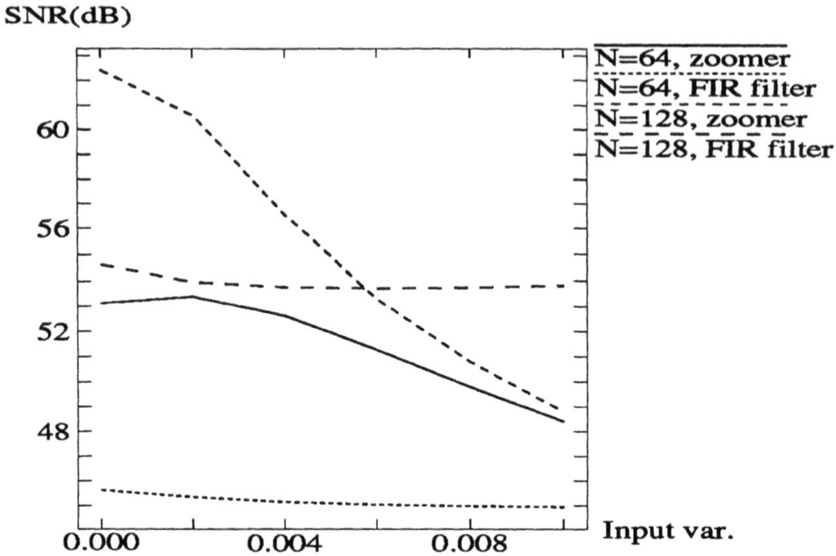


Figure 3.17: *Single-loop encoder: SNR for sample sizes 64 and 128 as a function of input variation; the input is assumed to vary linearly with time. The variation is normalized by b.*

$(N-1)sb$. We see that the SNR decreases fairly rapidly as the input variation increases. At a sample size of 128, the input variation corresponding to an SNR loss of 2 dB is about 0.2% of the dynamic range.

The method by which these results were obtained warrants comment. The expressions for SNR and worst-case resolution in equations (B.5) and (B.7) were derived under the assumption of constant input, and are thus strictly speaking invalid for non-zero slope. For small slopes, equation (B.4) for input variance remains approximately valid, and (B.5) and (B.7) can be used with the interpretation that the slope $s$ is constant and $X_{\text{mid}}$ is uniformly distributed on $D$.

In practice the assumption of known slope is not realistic, and the relative inadequacy of the zoomer in the situation of unknown slope can be ascribed to the fact that it is using a constant or zero'th order model on a sloping or first-order input. It is possible to devise a more sophisticated decoder which simultaneously estimates the average input $X_{\text{mid}}$ and the slope $s$, that is, a first-order decoder.[7] To find such a decoder, note that from (3.31), the optimal decoder obtains a series of bounds of the form

$$X_{\text{mid}} + \frac{n-N}{2}sb > \frac{1}{n}S_n \quad \text{if} \quad Q(U_n) = +b$$

[7] Theoretically, the input could be approximated or modeled by polynomials of arbitrary degree $P$; the dimension of the corresponding LP problem would be $P+1$. Making $P$ too large would however result in model overfitting.

$$X_{\text{mid}} + \frac{n-N}{2} sb \le \frac{1}{n} S_n \quad \text{if} \quad Q(U_n) = -b$$

Finding optimal joint bounds on $X_{\text{mid}}$ and $s$ is thus a two-dimensional Linear Programming (LP) problem. In that light, the original zoomer can be said to solve a one-dimensional LP problem which is much simpler than higher-dimensional problems. The reason is that in one dimension, two bounds always suffice to specify the allowable parameter region, whereas in higher dimensions, the number of hyperplanes necessary to bound the parameter region may be on the order of the number of linear constraints.

We now describe a simple way of simultaneously estimating the average input and the slope of a first-order or linearly varying signal; the method is based on the well known simplex algorithm of LP. We assume that the average input is bounded to the interval $(-Kb, +Kb)$, the slope is bounded to $(-Sb, +Sb)$, and that $K$ and $S$ are known to the decoder. Our proposed decoder solves the two-dimensional linear programming problem in $X_{\text{mid}}$ and $s$ twice using the simplex algorithm: Once with an objective function maximizing $X_{\text{mid}}$ resulting in $(X_{\text{mid1}}, s_1)$, and once with an objective function minimizing $X_{\text{mid}}$ resulting in $(X_{\text{mid2}}, s_2)$. The decoder then uses the average $(X_{\text{mid1}} + X_{\text{mid2}})/2$ as its estimate of $X_{\text{mid}}$. The decoder can also provide an estimate of the slope $s$ by averaging the optimization results, $s = (s_1 + s_2)/2$. This joint estimate of $(X_{\text{mid}}, s)$ is guaranteed to satisfy the LP constraints because the feasible set for any linear programming problem is convex, and the average of any two points in a convex set is also a member of the set.

The described approach is suboptimal in the sense that the optimal decoder would make use of a joint probability distribution function for $X_{\text{mid}}$ and $s$ to estimate the center of probability mass in parameter space. The simplified approach does, however, specialize correctly for constant inputs with $S = 0$ and $X_{\text{mid}}$ uniformly distributed on $(-Kb, +Kb)$.

Figure 3.18 shows the performance of the original zoomer, the first-order decoder described above, and the linear filter as a function of the maximum total linear input variation $(N-1)Sb$. For purposes of generating these plots, we assume that the average input and the slope are independent random variables which are uniformly distributed on their ranges. We observe that at a sample size of 128, the first-order decoder only experiences about a 1 dB decrease in SNR performance even for input variations in excess of 3% of the dynamic range, whereas the original zoomer loses about 20 dB at this level of maximum input variation. The first-order decoder has about the same sensitivity to slope as the linear decoder, but unlike the linear decoder, it can also produce reliable estimates of the slope.

### 3.3.2 Double-Loop Modulator

This section considers the effects of non-idealities on the double-loop zoomer algorithm. Throughout we compare the zoomer to the linear $N$-tap decoder with a $sinc^3$ transfer function.

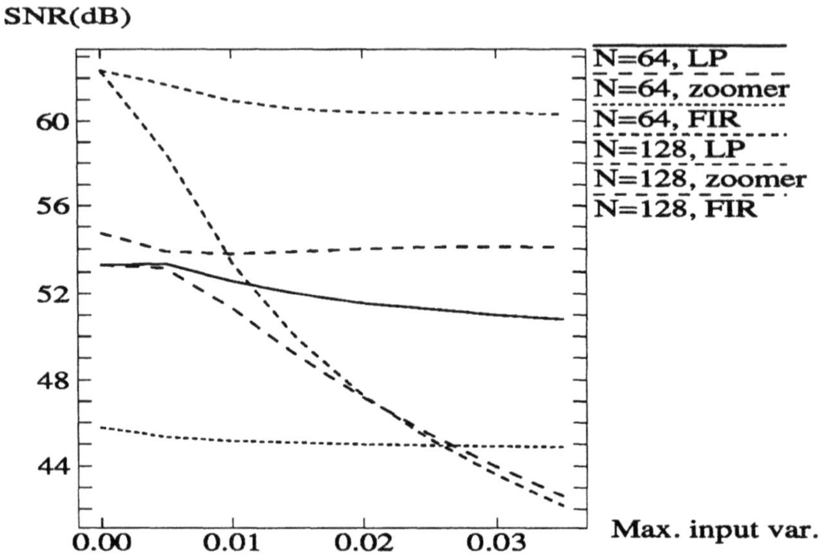


Figure 3.18: *Single-loop encoder: SNR for sample sizes 64 and 128 as a function of the maximum total linear input variation $(N-1)Sb$. Curves are shown for the original zoomer and the first order zoomer. The variation is normalized by $b$.*

#### 3.3.2.1 Non-Unity Gains

Denoting the gains of the first and second integrator by $g_1$ and $g_2$, respectively, the difference equations for the double-loop encoder can be written

$$\begin{aligned} U_n &= U_{n-1} + g_2(V_{n-1} - Q(U_{n-1})) \\ V_n &= V_{n-1} + g_1(X_n - Q(U_n)), \quad n \geq 1 \end{aligned}$$

and (3.11) becomes

$$U_n = g_2 \left( g_1 \sum_{i=1}^{n-1} (n-i) X_i - Q(U_0) - g_1 \sum_{i=1}^{n-1} (n-i+1) Q(U_i) - \right. \\ \left. (1-g_1) \sum_{i=1}^{n-1} (n-i) Q(U_i) \right)$$

We see that the gain of the second integrator has no effect on the sign of $U_n$, and hence it does not affect the performance of any decoder. The zoomer algorithm in the presence of non-unity gain remains unchanged, except that (3.16) becomes

$$\bar{X}_n = \frac{Q(U_0) + g_1 W_n + (1-g_1) S_n}{\frac{1}{2} g_1 n(n-1)} \tag{3.32}$$

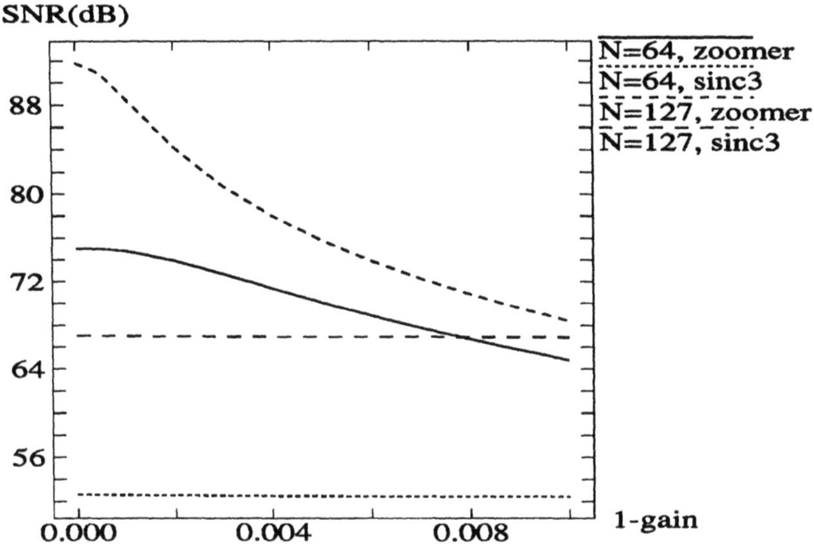


Figure 3.19: *Double-loop encoder: SNR for sample sizes 64 and 127 as a function of the first integrator gain $g_1$.*

with $S_n$ and $W_n$ given by (3.12) and (3.13). If the gain is known, it can thus be compensated for; in this case, performance is essentially identical to that of the original zoomer described in section 3.2.2.1.

If the gain is unknown, it can be adaptively estimated from (3.32) in the following way: Initially, we assume $g_1$ to be 1. Each time a non-codeword appears due to non-unity gain, the lower and upper zoomer bounds will be inconsistent; let us denote these bounds by $\bar{X}_n$ and $\bar{X}_m$, respectively. We can modify $g_1$ such that the bounds at time instants $n$ and $m$ become identical; that is, choose $g_1$ as close as possible to its nominal value of 1 while keeping the bounds barely consistent. To be specific, $g_1$ is chosen according to

$$g_1 = \frac{m(m-1)(Q(U_0)+S_n) - n(n-1)(Q(U_0)+S_m)}{n(n-1)(W_m - S_m) - m(m-1)(W_n - S_n)} \tag{3.33}$$

This updating process is repeated every time a set of incompatible bounds is encountered. The estimate of $X$ in the particular cycle where $g_1$ is updated will not be so accurate, but the accuracy of subsequent estimates will improve. Once $g_1$ is well estimated, the performance is essentially identical to that of the original zoomer with an ideal decoder.

We now assume that the integrator gains are unknown, and that the adaptive estimation scheme of equation (3.33) is not used. Figure 3.19 shows SNR curves for the original zoomer when the first integrator gain is non-unity. We see that at a sample size of 127,[8] an SNR degradation of 2 dB results for a gain which deviates about 0.1% from its nominal value.

#### 3.3.2.2 Leaky Integrators

We assume that the two integrators have the same leak factor $\alpha$, that is, the operational amplifiers have the same gain. The difference equations then become

$$\begin{aligned} U_n &= \alpha U_{n-1} + V_{n-1} - Q(U_{n-1}) \\ V_n &= \alpha V_{n-1} + X_n - Q(U_n), \quad n \geq 1 \end{aligned}$$

Let us define

$$S_n = \sum_{i=1}^{n-1} \alpha^{n-i-1} Q(U_i), \quad n \geq 2$$

$$W_n = \sum_{i=1}^{n-1} \alpha^{n-i-1} (n-i+1) Q(U_i), \quad n \geq 2$$

Note that if we define $S_1 = W_1 \triangleq 0$,

$$S_n = \alpha S_{n-1} + Q(U_{n-1}), \quad n \geq 2$$

$$W_n = \alpha W_{n-1} + S_n + Q(U_{n-1}), \quad n \geq 2$$

These recursive relations mimic the effect of the two integrators. With leaky integrators, (3.11) changes into

$$U_n = \frac{(n-1)\alpha^n - n\alpha^{n-1} + 1}{(1-\alpha)^2} X - \alpha^{n-1} Q(U_0) - W_n$$

and the zoomer bound (3.16) becomes

$$\bar{X}_n = \frac{(1-\alpha)^2}{(n-1)\alpha^n - n\alpha^{n-1} + 1} \left[\alpha^{n-1} Q(U_0) + W_n\right] \tag{3.34}$$

From the above discussion we see that if the integrator leaks are known, they can be compensated for, and in this case, the performance is essentially identical to that of the original zoomer considered in section 3.2.2.2.

We now assume that the integrator leaks are unknown. Figure 3.20 shows performance curves for the original zoomer in the presence of leak in both integrators. We see that op-amp gains on the order of 50,000 result in an SNR loss of 2 dB.

It is commonly held that non-idealities near the input summing node have relatively more impact on performance than non-idealities further inside the

[8] A sample size of 127 rather than 128 is chosen because we are comparing the zoomer with a $sinc^3$ filter, and this filter can only have impulse responses of length $3m+1$ with integer $m$.

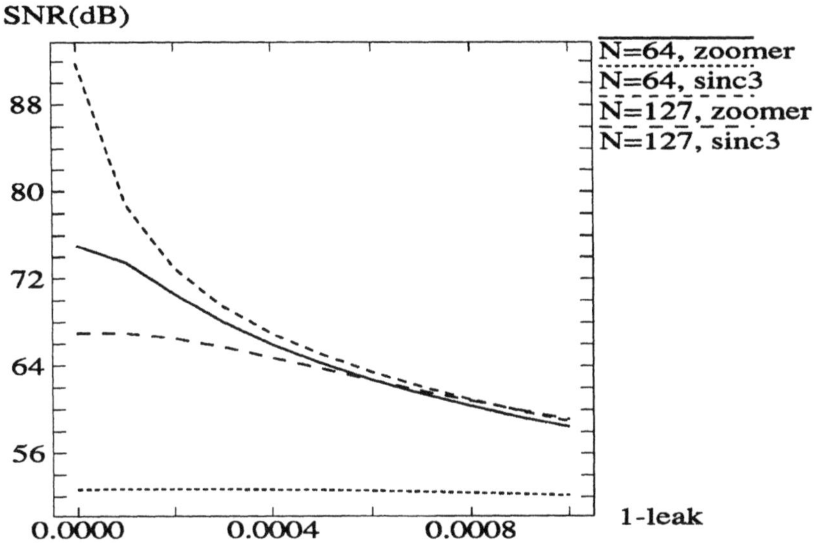


Figure 3.20: *Double-loop encoder: SNR for sample sizes 64 and 127 as a function of the integrator leak* $\alpha$. *Same leak is assumed for both integrators.*

loop. A qualitative argument in support of this observation is that non-idealities further inside the loop are better suppressed by the negative feedback. To quantify this, we show simulations on the effect of leak in the second integrator, assuming that the first integrator is ideal. In this case, the difference equations are

$$\begin{aligned} U_n &= \alpha U_{n-1} + V_{n-1} - Q(U_{n-1}) \\ V_n &= V_{n-1} + X_n - Q(U_n), \quad n \geq 1 \end{aligned}$$

We can define the running sums

$$S_n = \sum_{i=1}^{n-1} \alpha^{n-i-1} Q(U_i), \quad n \geq 2$$

$$W_n = \sum_{i=1}^{n-1} \frac{1-(2\alpha-1)\alpha^{n-i-1}}{1-\alpha} Q(U_i), \quad n \geq 2$$

If we define $S_1 = W_1 \triangleq 0$, we have the recursions

$$S_n = \alpha S_{n-1} + Q(U_{n-1}), \quad n \geq 2$$

$$W_n = \alpha S_{n-1} + S_n + Q(U_{n-1}), \quad n \geq 2$$

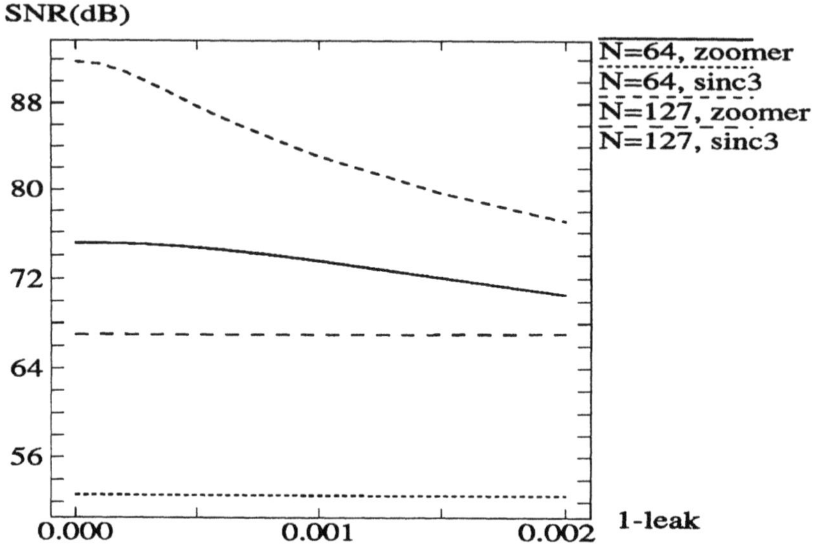


Figure 3.21: *Double-loop encoder: SNR for sample sizes 64 and 127 as a function of the second integrator leak* $\alpha$. *First integrator is assumed ideal.*

The zoomer bound becomes

$$\bar{X}_n = \frac{(1-\alpha)^2}{\alpha^n - n\alpha + (n-1)} \left[\alpha^{n-1} Q(U_0) + W_n\right]$$

This bound can be compared with (3.34), in which both the integrator leaks are assumed to equal $\alpha$. Figure 3.21 shows the isolated effect of the second integrator leak. Comparing with figure 3.20, we observe that the sensitivity towards the second integrator leak is on the order of 10 times smaller than that towards the first integrator leak; therefore the degradation in figure 3.20 is dominated by the first leak.

#### 3.3.2.3 Non-Zero Initial States

We assume that the initial state offsets are the same for both state variables, $U_0 = V_0 = \delta$. It can be shown that the state $U$ at time $n$ can then be written as

$$U_n = U_0 + nV_0 + \sum_{i=0}^{n-1}(n-i)X_i - Q(U_0) - \sum_{i=1}^{n-1}(n-i+1)Q(U_i)$$

The zoomer bound (3.16) changes into

$$\bar{X}_n = \frac{-(n+1)\delta + Q(U_0) + W_n}{\frac{1}{2}n(n-1)} \tag{3.35}$$

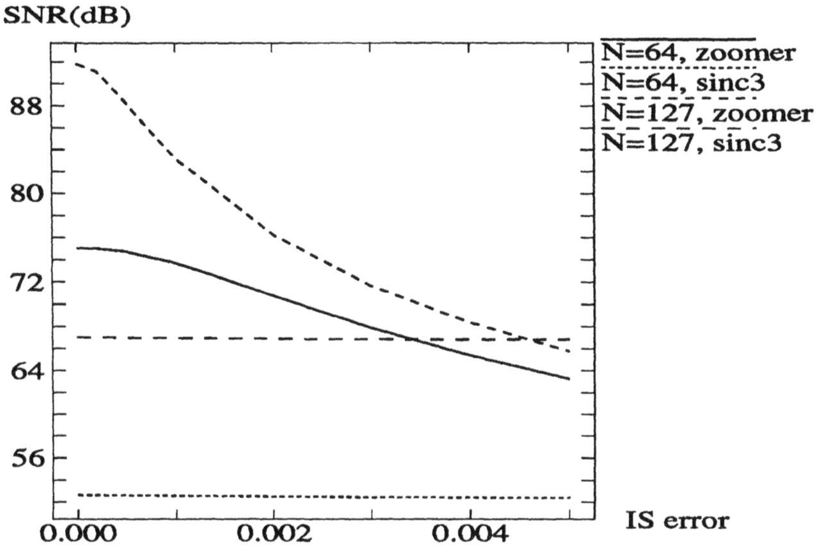


Figure 3.22: *Double-loop encoder: SNR for sample sizes 64 and 127 as a function of the initial state offset $\delta$. Same offset is assumed for both integrators. The offset is normalized by b.*

with $W_n$ given by (3.13). Note that $V_0$ contributes $n\delta$, while $U_0$ only contributes $\delta$. This indicates that more attention should be paid to the initial state of the first integrator. This result is in line with the general belief mentioned above that non-idealities closer to the input summing node have more impact on performance than non-idealities further inside the loop.

If the initial integrator states are known, they can be compensated for as shown in (3.35). We now assume that these states are unknown; figure 3.22 shows their effect. We see that for a sample size of 127, an SNR loss of 2 dB results when $\delta$ equals about 0.05% of the dynamic range.

#### 3.3.2.4 Quantizer Offset

If the 1-bit quantizer is characterized by (B.9), that is, by the offset $q_0$, the zoomer bound (3.16) changes into

$$\bar{X}_n = \frac{q_0 + Q(U_0) + W_n}{\frac{1}{2}n(n-1)} \tag{3.36}$$

with $W_n$ given by (3.13). Thus, a DC offset of $q_0$ is equivalent to a non-zero initial state of $U_0 = -q_0$. If the DC offset is known, it can be compensated for using (3.36).

We now assume that the DC offset is unknown; figure 3.23 shows its effect on the performance of the original zoomer. Since DC offset and non-zero initial state $U_0$ are equivalent, and non-zero initial state $V_0$ is known from section

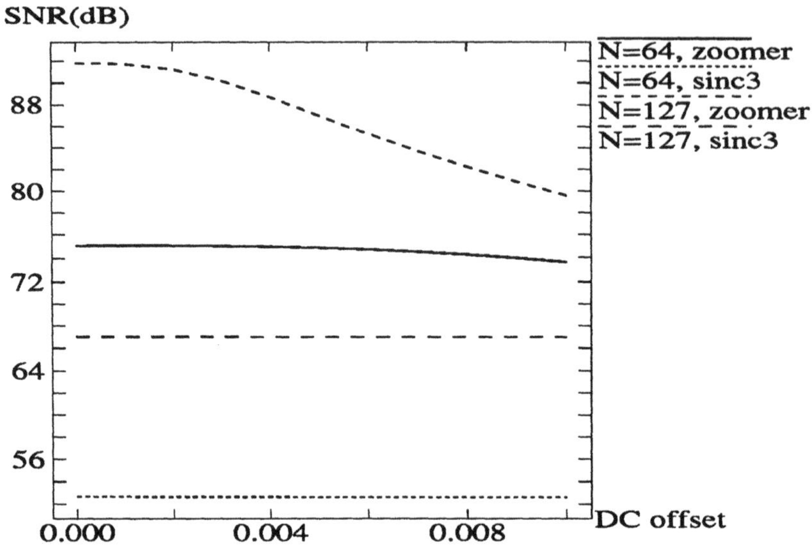


Figure 3.23: *Double-loop encoder: SNR for sample sizes 64 and 127 as a function of quantizer DC offset* $q_0$. *Same offset is assumed for both integrators. The offset is normalized by b.*

3.3.2.3 to degrade performance more seriously than non-zero $U_0$, we expect the sensitivity towards DC offset to be less than that towards initial states shown in figure 3.22. This is confirmed by figure 3.23. For a sample size of 127, an offset of 0.5% results in an SNR loss of about 2 dB.

#### 3.3.2.5 Random Noise

Consider noisy inputs of the form (B.10) specified in appendix B.3. If the noise sequence were known, the modified zoomer algorithm would involve replacing (3.16) with

$$\bar{X}_n = -\frac{1}{\frac{1}{2}n(n-1)}\sum_{i=1}^{n-1}(n-i)N_i + \frac{Q(U_0)+W_n}{\frac{1}{2}n(n-1)}$$

The variance of the right hand side is

$$\frac{4}{n^2(n-1)^2}\sum_{i=1}^{n-1}\frac{(Mb)^2}{3}i^2 = \frac{2(2n-1)(Mb)^2}{9n(n-1)}$$

Since the noise sequence is unknown in practice, one possible countermeasure against it is to loosen the bound in (3.16) by two standard deviations.[9] This

[9] Simulations indicate that it is better to use two standard deviations than one, but no claim is made that two is optimal.

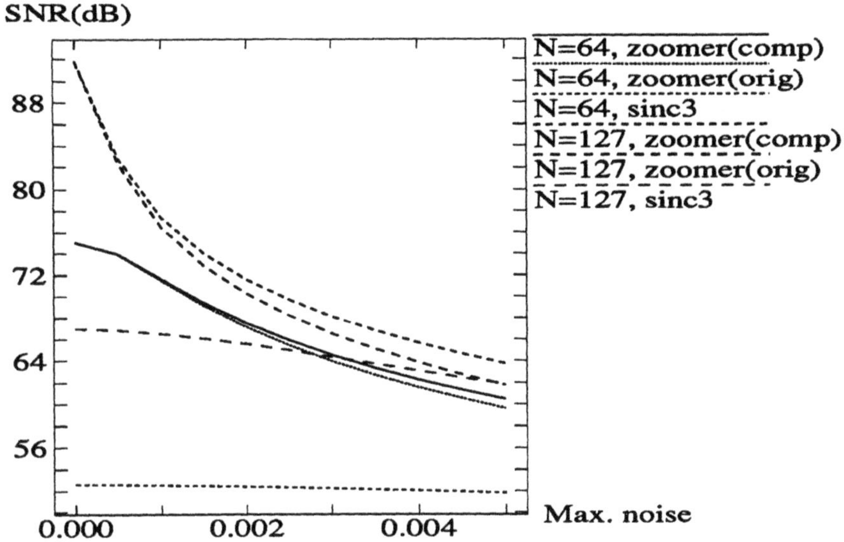


Figure 3.24: *Double-loop encoder: SNR for sample sizes 64 and 127 as a function of the maximum normalized noise level M. Curves are shown for the original zoomer and the compensated zoomer which uses (3.37).*

changes (3.16) into

$$\bar{X}_n = \frac{Q(U_0) + W_n}{\frac{1}{2}n(n-1)} - 2\sqrt{\frac{2(2n-1)}{n(n-1)}\frac{MQ(U_n)}{3}} \tag{3.37}$$

Note that the sign of the noise correction is opposite that of $Q(U_n)$, so that the bound is loosened. Knowledge of the maximum noise level $Mb$ is required.

Figure 3.24 shows the effect of random noise on the input. At a sample size of 127, an SNR loss of 2 dB results for a maximum noise level of about 0.02%.

#### 3.3.2.6 Time-Varying Input

To first order, the input signal can be approximated by

$$X_i = X_{\text{mid}} + \left(i - \frac{N-1}{2}\right) sb, \quad 0 \le i \le N-1 \tag{3.38}$$

where $s$ is the slope, and $X_{\text{mid}}$ is the average input. It is found that

$$\sum_{i=1}^{n-1}(n-i)X_i = \frac{1}{2}n(n-1)\left[X_{\text{mid}} + \frac{2n-3N+5}{6}sb\right] \tag{3.39}$$

If the slope is known and we wish to estimate $X_{\text{mid}}$, the zoomer bound thus becomes

$$\bar{X}_n = \frac{Q(U_0) + W_n}{\frac{1}{2}n(n-1)} + \frac{3N-2n-5}{6}sb$$

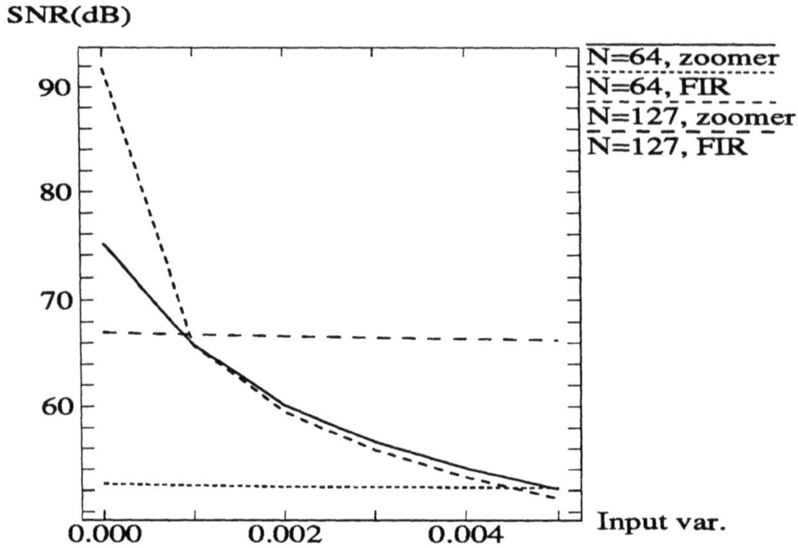


Figure 3.25: *Double-loop encoder: SNR for sample sizes 64 and 127 as a function of input variation; the input is assumed to vary linearly with time. The variation is normalized by b.*

If the slope is known, it can be compensated for using this equation. In practice the slope is unknown, and $X_{\text{mid}}$ and $s$ must be estimated simultaneously. Using (3.39), this leads to a two-dimensional linear programming problem as for the single-loop encoder.

We now assume that the slope is unknown. Figure 3.25 illustrates the performance of the original zoomer as a function of the total linear variation, $(N-1)sb$. The same caveats apply as in section 3.2.1.2. We see that an SNR degradation of 2 dB results for a total input variation of 0.01% of the dynamic range.

### 3.3.3 Two-Stage Modulator

This section considers the effects of non-idealities on the two-stage zoomer algorithm. Throughout we compare the zoomer to the linear $N$-tap decoder with a $sinc^3$ transfer function.

#### 3.3.3.1 Non-Unity Gains

Denoting the integrator gains of the first and second stage by $g_1$ and $g_2$, respectively, the difference equations can be written

$$\begin{aligned} U_n &= U_{n-1} + g_1(X_{n-1} - Q(U_{n-1})) \\ V_n &= V_{n-1} + g_2(-U_{n-1} + Q(U_{n-1} - Q(U_{n-1})), \quad n \geq 1 \end{aligned}$$

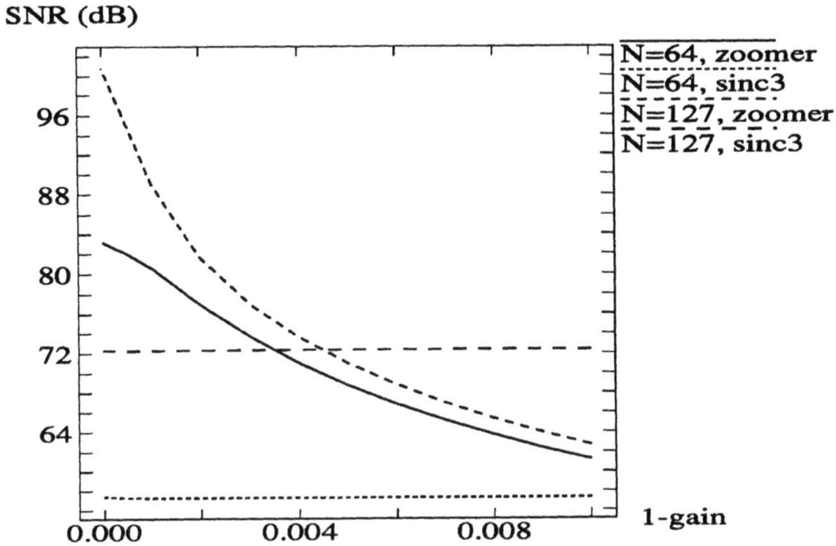


Figure 3.26: *Two-stage encoder: SNR for sample sizes 64 and 127 as a function of the first integrator gain $g_1$.*

Solving these with the definitions of Section 3.2.3.1 yields the zoomer bounds

$$\bar{X}_n^{(1)} = \frac{1}{n} S_n, \quad \bar{X}_n^{(2)} = \frac{g_1 W_n + (1-g_1) S_n - T_n}{\frac{1}{2} n(n-1) g_1} \tag{3.40}$$

We see that $g_2$ does not appear in the equations. If the first integrator gain is known, it can be compensated for using (3.40).

We now assume that the first integrator gain is unknown, and that an adaptive estimation scheme analogous to (3.33) for the double-loop case is not used. Figure 3.26 shows the SNR curves for the original zoomer with non-unity $g_1$. We see that for a sample size of 127, an SNR loss of 2 dB results for a gain $g_1$ which deviates about 0.02% from its nominal value.

#### 3.3.3.2 Leaky Integrators

We assume that the two integrators have the same leak factor $\alpha$, that is, that the op-amps have the same gain. The difference equations then become

$$\begin{aligned} U_n &= \alpha U_{n-1} + X_{n-1} - Q(U_{n-1}) \\ V_n &= \alpha V_{n-1} + Q(U_{n-1}) - U_{n-1} - Q(V_{n-1}), \quad n \geq 1 \end{aligned}$$

Defining $S_0 = T_0 = W_0 \triangleq 0$ and the running sums

$$S_n = \sum_{i=0}^{n-1} \alpha^{n-i-1} Q(U_i), \quad n \geq 1$$

$$T_n = \sum_{i=0}^{n-1} \alpha^{n-i-1} Q(V_i), \quad n \geq 1$$

$$W_n = \sum_{i=0}^{n-1} [n - i - (1-\alpha)] \, \alpha^{n-i-2} Q(U_i), \quad n \geq 1$$

we then have the recursions

$$S_n = \alpha S_{n-1} + Q(U_{n-1}), \quad n \geq 1$$

$$T_n = \alpha T_{n-1} + Q(V_{n-1}), \quad n \geq 1$$

$$W_n = \alpha W_{n-1} + S_n + (1-\alpha) S_{n-1}, \quad n \geq 1$$

The zoomer bounds change into

$$\bar{X}_n^{(1)} = \frac{1-\alpha}{1-\alpha^n} S_n, \quad \bar{X}_n^{(2)} = \frac{(1-\alpha)^2}{1 + (n-1)\alpha - n\alpha^{n-1}} [W_n - T_n]$$

Similarly to the double-loop case, we consider the effect of leak in the second integrator separately. The difference equations for this situation are

$$\begin{aligned} U_n &= U_{n-1} + X_{n-1} - Q(U_{n-1}) \\ V_n &= \alpha V_{n-1} + Q(U_{n-1}) - U_{n-1} - Q(V_{n-1}), \qquad n \geq 1 \end{aligned}$$

Defining $S_0 = T_0 = W_0 \triangleq 0$ and the running sums

$$S_n = \sum_{i=0}^{n-1} Q(U_i), \quad n \geq 1$$

$$T_n = \sum_{i=0}^{n-1} \alpha^{n-i-1} Q(V_i), \quad n \geq 1$$

$$W_n = \sum_{i=0}^{n-1} \frac{1-\alpha^{n-i}}{1-\alpha} Q(U_i), \quad n \geq 1$$

we have the recursions

$$S_n = S_{n-1} + Q(U_{n-1}), \quad n \geq 1$$

$$T_n = \alpha T_{n-1} + Q(V_{n-1}), \quad n \geq 1$$

$$W_n = \alpha W_{n-1} + S_n, \quad n \geq 1$$

The zoomer bounds change into

$$\bar{X}_n^{(1)} = \frac{1}{n} S_n, \quad \bar{X}_n^{(2)} = \frac{(1-\alpha)^2}{\alpha^n - n\alpha + (n-1)} [W_n - T_n]$$

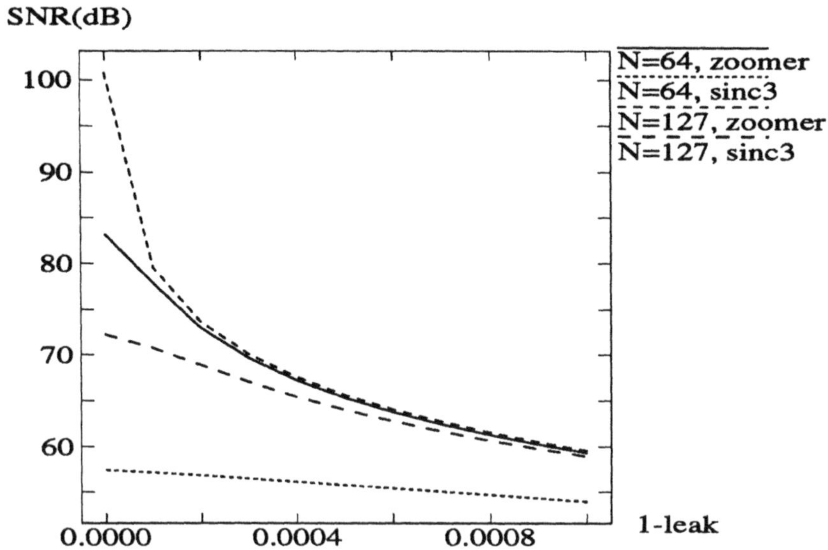


Figure 3.27: *Two-stage encoder: SNR for sample sizes 64 and 127 as a function of the integrator leak $\alpha$. Same leak is assumed for both integrators.*

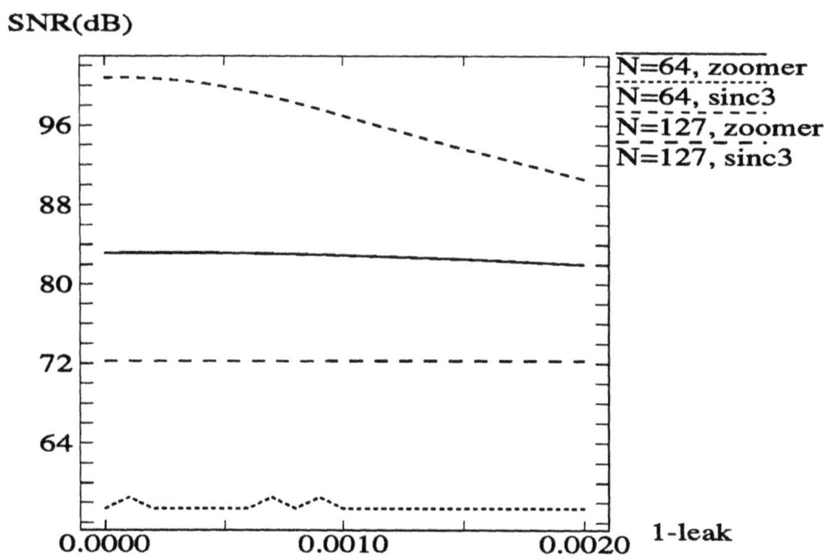


Figure 3.28: *Two-stage encoder: SNR for sample sizes 64 and 127 as a function of the second integrator leak $\alpha$. First integrator is assumed ideal.*

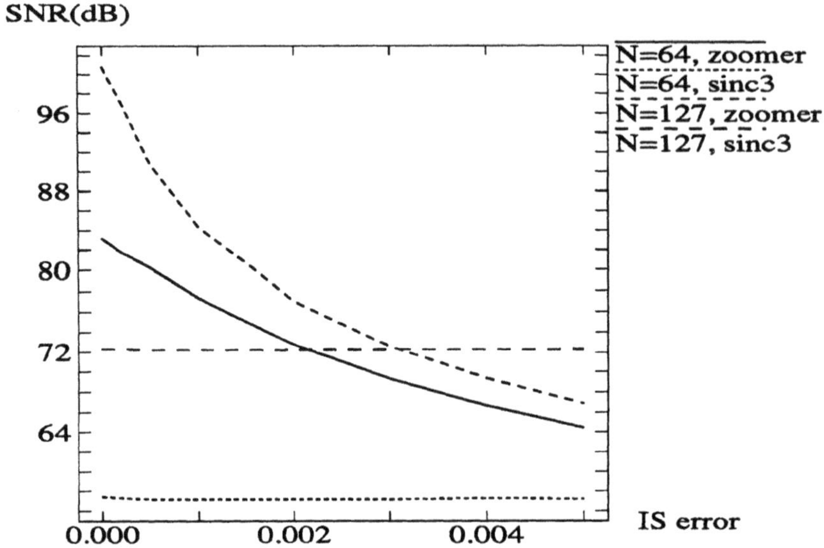


Figure 3.29: *Two-stage encoder: SNR for sample sizes 64 and 127 as a function of the initial state offset $\delta$. Same offset is assumed for both integrators. The offset is normalized by b.*

From the above discussion we see that if the integrator leaks are known, they can be compensated for.

We now assume that the leaks are unknown; figure 3.27 shows performance curves for the original zoomer as a function of the leak. We see that for a sample size of 127, op-amp gains on the order of 100,000 result in an SNR loss of 2 dB. Figure 3.28 shows the isolated effect of the second integrator leak on the performance of the ideal zoomer. Comparing with figure 3.27, we observe that the sensitivity towards the second integrator leak is about 50 times smaller than that towards the first integrator leak; the degradations in figure 3.27 are therefore dominated by the first leak.

#### 3.3.3.3 Non-Zero Initial States

For non-zero initial integrator states $U_0, V_0$, the zoomer bounds become

$$\bar{X}_n^{(1)} = \frac{-U_0 + S_n}{n}, \quad \bar{X}_n^{(2)} = \frac{-nU_0 + V_0 + W_n - T_n}{\frac{1}{2}n(n-1)} \tag{3.41}$$

with $S_n, T_n, W_n$ defined by (3.19), (3.20), (3.21). Note that in (3.41), $U_0$ is more important than $V_0$, again confirming that non-idealities closer to the input summing node have greater impact on performance. If the initial states are known, they can be compensated for using (3.41).

We now assume that the initial states are unknown; for simplicity, we assume that $U_0 = V_0 = \delta$. Figure 3.29 shows the effect of non-zero initial states on

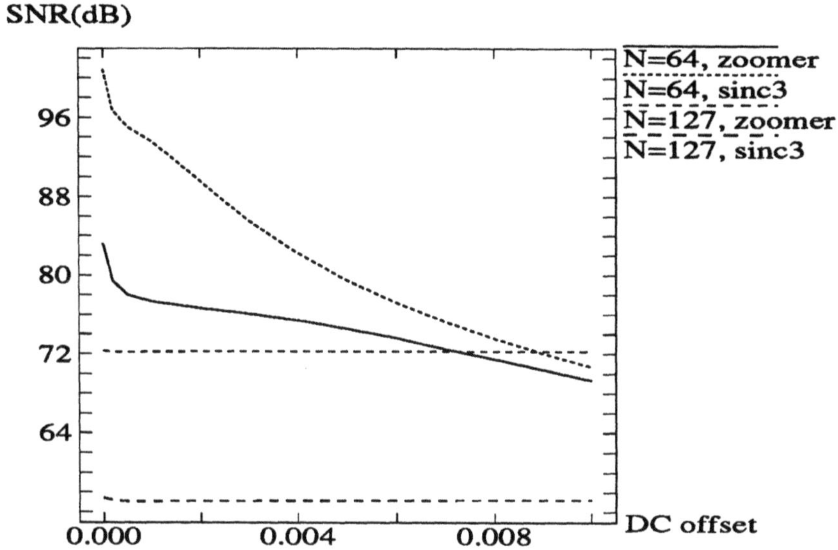


Figure 3.30: *Two-stage encoder: SNR for sample sizes 64 and 127 as a function of quantizer DC offset* $q_0$. *Same offset is assumed for the two quantizers. The offset is normalized by* $b$.

the performance of the original zoomer. We see that at a sample size of 127, an SNR loss of 2 dB results when $\delta$ deviates from zero by about 0.01% of the dynamic range.

#### 3.3.3.4 Quantizer Offset

Assuming that both quantizers are characterized by the same DC offset $q_0$, the zoomer bounds become

$$\bar{X}_n^{(1)} = \frac{q_0 + S_n}{n}, \quad \bar{X}_n^{(2)} = \frac{-q_0 + W_n - T_n}{\frac{1}{2}n(n-1)} \tag{3.42}$$

Note the similarities and differences between the equations (3.42) and equations (3.41) for non-zero initial states. The analogies with non-zero initial states indicate that an offset in the first quantizer is comparatively more detrimental to performance. If the DC offsets are known, they can be compensated for using (3.42).

We now assume that the DC offsets are unknown, but equal; figure 3.30 shows their effect on the performance of the original zoomer. The sensitivity of the modulator towards DC offset is somewhat smaller than that towards non-zero initial states shown in figure 3.29. At a sample size of 127, an SNR loss of 2 dB results for offsets on the order of 0.01% of the dynamic range.

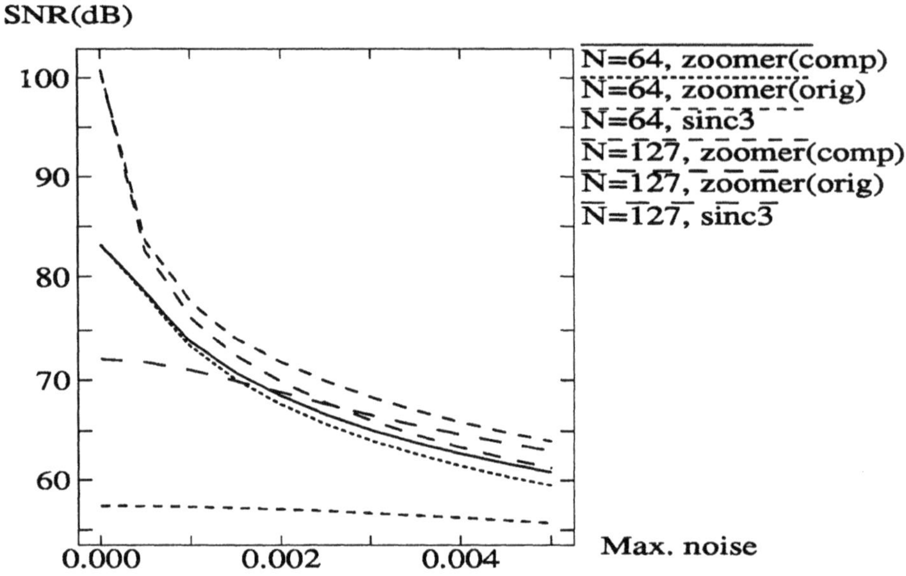


Figure 3.31: *Two-stage encoder: SNR for sample sizes 64 and 127 as a function of the maximum normalized noise level M. Curves are shown for the original zoomer and the compensated zoomer which uses (3.43), (3.44).*

#### 3.3.3.5 Random Noise

Following the derivations of previous sections, a simple countermeasure against white uniform noise in the interval $(-Mb, +Mb)$ is to change the zoomer bounds to

$$\bar{X}_n^{(1)} = \frac{1}{n}S_n - 2\frac{MQ(U_n)}{\sqrt{3n}} \tag{3.43}$$

$$\bar{X}_n^{(2)} = \frac{W_n - T_n}{\frac{1}{2}n(n-1)} + 2\sqrt{\frac{2(2n-1)}{n(n-1)}}\frac{MQ(V_n)}{3} \tag{3.44}$$

In (3.43), the sign of the noise correction is chosen opposite that of $Q(U_n)$, whereas in (3.44), it is chosen to be the same as that of $Q(V_n)$. In both cases, this leads to weaker bounds. Figure 3.31 shows the effect of random noise on the performance of the original zoomer. At a sample size of 127, a maximum noise level of 0.005% results in an SNR loss of about 2 dB.

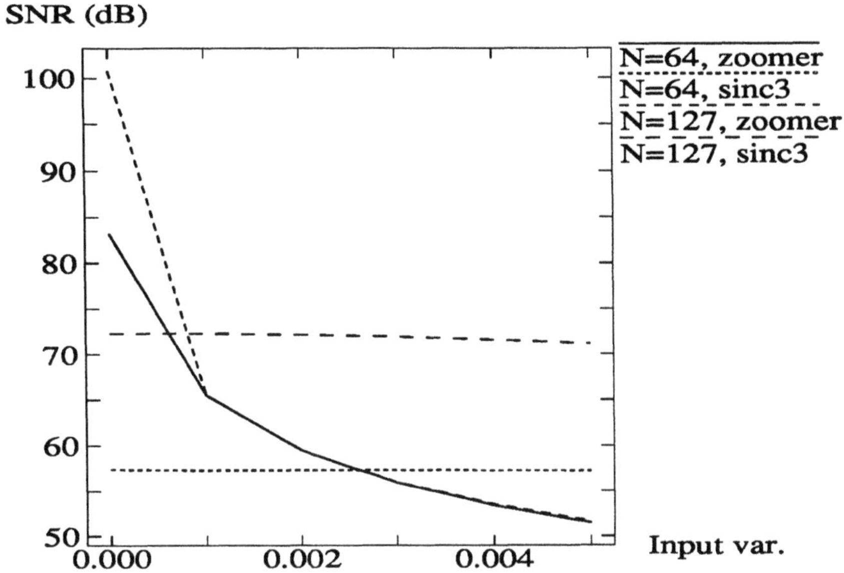


Figure 3.32: *Two-stage encoder: SNR for sample sizes 64 and 127 as a function of input variation; the input is assumed to vary linearly with time. The variation is normalized by b.*

#### 3.3.3.6 Time-Varying Input

Consider a time-varying input signal of the form (3.38). The first zoomer bound changes into

$$\bar{X}_n^{(1)} = \frac{1}{n} + \frac{N-n}{2}sb$$

We find that

$$\sum_{i=0}^{n-2}(n-i)X_i = \frac{1}{2}n(n-1)\left[X_{\text{mid}} + \frac{2n-3n-1}{6}sb\right]$$

The second zoomer bound thus changes into

$$\bar{X}_n^{(2)} = \frac{W_n - T_n}{\frac{1}{2}n(n-1)} + \frac{3N-2n+1}{6}sb$$

These equations show that if the slope can be considered known, it can be compensated for. We now assume the slope to be unknown; figure 3.32 illustrates the performance of the original zoomer as a function of the total linear variation, $(N-1)sb$. The same caveats apply as in Section 3.2.1.2. We see that the input variation should not exceed about 0.005%.

## 3.4 Summary

In this chapter we introduced a general technique for optimal decoding of the output of ideal $\Sigma\Delta$ encoders, under the assumptions of constant input and known initial integrator states. The technique is based on deriving a succession of upper and lower bounds on the input interval generating a given output sequence. The optimal decoder is nonlinear, as might be expected from the nonlinear nature of the encoder. Our results indicate that under ideal circumstances, substantial improvements in SNR and worst-case resolution can be achieved; these results are summarized in table 3.1. The improvements can be exploited as enhanced performance at the same data acquisition time, or alternatively as substantial reductions in data acquisition time for the same performance. We also presented simulation results for performance in the presence of various circuit imperfections; these results are summarized in table 3.2. It is interesting to compare the sensitivities of the zoomers for the encoder structures considered. In general, the single-loop zoomer is the least sensitive at a given sample size. The double-loop and two-stage zoomers both achieve better trade-offs with sample size than does the single-loop zoomer, but they are also more sensitive to non-idealities at a given sample size. The ideal performance of the double-loop and two-stage zoomers is roughly comparable, but we see that in general, the two-stage zoomer is about 2–5 times more sensitive towards circuit imperfections.

# Chapter 4

# An SVD-Based Method for Band-Limiting Finite Extent Signals

In this chapter[1] we present a novel Singular Value Decomposition (SVD) based method for band-limiting oversampled discrete-time finite extent signals. As we will see in chapter 5, this general signal processing tool can be used in a reconstruction algorithm for band-limited signals from their $\Sigma\Delta$ encoded binary signals. The presented band limitation method has maximum energy concentration as defined in the chapter, its dimension agrees asymptotically with Slepian's dimension result, and the method specializes correctly to the Discrete-Time Fourier Transform (DTFT) as the sample size tends to infinity. For a specific example, we show that the SVD method gains more than 20 dB in Signal-to-Noise Ratio (SNR) over one based on the Discrete Fourier Transform (DFT). We propose an efficient computational method, based on the Lanczos algorithm of numerical analysis, for computing only those singular vectors that are needed for the SVD-based band limitation. The SVD is signal-independent, only needs to be done once and can be stored. The SVD-based band limitation is comparable in speed to the Fast Fourier Transform (FFT) for sample sizes on the order of 4096.

## 4.1 Introduction

The problem of band-limiting a finite-extent discrete-time signal is seemingly well understood. The standard solution is to take the Discrete Fourier Transform (DFT) of the signal, set the out-of-band coefficients to zero, and take the inverse DFT. This solution relies on a DFT-based definition of band limitation,

[1] Based on [94], ©IEEE.

which is natural because of the relationship between the DFT and the Discrete-Time Fourier Transform (DTFT).[2] As a further advantage, the DFT has a fast implementation, namely the Fast Fourier Transform (FFT) [95]. However, the DFT-based method relies on a periodicity assumption which can produce misleading results. For instance, an infinite-extent purely sinusoidal signal can have a band-limited DTFT, whereas a finite set of samples of the same signal only has a band-limited DFT if the frequency happens to be a bin frequency for the DFT.

In this chapter we present an alternative approach to band-limiting finite-extent signals which is aimed at alleviating the sensitive frequency dependence of the DFT. Our approach is inspired by existing literature on band-limited extrapolation of discrete-time signals [96], [98], [103], [109], and redefines band limitation in terms of the singular vectors of a matrix which has been studied in detail by Slepian [96]. These singular vectors, for which no closed-form expression exists, are known as the Discrete Prolate Spheroidal Sequences (DPSSs). The DPSSs enjoy an optimal energy concentration property [96] which is preserved in our band limitation method. Among many other signal processing areas, Singular Value Decomposition (SVD) based methods have previously been applied to band-limited extrapolation [96], [98], [103], [109], the design of digital filters with maximal energy concentration in pass-band [99], [100], and image restoration [101], [102].

Although our analysis and techniques are generally applicable, they were developed for an application to nonlinear, iterative signal reconstruction from binary encoded sequences such as those produced by $\Sigma\Delta$ modulators. This particular problem is characterized by two assumptions: First, the OverSampling Ratio (OSR) is large, where the OSR is defined as the ratio between the sampling frequency and the Nyquist frequency of the class of considered signals. Second, only a small or moderate number of samples are available, so that consecutive blocks of samples cannot be overlapped to improve performance.

The chapter is organized as follows. In section 4.2 we propose our SVD-based method for band limitation. In section 4.3 we describe a relatively efficient numerical method for performing the required, signal-independent SVD. The SVD required for our band limitation is signal independent and only needs to be done once for all signals with the same sample size and OSR. In section 4.4 we discuss the computational complexity of the DFT-based and SVD-based methods. As we will see, the actual band limitation is comparable in speed to the FFT for a sample size of 4K (1K = 1024) and an OSR of 48. In section 4.5 we present analytical and simulation results on the performance of the DFT-based and SVD-based methods as well as linear filtering. Section 4.6 contains conclusions.

[2] The DTFT is defined for infinite-extent discrete-time signals.

## 4.2 SVD-Based Band Limitation

This section describes our proposed method for band limitation. In section 4.2.1 we present the mathematical background for the SVD-based method, drawing on previous work by several researchers on band-limited extrapolation [96], [98], [103], [109]. In section 4.2.2 we provide an alternative interpretation of the band-limited extrapolation method in [103], [109]. In section 4.2.3 we propose our band limitation method and state its properties based on Slepian's dimension result [96] and an optimal energy concentration result. In section 4.2.4 we point out a relation between the SVD-based method and the so-called regularized MNLS extrapolation method [98].

### 4.2.1 Mathematical Background

In this section we introduce notation and formulate our band limitation problem as well as a related band-limited extrapolation problem.

Consider an infinite-extent sequence $\mathbf{x} = \{x_n\}$ and its DTFT,

$$X(e^{j\omega}) = \sum_{n=-\infty}^{\infty} x_n e^{-j\omega n}, \quad -\pi < \omega \leq \pi$$

The DTFT is said to be band-limited to the frequency range $\Omega = (-\alpha, +\alpha)$, $0 < \alpha < \pi$, if and only if $X(e^{j\omega}) = 0$ for all $\omega \notin \Omega$. We define the oversampling ratio (OSR) to be $\pi/\alpha$. We also define an infinite-extent noisy version $\mathbf{s} = \{s_n\}$ of $\mathbf{x}$. No assumption of white or uncorrelated noise is made at this point. The noise on $\mathbf{s}$ outside of $\Omega$ can be rejected using an ideal low-pass filter. The ideal low-pass operation can also be viewed as multiplying $\mathbf{s}$ by an infinite-dimensional matrix[3] $\mathbf{L} = \{\ell_{nm}\}$ with elements [98], [103]

$$\ell_{mn} = \begin{cases} \dfrac{\sin \alpha(m-n)}{\pi(m-n)} & \text{for } m \neq n \\ \dfrac{\alpha}{\pi} & \text{for } m = n \end{cases}, \quad m, n \in \mathcal{Z} = \{\cdots, -1, 0, 1, \cdots\}$$

We use the subscript $\Lambda$ to denote the time limitation of a vector to an index set $\Lambda = \{0, 1, \cdots, N-1\}$. Defining an $N \times \infty$ time limitation matrix $\mathbf{T} = \{t_{mn}\}$ as

$$t_{mn} = \begin{cases} 1 & \text{for } m = n \\ 0 & \text{for } m \neq n \end{cases}, \quad m \in \Lambda, n \in \mathcal{Z}$$

we thus have $\mathbf{s}_\Lambda = \mathbf{T}\mathbf{s}$ and $\mathbf{x}_\Lambda = \mathbf{T}\mathbf{x}$.

The simplest version of our band limitation problem is to estimate $\mathbf{x}_\Lambda$ from the set of observed samples $\mathbf{s}_\Lambda$. This can be related to the well-known band-limited extrapolation problem [98], [103] of finding an infinite-extent sequence

[3] The use of infinite-dimensional matrices is purely a notational convenience.

$\tilde{\mathbf{x}}$ such that $\tilde{\mathbf{x}}$ is band-limited and is as close as possible to the observed vector $\mathbf{s}_\Lambda$:

$$\tilde{\mathbf{x}} = \mathbf{L}\tilde{\mathbf{x}} \quad \text{and} \quad \|\mathbf{s}_\Lambda - \mathbf{T}\tilde{\mathbf{x}}\|_\Lambda^2 = \|\mathbf{T}(\mathbf{s} - \tilde{\mathbf{x}})\|_\Lambda^2 \quad \text{is minimized} \tag{4.1}$$

In (4.1) we have used the time-limited 2-norm defined for an arbitrary index set $D$,

$$\|\mathbf{s}\|_D^2 \stackrel{\Delta}{=} \sum_{n \in D} s_n^2$$

Having solved the related problem (4.1), we can consider the time limitation $\tilde{\mathbf{x}}_\Lambda = \mathbf{T}\tilde{\mathbf{x}}$ to be the band limitation of $\mathbf{s}_\Lambda$.

As might be expected from the sampling theorem [24], there are in fact infinitely many infinite-extent sequences $\tilde{\mathbf{x}}$ which have band-limited DTFTs, that is, $\tilde{\mathbf{x}} = \mathbf{L}\tilde{\mathbf{x}}$, and which also pass through any finite collection of samples $\mathbf{s}_\Lambda$, that is, $\mathbf{s}_\Lambda = \mathbf{T}\tilde{\mathbf{x}}$ [103]. Therefore the solution to (4.1) is not unique, and the optimal estimate of $\mathbf{x}_\Lambda$ is the trivial solution $\tilde{\mathbf{x}}_\Lambda = \mathbf{s}_\Lambda$. In the context of band-limited extrapolation, the standard way to obtain a unique solution to (4.1) is to choose the minimum-energy one [98], [103]. This is empirically justified because there exist solutions to (4.1) for any $\mathbf{s}_\Lambda$ which assume arbitrarily large values outside of $\Lambda$, whereas in practice, we are interested in extrapolates with similar behavior over $\Lambda$ and outside of $\Lambda$. One can therefore replace (4.1) by the following least-squares band-limited extrapolation problem: Among all the infinite-extent sequences which pass through $\mathbf{s}_\Lambda$, choose the one with minimum energy, that is:

$$\min \|\mathbf{s}_\Lambda - \mathbf{T}\mathbf{L}\tilde{\mathbf{x}}\|_\Lambda^2 \quad \text{such that} \quad \|\tilde{\mathbf{x}}\|_\mathcal{Z}^2 \quad \text{is minimized} \tag{4.2}$$

Note that there is no explicit requirement for $\tilde{\mathbf{x}}$ to be band-limited, $\tilde{\mathbf{x}} = \mathbf{L}\tilde{\mathbf{x}}$ as in (4.1). However, it is easily shown that due to the minimization of the energy of $\tilde{\mathbf{x}}$, the solution to (4.2) is implicitly band-limited. Rephrasing our band limitation problem in terms of (4.2), we can consider the time limitation $\mathbf{b} \stackrel{\Delta}{=} \mathbf{T}\tilde{\mathbf{x}}$ to be the band-limited estimate of $\mathbf{x}_\Lambda$.

The solution to (4.2) can be expressed in terms of the $N$ orthogonal infinite-extent Discrete Prolate Spheroidal Sequences (DPSSs) [96] which we will denote by $\{\mathbf{U}_n\}$. For any finite $N$, all the $N$ DPSSs $\{\mathbf{U}_n\}$ depend only on $N$ and the OSR, and have DTFTs that are band-limited to $\Omega$. We will also use the truncated DPSSs $\{\mathbf{u}_n\}$ defined by $\mathbf{u}_n = \mathbf{T}\mathbf{U}_n$; with suitable normalization, these are orthonormal over $\Lambda$ [96]. The truncated DPSSs are the left singular vectors of the matrix $\mathbf{A} = \mathbf{T}\mathbf{L}$, or equivalently, the eigenvectors of the $N \times N$ matrix $\mathbf{L}_\Lambda \stackrel{\Delta}{=} \mathbf{A}\mathbf{A}^T = \mathbf{T}\mathbf{L}\mathbf{T}^T$ [104]. Note that $\mathbf{L}_\Lambda$ equals the $\infty \times \infty$ matrix $\mathbf{L}$ truncated in both dimensions to the index set $\Lambda$. We denote the ordered singular values of $\mathbf{L}_\Lambda$ by $\{\sigma_0, \cdots, \sigma_{N-1}\}$; they satisfy $1 > \sigma_0 > \cdots > \sigma_{N-1} > 0$ [96]. We see that $\mathbf{L}_\Lambda$ has full rank.

Returning to (4.2), the band-limited extrapolation solution is [96]

$$\tilde{\mathbf{x}} = \sum_{n=0}^{N-1} (\mathbf{u}_n^T \mathbf{s}_\Lambda) \cdot \mathbf{U}_n \tag{4.3}$$

Truncating $\tilde{\mathbf{x}}$ to $\Lambda$, we again obtain the trivial band limitation solution

$$\mathbf{b} = \mathbf{T}\mathbf{L}\tilde{\mathbf{x}} = \sum_{n=0}^{N-1} (\mathbf{u}_n^T \mathbf{s}_\Lambda) \cdot \mathbf{u}_n = \mathbf{s}_\Lambda \tag{4.4}$$

This is also clear from (4.2) itself, because the least squares constraint only singles out the extrapolate with minimum energy that goes through the given samples $\mathbf{s}_\Lambda$. Thus, even with the energy minimization, any finite-extent sequence can be considered band-limited. We resolve this conflict with intuition in the next sections.

### 4.2.2 Dimension Considerations

It has been proposed to replace (4.3) by the truncation [103], [109]

$$\tilde{\mathbf{x}} = \sum_{n=0}^{r-1} (\mathbf{u}_n^T \mathbf{s}_\Lambda) \cdot \mathbf{U}_n, \qquad r \approx N/\mathrm{OSR} \tag{4.5}$$

This change is often suggested as a way to improve the numerical stability in computing (4.3) [103], [109]. Specifically, it is shown in [96] that approximately $N/\mathrm{OSR}$ of the singular values $\{\sigma_n\}$ are very close to 1, and the rest are very close to 0. Thus, (4.5) amounts to discarding the part of $\mathbf{L}_\Lambda$ that does not have full numerical rank.[4] Note that $\tilde{\mathbf{x}}$ as defined in (4.5) no longer solves the band-limited extrapolation problem (4.2).

In this section we make the alternative interpretation that a dimension consideration suggests the use of (4.5). We base this claim on the behavior of (4.3) as $N \to \infty$: Due to the summation limits in (4.3), the right-hand side tends to $\mathbf{s}_\Lambda$, and this band-limited extrapolation does not agree with the desired result $\tilde{\mathbf{x}} = \mathbf{L}\mathbf{s}_\Lambda$ obtained with the DTFT. The disagreement can be removed by changing (4.3) to (4.5), because of Slepian's dimension result which is summarized as follows in [96]: *"For large N the set of sequences of bandwidth W that are confined to an index set of length about N has dimension approximately 2WN."* In our case, the product $2WN$ equals $N/\mathrm{OSR} = N\alpha/\pi$, so $r = N/\mathrm{OSR}$ orthonormal vectors are asymptotically necessary to span the space of finite-extent sequences on $\Lambda$ that are band-limited to $\Omega$. By choosing $r \approx N/\mathrm{OSR}$, we obtain the correct asymptotic specialization at the cost of not exactly solving the extrapolation problem (4.2).

### 4.2.3 Proposed Band Limitation Method

In this section we propose our band limitation method and describe two important properties, namely optimal energy concentration and correct specialization to the DTFT.

[4] The concept of numerical rank is defined in [104] and roughly refers to the fact that matrices with progressively larger condition numbers become harder to distinguish from singular matrices on finite-precision computers.

The re-interpretation of (4.5) in the previous section suggests the use of (4.5) as the basis of a band limitation method. Equation (4.5) can be viewed as connecting the well-known problem of band-limited extrapolation to the present problem of band limitation. Based on (4.5), we therefore propose the band limitation method

$$\mathbf{b} = \mathbf{T}\tilde{\mathbf{x}} = \sum_{n=0}^{r-1} (\mathbf{u}_n^T \mathbf{s}_\Lambda) \cdot \mathbf{u}_n, \qquad r \approx N/\mathrm{OSR} \tag{4.6}$$

that is, a projection onto the $r$ truncated DPSSs with the largest singular values, or the "largest eigenvectors" for short. The band limitation in (4.6) is not a solution to a specific minimization problem, but it has two desirable properties described below. Before that, however, we discuss possible definitions of band limitation for finite-extent sequences.

Equation (4.6) is a special case of the most general linear band limitation method with dimension $r$,

$$\mathbf{b}_{\text{general}} = \sum_{n=0}^{r-1} (\mathbf{a}_n^T \mathbf{s}_\Lambda) \cdot \mathbf{a}_n \tag{4.7}$$

where $\{\mathbf{a}_0, \cdots, \mathbf{a}_{r-1}\}$ is an arbitrary set of orthonormal $N$-dimensional "baseband" vectors. DFT-based band limitation is another special case of (4.7) with complex exponentials as the baseband vectors. We are free to view (4.7) as *defining* a class of band-limited finite-extent sequences for each choice of $\{\mathbf{a}_n\}$, namely the span of $\{\mathbf{a}_n\}$. There is no immediate reason to prefer one choice of vectors $\{\mathbf{a}_n\}$ over another. However, for consistency with the established notion of band-limitedness for infinite-extent sequences, we will require that the band limitation method (4.7) specializes correctly to the DTFT as the sample size tends to infinity.

The specific band limitation method (4.6) does indeed specialize to the DTFT as $N \to \infty$. This desirable property follows since for $N = \infty$, the matrix $\mathbf{L}_\Lambda$ becomes the ideal low-pass matrix $\mathbf{L}$ which has the complex exponentials $\exp(j\omega n)$ as its eigenvectors. For $N = \infty$ we in fact have the following properties: (a) The DPSSs and the truncated DPSSs are identical and are equal to complex exponentials. (b) The fraction 1/OSR of the infinite-extent eigenvectors are band-limited to the frequency range $\Omega$, and correspond to the eigenvalue 1 of the infinite-extent matrix $\mathbf{L}$. (c) The remaining eigenvectors contain strictly high-frequency energy, and correspond to the eigenvalue 0 of $\mathbf{L}$. Note that although the matrix $\mathbf{L}_\Lambda$ is non-singular and the DPSSs are band-limited for finite $N$, the infinite-extent matrix $\mathbf{L}$ does not depend on $N$ and is always singular. Only the fraction 1/OSR of its eigenvectors are band-limited.

The second desirable property of the band limitation method (4.6) is that within the class of methods (4.7), choosing the baseband vectors $\{\mathbf{a}_n\}$ to be the truncated DPSSs is optimal in an energy concentration sense. This result is based on the fact that among all band-limited infinite-extent sequences, the first DPSS, $\mathbf{U}_0$, is the one with the largest possible fraction, namely $\sigma_0^2$, of its energy on the set $\Lambda$. The second DPSS, $\mathbf{U}_1$, is the band-limited sequence which is orthogonal to $\mathbf{U}_0$ and has the largest fraction of its energy, namely $\sigma_1^2$, on $\Lambda$, and so on [96]. We define the energy concentration of the general band limitation method (4.7) as a function of $\mathbf{s}_\Lambda$ to be

$$C_{\text{general}}(\mathbf{s}_\Lambda) \triangleq \frac{\| \mathbf{b}_{\text{general}} \|_\Lambda^2}{\| \mathbf{b}_{\text{BLX,general}} \|_\mathcal{Z}^2} \tag{4.8}$$

where $\mathbf{b}_{\text{BLX,general}}$ is the minimum-energy band-limited extrapolate of $\mathbf{b}_{\text{general}}$ given by (4.5). Maximizing $C_{\text{general}}(\mathbf{s}_\Lambda)$ is reasonable, since $\| \mathbf{b}_{\text{general}} \|_\Lambda^2$ should contain as much as possible of the energy of the true band limitation solution, $\mathbf{s}_\Lambda$ in (4.4), in relation to the energy of the infinite-extent band-limited extrapolate given by (4.5). We show in appendix C.1 that for (4.7),

$$\min_{\mathbf{s}_\Lambda} C_{\text{general}}(\mathbf{s}_\Lambda) \leq \sigma_{r-1}^2 \tag{4.9}$$

We also show in appendix C.1 that the SVD-based method (4.6) has optimal energy concentration in that it achieves the upper bound $\sigma_{r-1}^2$ in (4.9). Thus, choosing the vectors $\{\mathbf{a}_n\}$ to be the truncated DPSSs $\{\mathbf{u}_n\}$ indeed optimizes energy concentration.

### 4.2.4 Relation to Regularized Extrapolation

In this section we describe a connection between our proposed band limitation method (4.6) and the so-called regularized Minimum Norm Least Squares (MNLS) method for band-limited extrapolation mentioned in [98]. In the more general context of ill-conditioned linear least squares systems of equations, our SVD-based method is known as Truncated SVD (TSVD) [105], whereas the MNLS-based method corresponds to Tikhonov regularization [105]. The regularized MNLS method amounts to finding the signal $\tilde{\mathbf{x}}$ that minimizes

$$\| \mathbf{s}_\Lambda - \mathbf{TL}\tilde{\mathbf{x}} \|_\Lambda^2 + \epsilon \, \| \tilde{\mathbf{x}} \|_\mathcal{Z}^2, \quad \epsilon > 0 \text{ small}$$

rather than $\| \mathbf{s}_\Lambda - \mathbf{TL}\tilde{\mathbf{x}} \|_\Lambda^2$ as in (4.2). By time-limiting the solution $\tilde{\mathbf{x}}$, we obtain the band limitation [105]

$$\mathbf{b}_{\text{Tikhonov}} = \sum_{n=0}^{N-1} \frac{\sigma_n^2}{\sigma_n^2 + \epsilon^2} (\mathbf{u}_n^T \mathbf{s}_\Lambda) \cdot \mathbf{u}_n \tag{4.10}$$

As noted in [105], $\epsilon$ significantly reduces the components of $\mathbf{b}_{\text{Tikhonov}}$ along the singular vectors $\mathbf{u}_n$ whose singular values $\sigma_n$ are much smaller than $\epsilon$. If $\epsilon$ lies in a gap in the singular value spectrum, (4.10) is a good approximation to (4.6). In practice, the spectrum of $\mathbf{L}_\Lambda$ decreases rapidly from 1 towards 0. Therefore, if $\epsilon \ll 1$ as is usually the case [98], the number of singular values close to $\epsilon$ is small. We conclude that Tikhonov band limitation can be viewed as an approximation to TSVD, where the number of terms in the TSVD are related to $\epsilon$.

## 4.3 Implementation of the SVD

In this section we describe an efficient numerical implementation of the SVD required in section 4.2.1. The decomposition only needs to be done once for each sample size and OSR. The key observations regarding the matrix $\mathbf{L}_\Lambda$ are the following:

1. We only need about $r = N/\text{OSR}$ of the eigenvectors of $\mathbf{L}_\Lambda$.

2. $\mathbf{L}_\Lambda$ is a Toeplitz matrix, so we can multiply a vector by $\mathbf{L}_\Lambda$ in $O(2N \log 2N)$ operations using a $2N$-point FFT, rather than $O(N^2)$ operations calculating the product directly.

Observation 1 implies that the Lanczos algorithm is ideally suited for the eigen-decomposition [104], [106]. This algorithm generates a series of tridiagonal matrices, starting with a small matrix in the first iteration and ending with an $N \times N$ one in the last iteration. The final $N \times N$ matrix has the same eigen-decomposition as the original matrix $\mathbf{L}_\Lambda$, and the algorithm can take advantage of the fact that there exist efficient methods to decompose tridiagonal matrices. In practice, the smaller tridiagonal matrices generated in the early iterations of the Lanczos algorithm have eigenvalues and eigenvectors that are good approximations to the largest and smallest eigenvalues and corresponding eigenvectors of $\mathbf{L}_\Lambda$. In addition, the algorithm provides error bounds on its estimates which become tighter from one iteration to the next. Since our goal is to obtain the largest singular vectors, we can stop the iteration once the error bounds on the estimates of these vectors become sufficiently small. In addition, the iteration should not terminate until a number of iterations have failed to produce new eigenvalue estimates close to 1 [107].

As a rule of thumb, we have found in our simulations that by generating $2N/\text{OSR}$ eigenvectors, that is, generating the $(2N/\text{OSR}) \times (2N/\text{OSR})$ tridiagonal matrix, we obtain all the about $N/\text{OSR}$ large eigenvectors with sufficient accuracy. In doing so, we also inevitably obtain about $N/\text{OSR}$ small eigenvectors which we discard. The Lanczos algorithm is numerically sensitive, and must be implemented with a keen eye on the orthogonality of the eigenvector estimates as they are iteratively generated [107].

The Lanczos algorithm requires modest storage, because no storage for intermediate matrices is necessary. In addition, explicit storage of $\mathbf{L}_\Lambda$ is not necessary, as the only requirement is a subroutine that can multiply an arbitrary vector by $\mathbf{L}_\Lambda$. The $N \times N$ matrix $\mathbf{L}_\Lambda$ is completely specified by $N$ numbers, and the time required for multiplications by $\mathbf{L}_\Lambda$ is significantly reduced through observation 2 above. Thus the algorithm can be implemented so that it essentially only requires memory for the desired large eigenvectors, as well as for some of the small eigenvectors which also converge. The latter ones are discarded, and not stored permanently.

We have found two ways of ascertaining that all large eigenvectors are generated. One is to use the knowledge that there are approximately $N/\text{OSR}$ large eigenvectors. A further, empirically motivated check is to count the number of sign changes in the large eigenvectors viewed as time sequences from 0 to $N - 1$. If the eigenvectors are numbered in terms of decreasing eigenvalues starting from $n = 0$, the number of sign changes in the $n$th eigenvector appears to be $n$; therefore, a gap in the ranked list of sign changes reveals a missing eigenvector. We have found no proof of this sign change conjecture, but a similar observation can be proved for discrete prolate spheroidal wave functions [96] which are the continuous-time equivalents of the discrete-time DPSSs.

In practice, the sign change conjecture can break down because there are many large eigenvalues that are extremely close to 1 and thus to each other, and the Lanczos algorithm cannot distinguish between these. Therefore, the algorithm tends to just produce *some* orthonormal basis of the subspace spanned by the large eigenvectors, rather than the actual orthonormal basis of truncated DPSSs. For our application of projecting onto the corresponding subspace, this fact is unimportant. But for purposes of counting sign changes, the stated conjecture breaks down for the large eigenvectors as a result. However, in the transition region where eigenvalues start to decrease from 1 to 0, each numerically obtained eigenvalue is typically spaced apart by one or two orders of magnitude from its neighbors, until the eigenvalues become so small as to reach machine precision. These transitory eigenvectors can be clearly distinguished by the algorithm, and the sign change conjecture is found to hold for them. To ascertain that one has obtained all large eigenvectors, one can therefore compare their number to the number expected when counting backwards from the number of sign changes of the transitory eigenvectors.

## 4.4 Computational Properties of Band Limitation Methods

In this section we qualitatively compare the computational characteristics of the DFT, linear filtering and the SVD-based method introduced in section 4.2 for our band limitation problem. We also discuss the applicability of gradient and conjugate gradient algorithms.

The FFT implementation of the DFT, and hence the DFT-based method, has an execution time of $O(N \log N)$ and a memory requirement of $O(N)$.

Linear filtering for band limitation can be implemented by padding the $N$-sample segment $\mathbf{s}_\Lambda$ with zeros, filtering the result as an infinite-extent sequence, and truncating the result to $\Lambda$. In its direct implementation, linear filtering has an execution time of order $O(MN)$ and a memory requirement of order $O(N)$ for an $M$-tap Finite Impulse Response (FIR) filter. If $M$ is large, the execution time can be reduced to order $O(N \log N)$ by appropriate use of the FFT.

For the SVD-based method (4.6), we distinguish between performing the initial SVD using the Lanczos algorithm, and performing the SVD-based band limitation. On a SUN Sparc IPC, a Lanczos-based partial SVD of the $\mathbf{L}_\Lambda$-matrix takes $3\frac{1}{2}$ minutes when $N = 2048$ and OSR $= 48$, and 27 minutes when

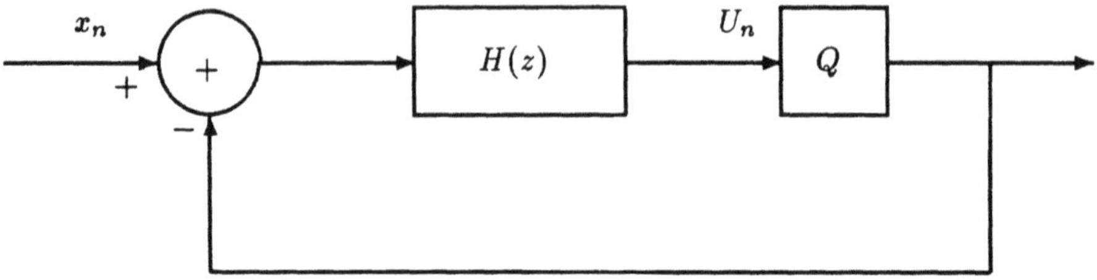


Figure 4.1: *Discrete-time model of the interpolative ΣΔ encoder.*

$N = 4096$ and OSR = 48. The disk storage requirement is $O(N^2/\text{OSR})$, or 3 Mbytes in the latter case. For comparison, a full eigen-decomposition even in the former case takes days and requires about 70 Mbytes of virtual memory.

Performing an SVD-based band limitation requires $O(N^2/\text{OSR})$ time, which is only somewhat slower than the DFT-based method for small to moderate $N$. For $N = 4096$ and OSR = 48, a DFT band limitation takes 0.4 seconds, whereas an SVD band limitation takes 1.2 seconds on a SUN Sparc IPC.

A different approach to the band limitation and band-limited extrapolation problems is that of approximately and iteratively solving a system of $N$ linear equations that follow from (4.2) [103]. Jain and Ranganath [103] propose the use of the gradient and conjugate gradient algorithms. These algorithms require modest storage, and the most computationally intensive steps are multiplications with the matrix $\mathbf{L}_\Lambda$ which can be done via an FFT because $\mathbf{L}_\Lambda$ is a Toeplitz matrix. Theoretically, both algorithms guarantee convergence to a unique solution, because $\mathbf{L}_\Lambda$ is positive definite. In practice, however, $\mathbf{L}_\Lambda$ is extremely ill-conditioned even for small values of $N$ such as 16. For $N = 16$ and OSR = 48, we have found by simulations that the solution obtained by the gradient algorithm is extremely sensitive to round-off in the calculations, although it does converge. Several variants of the conjugate gradient algorithm, including that in [103] and those in [104], always diverged for our application, for small as well as large values of $N$.

## 4.5 Results

This section presents performance results for the SVD-based method for band limitation from section 4.2.1, as well as for the DFT-based method and linear filtering. The sample size is 4K throughout. We consider two classes of input signals. The first is the class of pure sinusoids with baseband frequencies that are not necessarily bin frequencies for the DFT. The second class consists of binary encoded versions of signals in the first class. The encoder we use is the fourth order interpolative ΣΔ encoder described in [32], which is designed for an oversampling ratio of OSR = 48. The encoder is shown in figure 4.1, and the coefficients of the discrete-time low-pass filter $H(z)$ can be found in

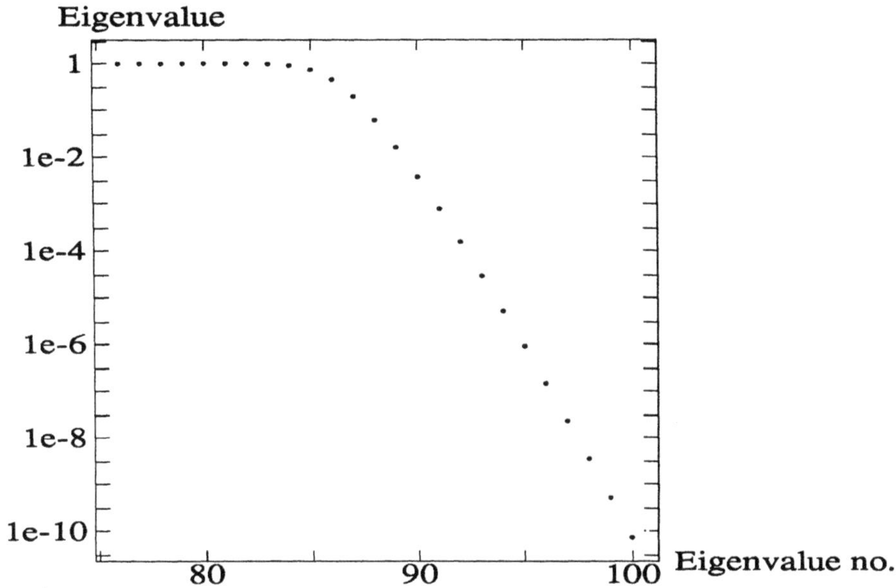


Figure 4.2: *Eigenvalues of the matrix* $\mathbf{L}_\Lambda$ *for* $N = 4096$ *and* $\text{OSR} = 48$. *Approximately* $N/\text{OSR} = 85$ *of the eigenvalues are close to 1.*

section 7.4.1. The noise introduced by the $\Sigma\Delta$ encoder is not white—in fact, it has a strong high-pass characteristic. Neither is the noise uncorrelated with the signal, but it can be treated approximately as such. A Signal-to-Noise Ratio (SNR) on the order of 90 dB can be achieved with certain operations on the binary sequence [32], [93], [108], and band limitation is an essential part of these operations. The SNR in approximating an actual signal $\mathbf{x} = \{x_0, x_1, \cdots, x_{N-1}\}$ with the estimate $\tilde{\mathbf{x}} = \{\tilde{x}_0, \tilde{x}_1, \cdots, \tilde{x}_{N-1}\}$ is defined by $10\log_{10}(E_{\text{signal}}/E_{\text{noise}})$, where the signal and noise energies are given by

$$E_{\text{signal}} = \sum_{n=0}^{N-1} x_n^2, \quad E_{\text{noise}} = \sum_{n=0}^{N-1} (x_n - \tilde{x}_n)^2$$

### 4.5.1 SVD-Based Method

For $N = 4096$ and $\text{OSR} = 48$, we expect about $N/\text{OSR} = 85$ of the eigenvalues of $\mathbf{L}_\Lambda$ to be close to 1, and the rest to be close to 0, as explained in section 4.2.1. Figure 4.2 shows the eigenvalues around index 85 in a logarithmic plot. The plot confirms the theory. The eigenvalues obtained numerically continue to decrease from 1 towards 0 until they reach machine precision, which is about $10^{-16}$ in our case.

The eigenvectors corresponding to large eigenvalues are expected to have DFTs with a low-pass character. Figure 4.3 shows the magnitudes of the DFTs

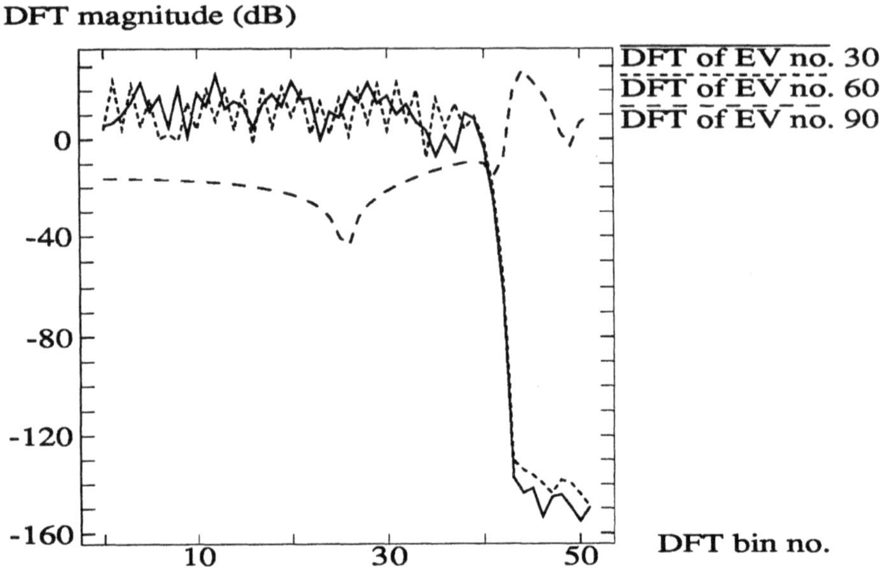


Figure 4.3: *DFTs of three eigenvectors of the matrix* $\mathbf{L}_\Lambda$ *for* $N = 4096$ *and* OSR $= 48$. *The two eigenvectors with indices 30 and 60 are low-pass, and the one with index 90 is high-pass. The baseband ends at bin frequency no. 43, and only a small portion of the whole spectrum is shown.*

of three eigenvectors, namely the ones with index 30, 60 and 90. The baseband extends up to bin number $N/(2\text{OSR}) = 43$, and only a portion of the whole spectrum with $N$ bins is shown.[5] As described in section 4.3, the large eigenvectors obtained numerically are not actually truncated DPSSs, but are still orthonormal and span the same space as the corresponding truncated DPSSs. The figure shows that the DFTs of the two eigenvectors with indices less than 85 have a low-pass character, whereas the last DFT has a high-pass character.

The problem of choosing the number $r$ of eigenvectors in (4.6) still remains, although asymptotically we must have $r \approx N/\text{OSR}$. In the context of band-limited extrapolation, it is well known that the parameter $r$ controls a trade-off between two properties of the extrapolate, namely the accuracy with which noise-free signals can be represented, and the noise sensitivity [101], [102], [109]. The same observation holds for our band limitation method (4.6): Larger values of $r$ means that the subspace onto which we project has a larger dimension, implying that more of a signal, but also more noise can be represented. A good choice of $r$ thus depends on the expected amount of noise. There exist formulas to aid the choice for band-limited extrapolation under various assumptions, such as when Karhunen-Loève expansions of both the random signal and its noise signal are available [102], [109]. In this paper we will choose the exact

[5] There are also $N/(2\text{OSR})$ baseband bins symmetrically around the middle bin, bin number $N/2$, with complex DFT coefficients that are complex conjugates of the shown ones.

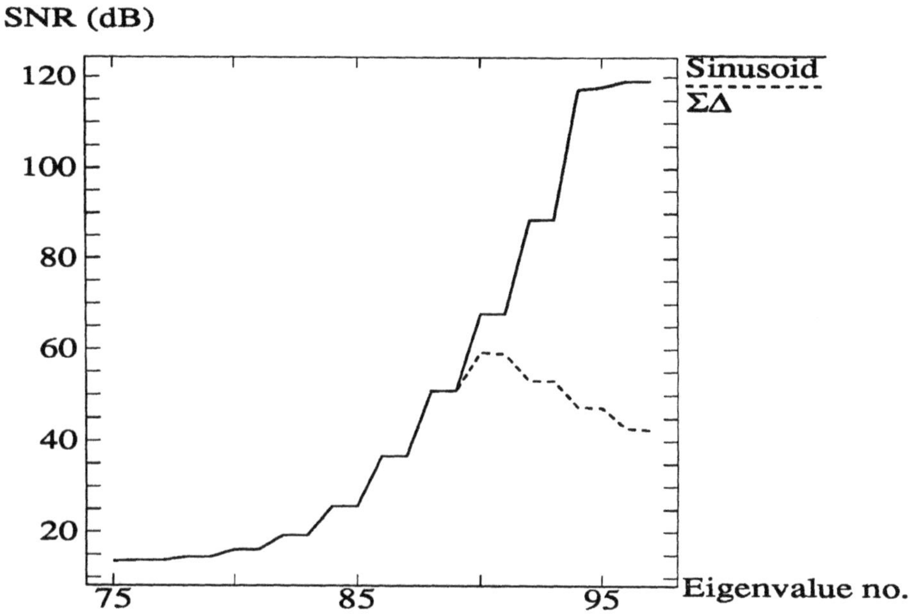


Figure 4.4: *SNR for the SVD-based method as a function of eigenvalue number $r$ of the matrix $\mathbf{L}_\Lambda$ for $N = 4096$ and OSR $= 48$. The input signals are a sinusoid and its $\Sigma\Delta$ encoded version. The input frequency is $\omega = \frac{9}{32}\sigma$, and the input amplitude is at -6 dB relative to the quantizer step size of the $\Sigma\Delta$ encoder.*

value of $r$ empirically.

Figure 4.4 shows SNR curves for the SVD-based method as a function of eigenvector number for indices around 85. Two signals are considered, namely a sinusoid and its $\Sigma\Delta$ encoded binary version. The frequency is $\alpha = \frac{9}{32}\sigma$, that is, a bin frequency safely below the baseband edge. As expected, the SNR for the sinusoid in this noise-free case is an increasing function of the number of singular vectors. For the very noisy $\Sigma\Delta$ encoded signal, the optimum value of $r$ is 90 or 91, and the corresponding SNR is 59 dB. The optimum value of $r$ depends on the input frequency, and should be chosen by taking the whole baseband range into account.

Figure 4.5 shows SNR curves as a function of input frequency for the SVD-based method with a fixed number $r = 91$ of eigenvectors. Again, the input sequences are baseband sinusoids and their $\Sigma\Delta$ encoded versions. Both bin and non-bin frequencies for a corresponding 4K DFT are considered. The figure shows that the SNR oscillates between about 58 and 65 dB for sinusoidal inputs, and between about 38 dB and 58 dB for $\Sigma\Delta$ encoded inputs. The curve for the $\Sigma\Delta$ encoded signals averages about 48 dB. Both curves exhibit some signal frequency dependence.

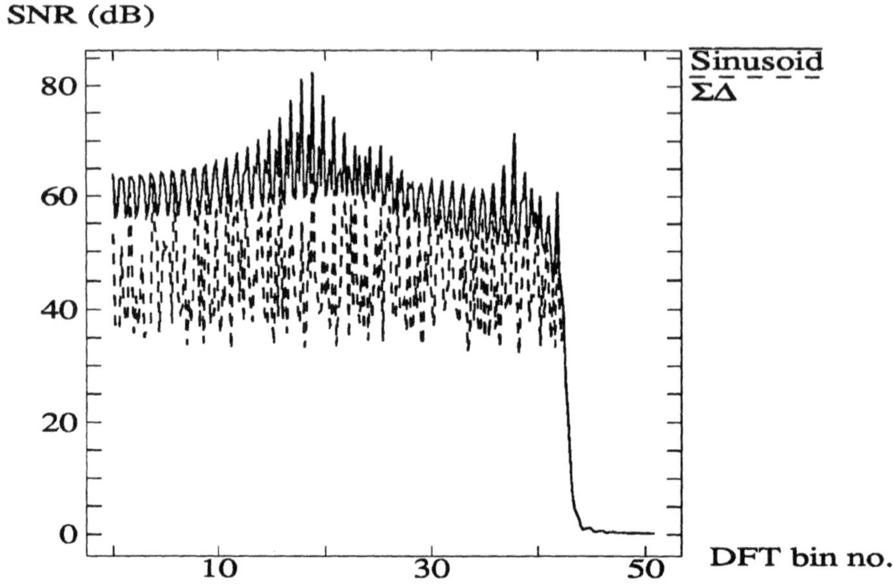


Figure 4.5: *SNR for the SVD method as a function of input frequency for a fixed number $r = 91$ of eigenvectors, $N = 4096$ and $\mathrm{OSR} = 48$. The input signals are a sinusoid and its $\Sigma\Delta$ encoded binary version. Frequencies are not limited to integer bins. The input amplitude is at -6 dB relative to the quantizer step size of the $\Sigma\Delta$ encoder.*

### 4.5.2 DFT-Based Method

Figures 4.6 and 4.7 are analogous to figure 4.5 and show SNR curves as a function of input frequency when the band limitation is done with the DFT. The input sequences used in figure 4.6 are baseband sinusoids, and the inputs used in figure 4.7 are their $\Sigma\Delta$ encoded versions. The figures show that the SNR depends heavily on the input frequency. For sinusoidal inputs, the SNR oscillates between infinity at bin frequencies and about 21 dB half way between bin frequencies; this should be contrasted with the SVD method for which the SNR oscillates between about 58 dB and 65 dB. For the $\Sigma\Delta$ encoded inputs, the SNR oscillates between about 21 dB and 75 dB, which should be contrasted with 38-58 dB for the SVD-based method. For both input types, the SVD-based method exhibits less frequency dependence than the DFT, but the largest SNR is smaller for the SVD-based method than for the DFT-based one.

In appendix C.2, we derive an approximation to the minimum SNR of the DFT-based method for sinusoidal baseband inputs. Specifically, we show that the signal frequency can be chosen half way between DFT bin frequencies such that

$$\mathrm{SNR}_{\min} \approx 10 \log_{10} \left( \frac{N\pi^2}{4\mathrm{OSR}} \right)$$

For $N = 4096$ and $\mathrm{OSR} = 48$, $\mathrm{SNR}_{\min} = 23$ dB which agrees well with 21 dB as

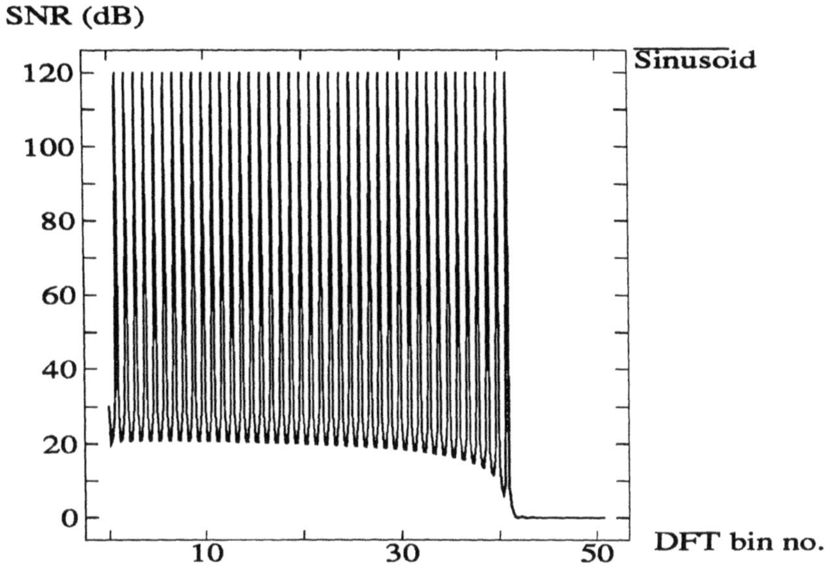


Figure 4.6: *SNR for the DFT method as a function of input frequency for* $N$ = 4096 *and* OSR = 48. *Infinite SNR is represented as 120 dB. The input signal is a sinusoid. Frequencies are not limited to integer bins.*

observed above. This minimum SNR only increases by only 3 dB/octave with the sample size.

The standard remedy against so-called spectral leakage of non-bin frequencies is windowing [24], [110]. However, the price for suppressing artificial high-frequency (HF) noise is the introduction of significant signal frequency error. As an illustration, consider a sinusoid at a bin frequency. It is shown in appendix C.3 that applying the $N$-point Hanning window [110] results in an SNR of 4.8 dB independent of OSR and sample size. This very low number results because the SNR metric includes all 2-norm error between the actual signal and the estimate, and windowing smears the signal spectrum.

### 4.5.3 Linear Filtering

In this section we apply an FIR filter to the sinusoidal and $\Sigma\Delta$ encoded 4K sequences used above. Since the sequences are padded with zeroes, filtering results in large edge effects. Edge effects can be reduced by reducing the filter length, but this also degrades the filter transfer function. Infinite Impulse Response (IIR) filters also suffer from the edge effect problem. In addition, IIR filters have non-constant group delay over baseband, and the resulting frequency-dependent time shift further distorts the signal components.

The FIR filter used has 215 taps, a stop-band attenuation of 118 dB, a pass-band ripple of 0.36 dB and a transition band of the same width as the

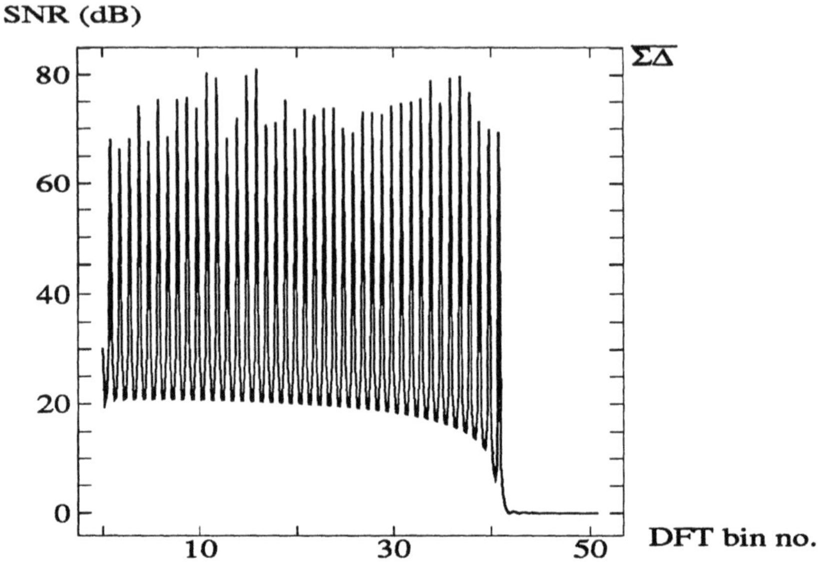


Figure 4.7: *SNR for the DFT method as a function of input frequency. Parameters are as in figure 4.6, except that the input signal is a $\Sigma\Delta$ encoded binary version of a sinusoid. The input amplitude is at -6 dB relative to the quantizer step size of the $\Sigma\Delta$ encoder.*

baseband. The filter was designed using the Parks-McClellan algorithm [111].[6] Figure 4.8 shows the SNR curves as a function of frequency for the FIR filter. The performance is seen to be almost the same for sinusoidal and $\Sigma\Delta$ encoded inputs. The frequency dependence of the SNR is fairly small, but the overall level of about 26 dB is more than 20 dB lower than the 48 dB level for the SVD method with $\Sigma\Delta$ encoded inputs.

## 4.6 Summary

In this chapter, we addressed the problem of band-limiting finite-extent sequences. We related the problem to that of band-limited extrapolation, and derived an SVD-based method for band limitation. In doing so, we redefined band limitation in terms of the truncated DPSSs. We stress the interpretation that the SVD-based method is chosen to have the correct dimension specified by Slepian's result [96], to specialize correctly to the DTFT as the sample size tends to infinity, and to have maximum energy concentration. As a result, the method cannot exactly solve the band limitation problem for finite $N$.

We presented a relatively efficient method for performing SVD-based band limitation of sequences with up to 4K or 8K samples, based on the Lanczos

[6]The specifications are taken from [112] where such a filter is used as the first stage in a two-stage low-pass filter for the same fourth order $\Sigma\Delta$ encoder that we use.

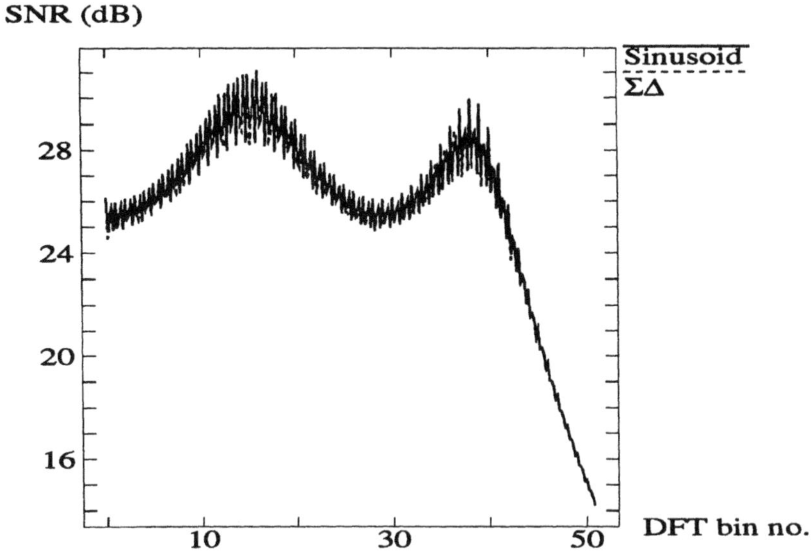


Figure 4.8: *SNR for an FIR filter as a function of input frequency, for* $N = 4096$ *and* $\mathrm{OSR} = 48$. *The input signals are a sinusoid and its* $\Sigma\Delta$ *encoded version. The input amplitude is at -6 dB relative to the quantizer step size of the* $\Sigma\Delta$ *encoder.*

algorithm. Finally, we presented a performance comparison between the SVD-based and DFT-based methods. The SVD-based method was found to result in SNR gains of up to 20 dB over the DFT for non-bin frequencies, but at bin frequencies the DFT is of course unsurpassed. The SVD-based method is less dependent on input frequency than the DFT-based one; its main drawback is the storage requirement for the singular vectors.

# Chapter 5

# Optimal Decoding for ONS Decoders with Band-Limited Inputs

In this chapter[1] we consider the problem of reconstructing a band-limited input signal from its $\Sigma\Delta$ encoded binary version. The proposed nonlinear decoding algorithm is based on the theory of Projections Onto Convex Sets (POCS), and alternates between a time-domain operation and a band limitation to find a signal invariant under both. The time-domain operation results in a quadratic programming problem. The band limitation can be based on a signal-independent singular value decomposition of a certain matrix. We show simulation results for the SNR performance of POCS-based decoding and linear decoding for a number of $\Sigma\Delta$ encoders, including the single-loop, double-loop and two-stage encoders and a specific fourth order interpolative encoder. We find improvements in SNR of up to 20-30 dB for the single-loop, double-loop and two-stage encoders depending on the oversampling ratio, and up to 10 dB for the fourth order encoder.

## 5.1 Introduction

Linearization is routinely used in the analysis and design of $\Sigma\Delta$ encoders because the resulting models are simple, and their shortcomings are perceived to be well known. In particular, the decoding of the binary encoder output stream has traditionally been based on linear filtering, and the optimality of this solution has not been questioned. In chapter 1 we pointed out that linear low-pass decoding is sub-optimal because it neglects that the encoder input resulted in the particular observed binary output stream. In chapter 3 we introduced an

[1] Based on [113], ©IEEE, [114], ©1991 IEEE, and [108], ©1992 IEEE.

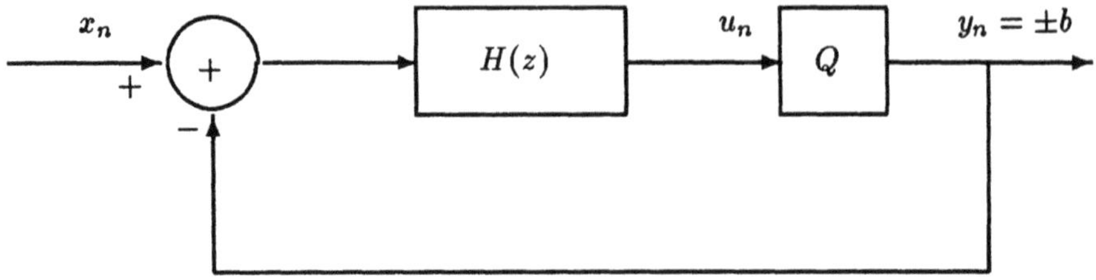


Figure 5.1: *Discrete-time model of the interpolative ΣΔ encoder.*

optimal nonlinear decoding algorithm for constant inputs, and demonstrated large gains over linear decoding.

In this chapter we consider nonlinear decoding for band-limited inputs [93], [108], [114], [115], [116]. We take a signal reconstruction point of view and derive a conceptually feasible, optimal nonlinear decoder. Simulations indicate that our decoder is superior to linear decoding by up to 20-30 dB for the single-loop, double-loop and two-stage encoders depending on the oversampling ratio, and by up to 10 dB for a specific fourth order interpolative encoder [32]. These results may serve as upper bounds on attainable performance. Our ultimate goal is a decoder which performs significantly better than linear decoding, is simple to implement and is as robust towards circuit imperfections as linear decoding.

The chapter is organized as follows. Section 5.2 describes the reconstruction algorithm. Section 5.3 presents numerical results for the considered encoders in the absence of circuit imperfections, and section 5.4 quantifies the effects of a number of non-idealities. Finally, section 5.5 contains a summary.

## 5.2 Reconstruction Algorithm

In this section we describe our proposed reconstruction algorithm in the context of the general interpolative ΣΔ encoder shown in figure 5.1. Although the double-loop and two-stage encoders are not interpolative, the algorithm easily extends to these architectures. Section 5.2.1 describes the POCS algorithm [82], [83] which iteratively enforces two constraints on an input signal estimate: First, that the input signal resulted in the observed output signal, and second, that it is band-limited. Section 5.2.2 shows that finding an input sequence which generates the observed output signal can be treated as a Quadratic Programming (QP) problem. Section 5.2.3 reviews our Singular Value Decomposition (SVD) based approach to band limitation. Section 5.2.4 describes modifications to the POCS algorithm for sliding-block decoding. Section 5.2.5 compares our proposed algorithm to that of Thao and Vetterli [117], [118], [119], [120].

### 5.2.1 Projections onto Convex Sets

In this section we briefly review the nonlinear POCS algorithm as it applies to our decoding problem. For any binary output signal $\mathbf{y} = \{y_0, \cdots, y_{N-1}\}$ of an interpolative $\Sigma\Delta$ encoder, we define the set $S_1$ of all input signals $\mathbf{x} = \{x_0, \cdots, x_{N-1}\}$ that result in $\mathbf{y}$ when applied to the encoder. We also define the set $S_2$ of all $N$-sample signals $\mathbf{x}$ that are band-limited; the precise meaning of this is discussed in section 5.2.3. To estimate the input signal optimally, we must find a signal $\hat{\mathbf{x}} \in S_1 \cap S_2$. This problem formulation makes the POCS algorithm a natural choice [82]. We show below that $S_1$ and $S_2$ are convex[2] as assumed by POCS. If we denote the orthogonal projections onto $S_1$ and $S_2$ by $P_1$ and $P_2$, respectively, then the theory of POCS states that an element $\hat{\mathbf{x}} \in S_1 \cap S_2$ can be found from any initial guess $\mathbf{x}_0$ by the iteration [82]

$$\mathbf{x}_{n+1} = (P_2 \circ P_1)\mathbf{x}_n, \quad n \geq 0, \quad \hat{\mathbf{x}} = P_1\hat{\mathbf{x}} = P_2\hat{\mathbf{x}} = \lim_{n\to\infty} \mathbf{x}_n$$

In the following sections we discuss the projections separately.

### 5.2.2 Time-Domain Projection

In this section we consider the time-domain projection $P_1$ of a signal $\mathbf{x}$ onto the space $S_1$ of sequences that generate a given output sequence $\mathbf{y}$. We derive a time-domain characterization of the $\Sigma\Delta$ encoder, and show that the projection $P_1$ results in a QP problem.

We denote the impulse response of $H(z)$ by $\{h_0, h_1, \cdots\}$; $h_0$ must be zero to avoid delay-free loops. We define the $N \times N$ lower-triangular Toeplitz matrix

$$\mathbf{H} = \begin{bmatrix} h_1 & 0 & \cdots & 0 & 0 \\ h_2 & h_1 & \cdots & 0 & 0 \\ \vdots & \vdots & & \vdots & \vdots \\ h_{N-1} & h_{N-2} & \cdots & h_1 & 0 \\ h_N & h_{N-1} & \cdots & h_2 & h_1 \end{bmatrix}$$

For an $M$th order encoder, we define an $(M+1)$-dimensional state vector $\mathbf{s}$,[3] which we assume to be known in this section. We also define an $N \times (M+1)$ zero-input response matrix $\mathbf{Z}$ so that if the filter $H(z)$ is driven open-loop with initial state $\mathbf{s}$ and zero input over $N$ samples, the filter output is $\mathbf{Zs}$. We can then write the quantizer input vector $\mathbf{u} = \{u_0, u_1, \cdots, u_{N-1}\}$ as

$$\mathbf{u} = \mathbf{H}(\mathbf{x} - \mathbf{y}) + \mathbf{Zs}$$

Our only knowledge of any particular quantizer input $u_n$ is in the form of its sign, $Q(u_n)$. This knowledge provides us with a series of bounds on linear

---

[2] A set $S$ is convex if for all $\mathbf{a}, \mathbf{b} \in S$, $\alpha\mathbf{a} + (1-\alpha)\mathbf{b} \in S$ for any $0 < \alpha < 1$.

[3] As we will see in chapter 7, the overall delay in $H(z)$ in general adds one to the system order. For the single-loop encoder, however, the filter delay and the overall delay can be combined into one.

combinations of sample values $\{x_n\}$. To cast this in matrix notation, we define an $N \times N$ diagonal matrix with $\pm 1$ entries on its diagonal,

$$\mathbf{Q} = -\frac{1}{b}\text{diag}(y_1, y_2, \cdots, y_N)$$

where $b$ is the quantizer step size. The bounds imposed by the quantizer outputs are then

$$\mathbf{QHx} \leq \mathbf{Q}(\mathbf{Hy} - \mathbf{Zs}) \tag{5.1}$$

where the inequality sign is to be taken coordinate-wise. Equation (5.1) establishes a time-domain description of the interpolative encoder.[4] We can show that the set $S_1$ of signals satisfying (5.1) is convex. Specifically, if we consider two sample vectors $\mathbf{x}_1$ and $\mathbf{x}_2$ that both satisfy (5.1), then for any $0 < \alpha < 1$,

$$\begin{aligned} & \mathbf{QH}[\alpha\mathbf{x}_1 + (1-\alpha)\mathbf{x}_2] \\ \leq\ & \alpha\mathbf{Q}(\mathbf{Hy} - \mathbf{Zs}) + (1-\alpha)\mathbf{Q}(\mathbf{Hy} - \mathbf{Zs}) = \mathbf{Q}(\mathbf{Hy} - \mathbf{Zs}) \end{aligned}$$

We will adopt the 2-norm as our performance and projection metric, that is, we define the Signal-to-Noise Ratio (SNR) to be $10\log_{10}(E_{\text{signal}}/E_{\text{noise}})$ where

$$E_{\text{signal}} = \sum_{n=0}^{N-1} x_n^2, \quad E_{\text{noise}} = \sum_{n=0}^{N-1} (x_n - \hat{x}_n)^2 \tag{5.2}$$

Projecting onto $S_1$ in the 2-norm is equivalent to finding the signal $\hat{\mathbf{x}}$ that satisfies (5.1) and minimizes the distance

$$d^2 = \|\hat{\mathbf{x}} - \mathbf{x}\|_2^2 \tag{5.3}$$

Equations (5.1) and (5.3) constitute a linearly constrained QP problem which can be solved with techniques available in the literature [121].

In practice, sample sizes $N$ on the order of thousands are desirable to increase the accuracy of the band limitation projection described in section 5.2.3 below. Because of the large computational complexity of the QP problem, we will derive an approximation to $P_1$ which takes advantage of the fact that $\mathbf{H}$ is lower-triangular. The idea is to solve a number of $L$-dimensional QP problems ($L < N$) rather than a single $N$-dimensional one, by dividing the signal $\mathbf{x}$ into $L$-sample QP sub-blocks. In the simplest set-up, the sub-blocks are not overlapping. The small QP problems are solved in chronological order, and no sub-block is allowed to change the estimates of previous sub-blocks. This

[4] Similar descriptions can be found for non-interpolative encoders. The main differences for such encoders are as follows: For the double-loop encoder, the impulse response matrix seen by the input $\mathbf{x}$ and the output $\mathbf{y}$ are different. For the two-stage encoder, we assume as in chapter 3 that we have access to both quantizer outputs; each quantizer gives a series of bounds, and we can enforce the two sets of bounds separately. For predictive coders such as the Delta encoder, the impulse response matrix is the identity matrix, and as a result the time-domain projection becomes trivial.

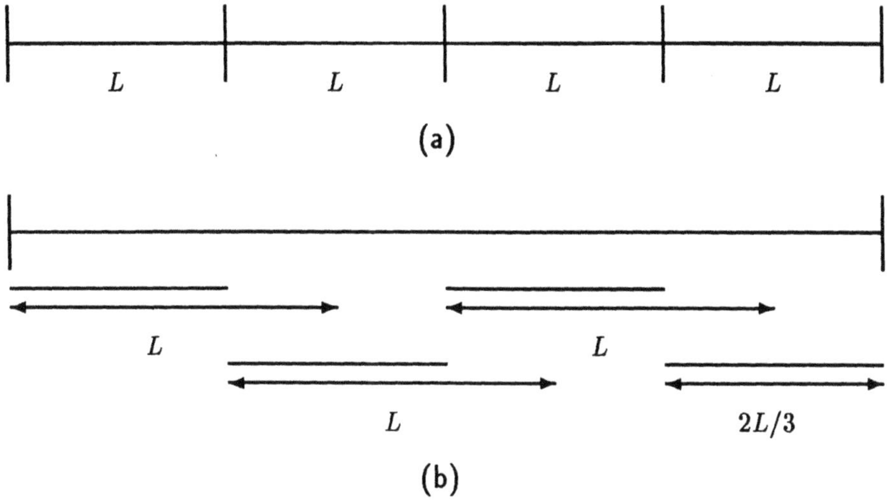


Figure 5.2: *Two ways of dividing an N-dimensional QP problem into smaller L-dimensional problems. (a) Non-overlapping blocks. (b) Blocks overlapping by 1/3; the marked parts of each block are used in the final solution estimate.*

approach is illustrated in figure 5.2(a) and is described mathematically in appendix D.1. The problem with this set-up is that large changes tend to be necessary at the beginning of each sub-block because **H** is lower-triangular, and so each bound violation must be corrected using only samples before the violation. Bound violations at the beginning of a sub-block therefore require larger sample modifications than do violations towards the end of a sub-block. In our preferred set-up the sub-blocks are partially overlapping, and thus the optimization of each sub-block takes into account a portion of the sub-block immediately following it. This approach is shown in figure 5.2(b) and is also described mathematically in appendix D.1. The overall QP estimate of an $N$-sample block is obtained by concatenating parts of the estimates resulting from each of the QP sub-blocks. Specifically, we use the estimate on that part of each sub-block which does not overlap with the immediately following sub-block. The overlapping portion of each sub-block with the next sub-block is used to improve upon the initial estimate of the QP solution for the following sub-block.

The choice of QP block size $L$ is a trade-off between the speed with which a solution is obtained, and the closeness of the solution to the true 2-norm solution. The closeness to the 2-norm is important because the frequency-domain projection will be done in the 2-norm, and the POCS algorithm only guarantees convergence if the projections are done in the same norm.

### 5.2.3 Frequency-Domain Projection

In this section we consider the frequency-domain projection $P_2$ of a signal $\mathbf{x} = \{x_0, \cdots, x_{N-1}\}$ onto the set $S_2$ of signals that are band-limited in a sense to be defined. This projection raises some general questions about band limitation that are discussed in chapter 4 and are summarized below.

The two standard techniques for band limitation are the Discrete Fourier Transform (DFT) and linear filtering. However, the DFT is only accurate when the signal frequencies are bin frequencies for the DFT, which cannot in general be assumed. We showed in chapter 4 that the frequency of a sinusoidal baseband signal can be chosen half way between DFT bin frequencies such that the SNR is as low as

$$\mathrm{SNR}_{\mathrm{min}} \approx 10 \log_{10} \left( \frac{N\pi^2}{4\mathrm{OSR}} \right)$$

For $N = 4096$ and $\mathrm{OSR} = 48$, we get $\mathrm{SNR}_{\mathrm{min}} = 23$ dB, and this minimum SNR only increases by 3 dB/octave with the sample size. Windowing can be used to suppress spectral leakage in the DFT, but this has the side-effect of smearing the signal spectrum significantly as shown in chapter 4.

Linear filtering for moderate-length block-oriented processing has the disadvantage that we must pad the block of samples with zeros to obtain a sequence on which a linear filter can operate, and the filtered result exhibits large edge effects due to this padding. In addition, a linear filter can only approach the ideal low-pass transfer function which has unity baseband gain and zero high-frequency gain, and the deviations from ideality result in errors. Finally, linear filtering is not a projection unless the filter transfer function only takes on the values 1 and 0.

Our preferred, general technique for band limitation is described in chapter 4 and is related to known algorithms for band-limited extrapolation [98]. In chapter 4 we also discuss the fact that any finite set of samples can be viewed as samples of infinitely many different infinite-extent sequences with band-limited Discrete Time Fourier Transforms (DTFTs), and thus it is not clear how to define band limitation for general finite-extent sequences. This conceptual problem is resolved by using energy and dimensional considerations. Our band-limitation technique is based on the SVD of an $N \times N$ symmetric Toeplitz matrix $\mathbf{L}$ of samples of the impulse response of an ideal low-pass filter, $\ell_{mn} = \sin[\sigma(m-n)]/[\pi(m-n)]$ where $\sigma = 1/\mathrm{OSR}$. The SVD leads to the truncated Discrete Prolate Spheroidal Sequences (DPSSs) of $N$-dimensional vectors $\{\mathbf{u}_0, \cdots, \mathbf{u}_{N-1}\}$ which have been studied extensively by Slepian [96] and others. The truncated DPSSs form an orthonormal basis of $\mathcal{R}^N$. Approximately $r = N/\mathrm{OSR}$ of the singular values of $\mathbf{L}$ are close to 1, while the remaining ones are close to 0. The truncated DPSSs for singular values close to 1 and 0 are analogous to the baseband and High-Frequency (HF) complex exponential basis functions, respectively, of the DFT. Our SVD-based algorithm projects signals

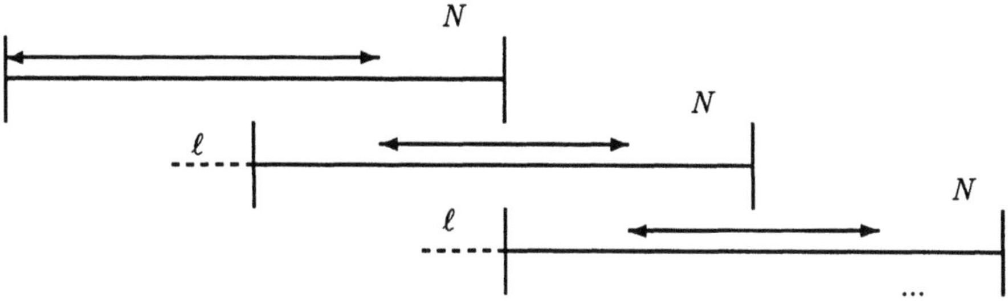


Figure 5.3: *50% overlap of the N-sample blocks on which the POCS iterations operate. The marked parts of each block are used in the overall solution. The dotted lines show an extension of the QP projection described in the text.*

onto the space $S_2$ spanned by the "baseband" DPSSs,

$$\hat{\mathbf{x}} = \sum_{n=0}^{r-1} (\mathbf{x}^T \mathbf{u}_n) \cdot \mathbf{u}_n, \quad r \approx N/\text{OSR} \tag{5.4}$$

The projection (5.4) is in the sense of the 2-norm, and the space $S_2$ is linear and thus convex. We consider the signals in the range space of $\{\mathbf{u}_0, \cdots, \mathbf{u}_{r-1}\}$ to be band-limited. The SVD required to obtain the vectors $\{\mathbf{u}_0, \cdots, \mathbf{u}_{r-1}\}$ only needs to be done once, and can be implemented efficiently using the Lanczos algorithm [106] and exploiting the Toeplitz form of $\mathbf{L}$. The computational complexity of the projection (5.4) is $O(N^2/\text{OSR})$ which is found to be comparable in speed to a Fast Fourier Transform (FFT) band-limitation for a sample size of 4K (1K = 1024).

### 5.2.4 Sliding Block Decoding

The POCS algorithm described in sections 5.2.1–5.2.3 is suitable for decoding blocks of $N$ samples when the initial encoder state is known. However, we find empirically that the algorithm suffers from larger estimation inaccuracy towards the edges of an $N$-sample block than on its middle. To reduce edge effects, we therefore also consider a sliding block set-up where the $N$-sample blocks overlap by 50% as shown in figure 5.3. Each block is decoded using the iterative POCS algorithm, and upon convergence, the initial state of the following block is estimated. The state estimation is done by simulating the behavior of an encoder which starts in the initial state of the current block and has the estimated input sequence as its input. The overall input estimate is the concatenation of the input estimates on the middle halves of the $N$-sample blocks. The SNR is calculated in the time domain over the middle half of each $N$-sample block, and the overall SNR is defined as the average of the SNRs on the contributing blocks.

The accuracy of the time-domain and frequency-domain projections can be improved upon in two ways by extending the idea of overlapping blocks. First, for all but the first $N$-sample block, the QP operation can be started a number $\ell$ samples before the beginning of the $N$-sample block and extend over $N+\ell$ samples. This implies that the QP operation on the previous $N$-sample block includes the first $\ell$ samples of the present $N$-sample block, and reduces the effects of early bound violations for the same reason that overlapping the QP sub-blocks reduces these effects. Second, the SVD projection tends to be most accurate on the middle of an $N$-sample block, so results from the part of the previous $N$-sample block that overlaps the current $N$-sample block can be used to improve upon the accuracy of the initial estimate for the current block. On samples where no better estimates are available from previous $N$-sample blocks, the initial estimate is chosen to be the binary output sequence of the encoder.

### 5.2.5 Comparison to Thao and Vetterli's Algorithm

In this section we compare our proposed algorithm [93], [108], [114], [115], [116] to the algorithm proposed by Thao and Vetterli [117], [118], [119], [120] which was developed independently. Both algorithms are POCS-based. Sections 5.2.5.1 and 5.2.5.2 discuss the time-domain and frequency-domain projections, and section 5.2.5.3 discusses the modes of operation for the two algorithms.

#### 5.2.5.1 Time-Domain Projection

Thao and Vetterli develop an approximate method for performing the time-domain projection which is valid for certain $n$th order multi-stage $\Sigma\Delta$ encoders as well as for other encoders. The idea is to only enforce a subset of the bounds imposed by the encoder outputs, and to recursively attempt to guess which bounds are active [121]. For orders $n$ above 1, the authors' algorithm only yields an approximate projection onto the convex subspace spanned by the particular bounds which are assumed active; the algorithm is thus approximate both in the way that the active constraints are chosen, and in the way that the resulting optimization problem is solved. The advantage of the authors' interesting approach over the more accurate approach of performing the full projection is a decrease in computational complexity. The authors do not comment on the achieved gain over conventional, efficiently implemented QP, or an approximate QP approach such as the one described in section 5.2.2 which uses all available bounds. Nor do they describe the loss of accuracy resulting from their approximations. The authors do not quantify the fraction of bounds that are typically used by their algorithm, but their most general algorithm described in [118] requires that the active bounds be spaced apart by at least $n$ samples. The authors do not state whether their algorithm permits a range of trade-offs between computational speed and accuracy. The authors' algorithm only works for a subset of $\Sigma\Delta$ encoders, most notably multi-stage encoders whose individual stages all have their open-loop filter poles at DC. The approach is thus not applicable to the general class of interpolative encoders with filter poles over signal baseband [32] such as the fourth order encoder considered in section 5.3. Because of the assumption of DC filter poles, it would be interesting to see simulation results for the sensitivity of the authors' algorithm towards unknown non-idealities in the encoder filters such as integrator leak.

#### 5.2.5.2 Frequency-Domain Projection

In order to have access to an ideal low-pass filtering for the frequency-domain projection, Thao and Vetterli assume that the input signal is periodic. They also assume that the period is equal to the number of samples $N$ to which the decoder has access [118]. In effect, they thus assume that the input frequency is known.

#### 5.2.5.3 Mode of Operation

With regard to a continuous mode of operation, Thao and Vetterli assume that the integrator states are reset to zero at the beginning of each conversion cycle. While our algorithm can also work in this mode, we found that increased accuracy is obtained by overlapping the blocks and propagating estimates of the encoder states, as described in section 5.2.4; this enables us to only accumulate estimation errors on the middle half of each POCS block. Another potential advantage of not periodically resetting the integrator states is that encoder transients are avoided.

## 5.3 Results

This section contains numerical results for single block and sliding block decoding for a number of $\Sigma\Delta$ encoders, including the single-loop, double-loop and two-stage encoders and the specific fourth order encoder given by [32]

$$H(z) = \frac{0.8653 - 2.2692z^{-1} + 2.0064z^{-2} - 0.59714z^{-3} + 0.000035z^{-4}}{1 - 3.99646z^{-1} + 5.992922z^{-2} - 3.996460866z^{-3} + 1.000000433z^{-4}} \cdot z^{-1}$$

For single block simulations, the SNR is calculated over the middle half of an $N$-sample block, and for sliding block simulations, the SNR is calculated as explained in section 5.2.4. We point out that our SNR measure includes both low-frequency and high-frequency signal errors.

Throughout the section, the input signal is a sinusoid whose amplitude is varied and whose frequency is $\alpha = 9/32$ times the largest possible signal frequency. This frequency is a bin frequency for the DFT to enable comparisons between the POCS decoders whose band limitations are based on the DFT and on the SVD (5.4), respectively. When we use a bin frequency, the DFT will perform better than the SVD, and so the shown DFT-based experimental results will be superior to the corresponding SVD-based ones. However, the

| Encoder | Decoder | OSR | Peak SNR (dB) DFT | SVD |
|---|---|---|---|---|
| Single-loop | POCS | 64 | 61 | 56 |
| | | 128 | 84 | 67 |
| | linear | 64 | 37 | 37 |
| | | 128 | 44 | 44 |
| Double-loop | POCS | 64 | 79 | 65 |
| | | 128 | 92 | 83 |
| | linear | 64 | 37 | 37 |
| | | 128 | 44 | 44 |
| Two-stage | POCS | 64 | 86 | 85 |
| | | 128 | 101 | 94 |
| | linear | 64 | 31 | 31 |
| | | 128 | 38 | 38 |
| 4th order | POCS | 48 | 93 | 83 |
| | linear | 48 | 84 | 64 |

Table 5.1: *Summary of peak SNR performance results for the single-loop, double-loop, two-stage and fourth order encoders; the encoder components are assumed ideal. Both the POCS and the linear decoders can be based on either the DFT or the SVD. The linear decoders are generally limited by signal frequency errors.*

SVD performs significantly better than the DFT on non-bin frequencies, and signal frequencies cannot in general be assumed to be bin frequencies.

For the shown results, the sample size for each block is 4K. For sliding block simulations, we consider three 4K blocks which overlap by 50% for a total of 8K samples. The overlap between QP sub-blocks is $L/3$ where $L = 192$ for the fourth order encoder and $L = 96$ for the single-loop, double-loop and two-stage encoders. The initial state vector $\mathbf{s}$ for the first block is assumed to be known and is set to zero. The number of singular vectors $r$ in (5.4) is 94, 74 and 39 for oversampling ratios OSR = 48, 64 and 128, respectively. The number of POCS iterations is 12.

Sections 5.3.1, 5.3.2, 5.3.3 and 5.3.4 contain results for the single-loop, double-loop, two-stage and fourth order encoders, respectively. The main results are summarized in table 5.1; we make comments on the dependence of the SNR upon the OSR in the individual sections below. In section 5.3.5 we compare our SNR metric and performance results to existing results in the literature.

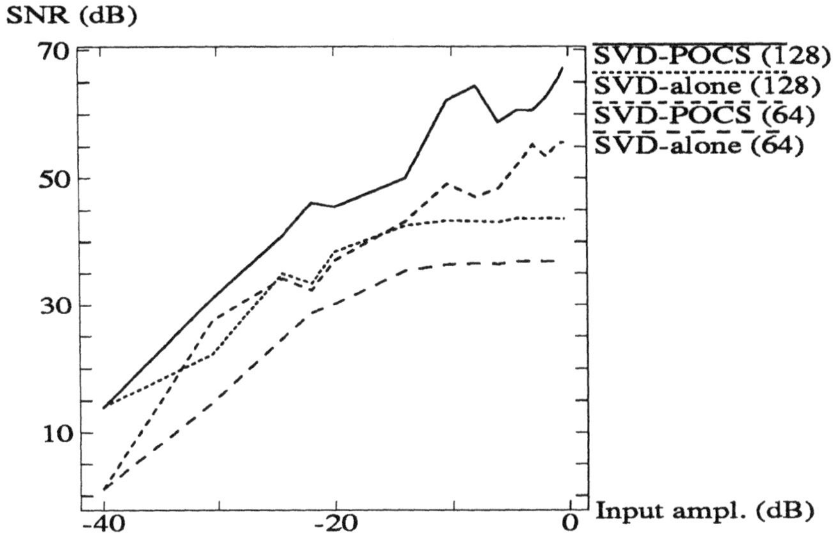


Figure 5.4: *SNR as a function of input amplitude for the single-loop encoder with single block decoding. The two decoders are an SVD-based low-pass decoder and the SVD-based POCS decoder.*

### 5.3.1 Single-Loop Modulator

We first consider single block decoding with the SVD-based band limitation (5.4) as the frequency domain projection, for the two oversampling ratios 64 and 128. Figure 5.4 shows SNR curves for the SVD-based POCS decoder, as well as for a low-pass decoder that uses (5.4). The figure shows that particularly for large input amplitudes, the POCS algorithm is clearly superior to the linear decoder. For OSR = 64, the peak SNRs are 56 dB and 37 dB, and for OSR = 128, the peak SNRs are 67 dB and 44 dB for the SVD-POCS and SVD-alone decoders, respectively.

We have carried out a set of simulations under an identical set-up to that of figure 5.4, except that we consider a sliding-block decoder. We find that the SNR curves for the sliding-block decoder are within 1-2 dB of the corresponding curves in figure 5.4, indicating that state uncertainty implies little loss in SNR, so the same conclusions hold as for figure 5.4.

For comparison, figure 5.5 shows SNR results for the POCS algorithm when using the DFT rather than the SVD-based method (5.4) as the band limitation. The figure is valid for single block decoding, but another set of simulations for sliding-block decoding shows that the difference between corresponding single block and sliding-block curves is 0.1–0.2 dB. For OSR = 64, the peak SNRs are 61 dB and 37 dB, and for OSR = 128, the peak SNRs are 84 dB and 44 dB for the DFT-POCS and DFT-alone decoders, respectively.

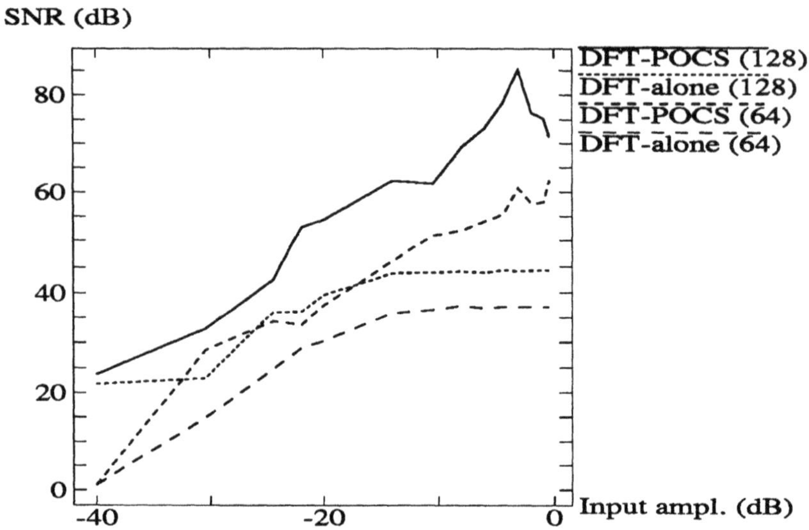


Figure 5.5: *SNR as a function of input amplitude for the single-loop encoder with single block decoding. The two decoders are a DFT-based low-pass decoder and the DFT-based POCS decoder.*

Comparing figures 5.4 and 5.5, the SVD-based POCS method is up to 20 dB superior to the DFT-alone and SVD-alone methods. We see that the SVD-based POCS method mostly loses 2-4 dB compared to the DFT-based POCS method for OSR = 64 and up to about 10 dB for OSR = 128. We find that the linear decoders are all severely limited by signal frequency errors, as commented on in section 5.3.5.1.

To demonstrate the convergence of our algorithm, figure 5.6 shows the energy modification performed by the frequency-domain projection as a function of POCS iteration number. The curve is valid for single-block decoding with the DFT-POCS algorithm and an input amplitude of 0.10. We see that the algorithm finds an estimated input which is within about $10^{-23}$ of satisfying both the time-domain and frequency-domain constraints; the small error may be attributed to machine inaccuracy in the DFT.

To assess the SNR dependence of the DFT-POCS algorithm upon the OSR, we have obtained simulation results for OSRs of 16, 32, 48, 64, 96 and 128. To combine the SNR performance over the range of input amplitudes into one indicator, we average the SNR over 14 fixed amplitudes between 0.01 and 0.95.[5] This average SNR should of course only be taken as a relative rather than absolute performance indicator, but it does permit comparison across OSRs. Figure 5.7 shows the trade-off between the average SNR and the OSR. We see

[5]The amplitudes are the same ones used to generate the plots in this section, specifically, 0.01, 0.03, 0.06, 0.08, 0.10, 0.20, 0.30, 0.40, 0.50, 0.60, 0.70, 0.80, 0.90, 0.95.

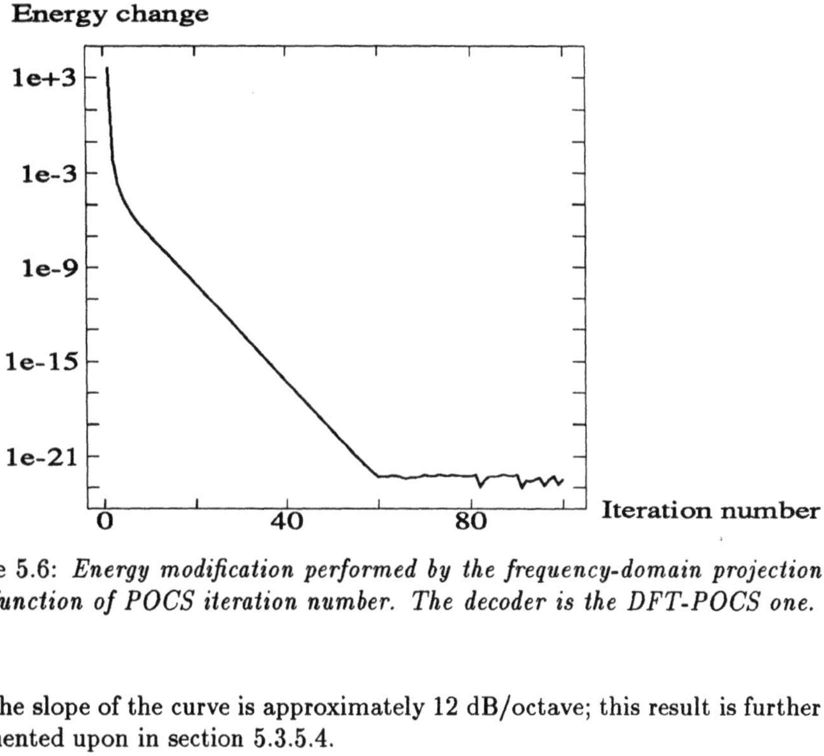


Figure 5.6: *Energy modification performed by the frequency-domain projection as a function of POCS iteration number. The decoder is the DFT-POCS one.*

that the slope of the curve is approximately 12 dB/octave; this result is further commented upon in section 5.3.5.4.

### 5.3.2 Double-Loop Modulator

For the double-loop encoder, we consider single block decoding with the SVD-based band limitation (5.4) as the frequency domain projection, for the two oversampling ratios 64 and 128. Figure 5.8 shows SNR curves for the SVD-based POCS decoder, as well as for a low-pass decoder that uses (5.4). The figure shows that particularly for large input amplitudes, the POCS algorithm is clearly superior to the linear decoder. For OSR = 64, the peak SNRs are 65 dB and 37 dB, and for OSR = 128, the peak SNRs are 83 dB and 44 dB for the SVD-POCS and SVD-alone decoders, respectively.

For comparison, figure 5.9 shows SNR results for the POCS algorithm when using the DFT rather than the SVD-based method (5.4) as the band limitation. The figure is valid for single block decoding, but another set of simulations for sliding-block decoding shows that the difference between corresponding single block and sliding-block curves is about 3 dB. For OSR = 64, the peak SNRs are 79 dB and 37 dB, and for OSR = 128, the peak SNRs are 92 dB and 44 dB for the DFT-POCS and DFT-alone decoders, respectively.

Comparing figures 5.8 and 5.9, the SVD-based POCS method is up to 40 dB superior to the DFT-alone and SVD-alone methods. We see that the SVD-based POCS method mostly loses 5-10 dB compared to the DFT-based POCS

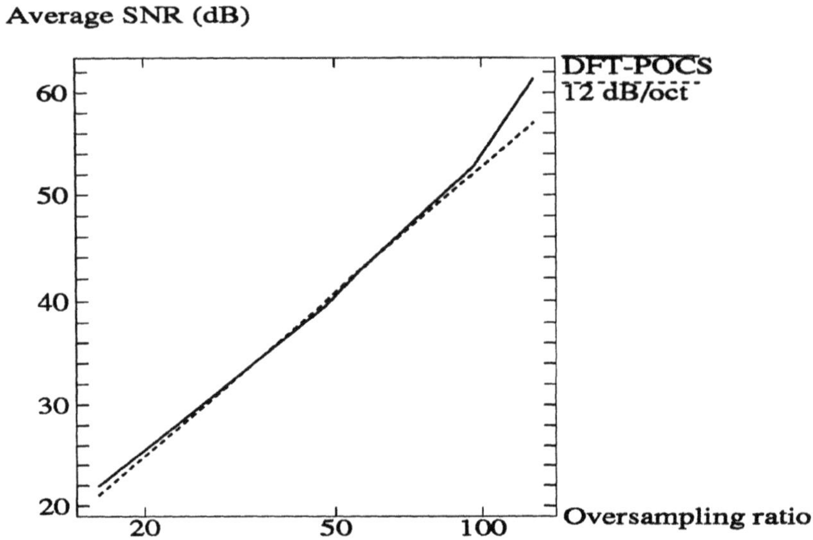


Figure 5.7: *Average SNR as a function of oversampling ratio for the single-loop encoder with single-block decoding and the DFT-POCS decoder. Also shown is a line with 12 dB/octave slope.*

method for OSR = 64 and up to 10-15 dB for OSR = 128. We find that the linear decoders are all severely limited by signal frequency errors.

As an example of the influence of the QP block size $L$, figure 5.10 shows the peak SNR of the DFT-POCS decoder as a function of $L$. This figure justifies our choice of $L = 96$ for the double-loop encoder.

To assess the SNR dependence of the DFT-POCS algorithm upon the OSR, we have obtained simulation results for OSRs of 16, 32, 48, 64, 96 and 128. To combine the SNR performance over the range of input amplitudes into one indicator, we average the SNR over 14 fixed amplitudes as in section 5.3.1. Figure 5.11 shows the trade-off between the average SNR and the OSR. We see that the slope of the curve is approximately 18 dB/octave on most parts of the curve, but the curve suffers a dip in average SNR around an OSR of 80, and the overall slope thus appears to be slightly less than 18 dB/octave. This result is further commented upon in section 5.3.5.4.

### 5.3.3 Two-Stage Modulator

For the two-stage encoder, we consider single block decoding with the SVD-based band limitation (5.4) as the frequency domain projection, for the two oversampling ratios 64 and 128. Figure 5.12 shows SNR curves for the SVD-based POCS decoder, as well as for a low-pass decoder that uses (5.4). The figure shows that particularly for large input amplitudes, the POCS algorithm

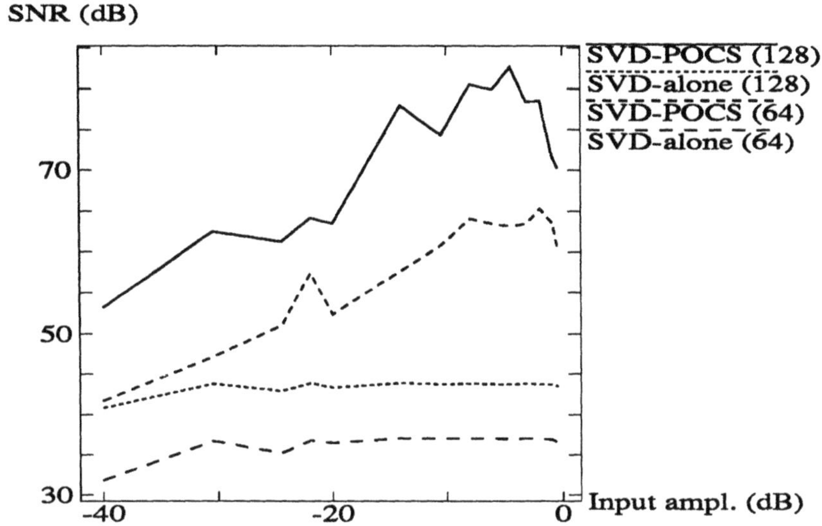


Figure 5.8: *SNR as a function of input amplitude for the double-loop encoder with single block decoding. The two decoders are an SVD-based low-pass decoder and the SVD-based POCS decoder.*

is clearly superior to the linear decoder. For OSR = 64, the peak SNRs are 85 dB and 31 dB, and for OSR = 128, the peak SNRs are 94 dB and 38 dB for the SVD-POCS and SVD-alone decoders, respectively.

For comparison, figure 5.13 shows SNR results for the POCS algorithm when using the DFT rather than the SVD-based method (5.4) as the band limitation. For OSR = 64, the peak SNRs are 86 dB and 31 dB, and for OSR = 128, the peak SNRs are 101 dB and 38 dB for the DFT-POCS and DFT-alone decoders, respectively.

Comparing figures 5.12 and 5.13, the SVD-based POCS method is up to 40 dB superior to the DFT-alone and SVD-alone methods. We see that the SVD-based POCS method mostly loses 5-10 dB compared to the DFT-based POCS method for OSR = 64 and OSR = 128. We find that the linear decoders are all limited by signal frequency errors.

To assess the SNR dependence of the DFT-POCS algorithm upon the OSR, we have obtained simulation results for OSRs of 16, 32, 48, 64, 96 and 128. To combine the SNR performance over the range of input amplitudes into one indicator, we average the SNR over 14 fixed amplitudes as in section 5.3.1. Figure 5.14 shows the trade-off between the average SNR and the OSR. We see that the slope of the curve is approximately 18 dB/octave up to an OSR of about 80, and then appears to drop slightly. This result is further commented upon in section 5.3.5.4.

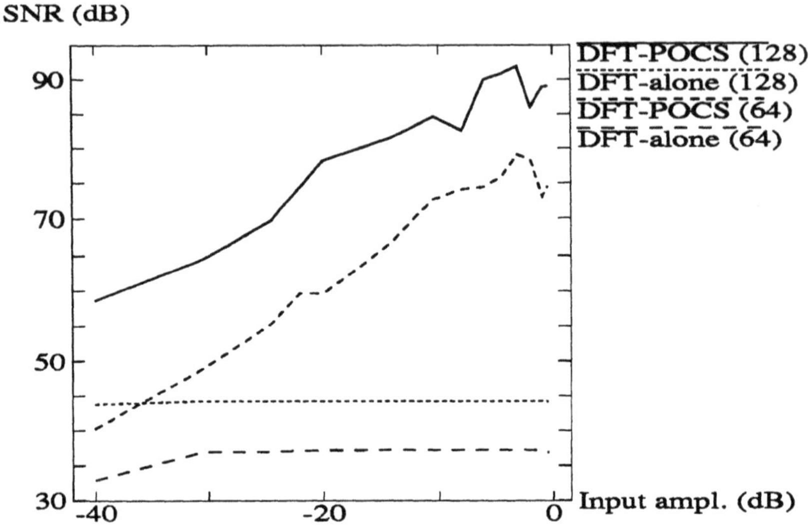


Figure 5.9: *SNR as a function of input amplitude for the double-loop encoder with single block decoding. The two decoders are a DFT-based low-pass decoder and the DFT-based POCS decoder.*

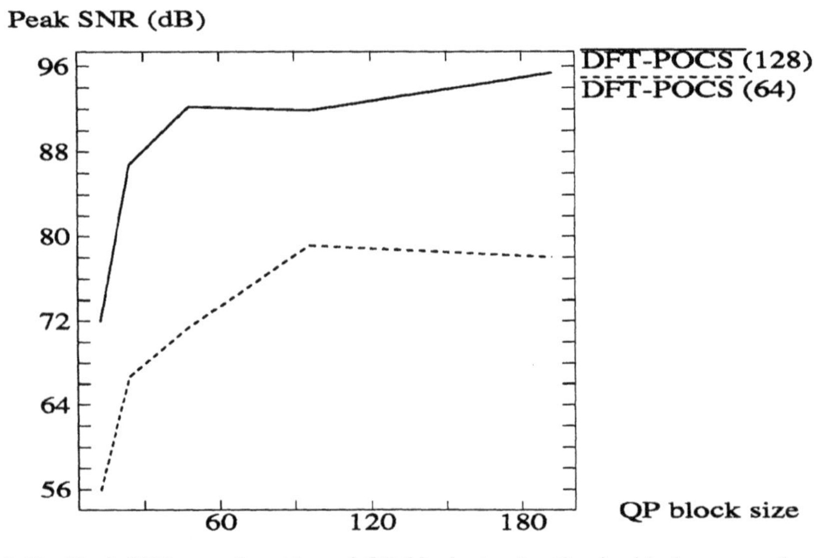


Figure 5.10: *Peak SNR as a function of QP block size for the double-loop encoder with single block DFT-POCS decoding.*

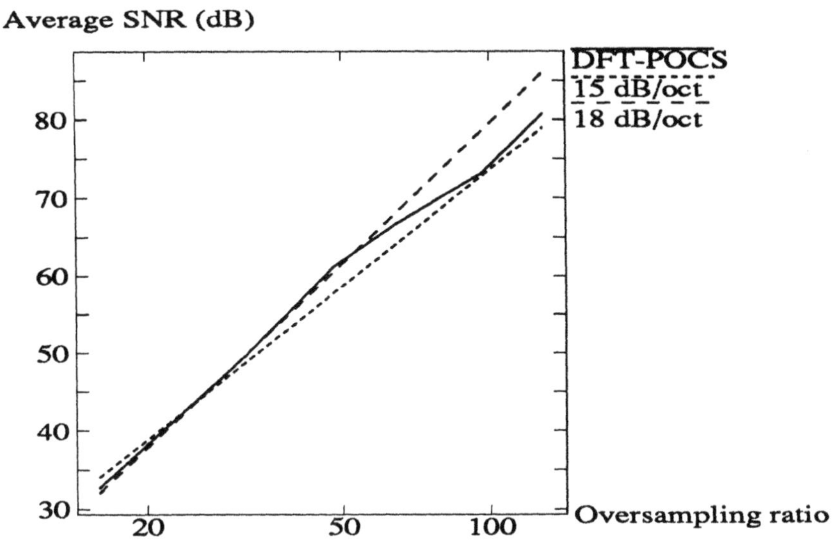


Figure 5.11: *Average SNR as a function of oversampling ratio for the double-loop encoder with single-block decoding and the DFT-POCS decoder. Also shown are lines with 15 dB/octave and 18 dB/octave slopes.*

### 5.3.4 Fourth Order Modulator

We first consider single block decoding with the SVD-based band limitation (5.4) as the frequency domain projection. Figure 5.15 shows SNR curves for the SVD-based POCS decoder, as well as for a low-pass decoder that only uses (5.4). The figure shows that the peak SNR is 83 dB for the SVD-POCS decoder and 64 dB for the SVD-alone decoder. The gain of the SVD-POCS decoder is 10-20 dB.

For comparison, figure 5.16 shows SNR results for single block decoding for the POCS algorithm when using the DFT rather than the SVD-based method (5.4) for band limitation. The peak SNRs are 93 dB for the DFT-POCS decoder and 84 dB for the DFT-alone decoder, and the SNR gain of the DFT-POCS decoder is 5-10 dB.

Other simulation results show that under the sliding block set-up, both the SVD-POCS and the DFT-POCS decoders lose 1-2 dB compared to the corresponding single block decoders. This means that state uncertainty results in little SNR loss.

The DFT-alone curve in figure 5.16 is generally 15-30 dB above the SVD-alone curve in figure 5.15. As explained above, we would expect the DFT-alone method to be superior because we are using a bin frequency signal as the input. However, the difference between the DFT-alone and SVD-alone curves could have been made smaller by a different choice of the number of singular vectors

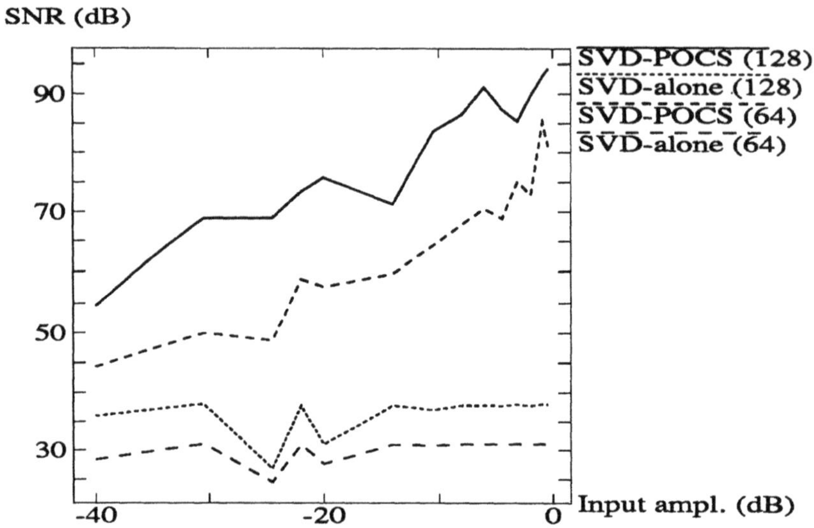


Figure 5.12: *SNR as a function of input amplitude for the two-stage encoder with single block decoding. The two decoders are an SVD-based low-pass decoder and the SVD-based POCS decoder.*

for the SVD band limitation (5.4). The number of singular vectors was chosen to optimize the SVD-POCS curves rather than the SVD-alone ones.

### 5.3.5 Comparison with Existing Results

In this section we compare the results in sections 5.3.1–5.3.4 to existing results in the literature. In section 5.3.5.1 we discuss the choice of SNR definition, and in section 5.3.5.2 we present simulation results to illustrate the performance of the POCS algorithm under different SNR definitions. In section 5.3.5.3 we show that the POCS algorithm may be capable of suppressing the spectral tones that plague linear decoders for low-order encoders. In section 5.3.5.4 we compare our results to those of Thao and Vetterli.

#### 5.3.5.1 SNR Definition

To enable comparisons with existing results, we change our definition of SNR from the current definition (5.2) which counts the 2-norm difference between the sinusoidal input signal and its decoded version as noise. Our argument for using (5.2) is that a decoder should ultimately be judged on the total amount of noise that it leaves on its input estimate—otherwise, any additional circuitry to further reduce the noise should be viewed as part of the decoder. Nonetheless, the changes that we consider are:

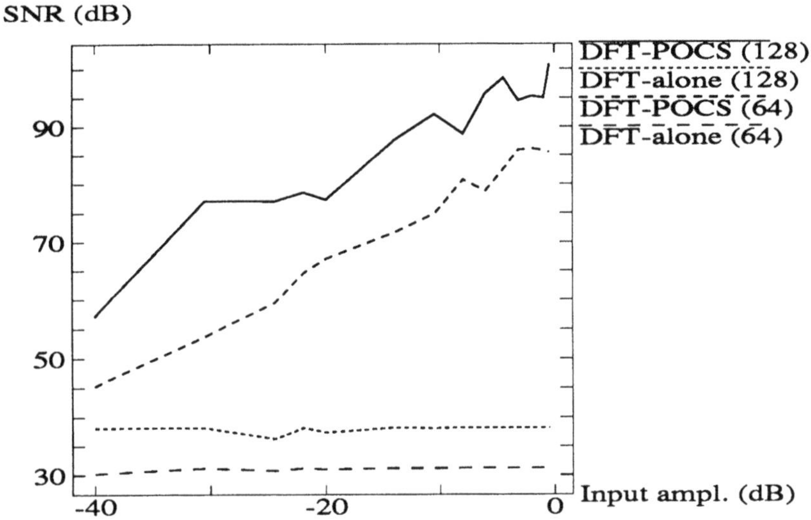


Figure 5.13: *SNR as a function of input amplitude for the two-stage encoder with single block decoding. The two decoders are a DFT-based low-pass decoder and the DFT-based POCS decoder.*

1. Exclusion of the noise contribution from so-called linear errors in the SNR calculation. Linear errors are the errors at the frequency of the sinusoidal input signal. This modification is in agreement with [32], [37], [122].

2. Using an $N$-point Hanning window on the difference signal between the input and the decoder output [24] before computing the noise power. This modification is in agreement with [32].

Exclusion of linear errors is generally justified by noting that they can be compensated for with simple scaling. However, there are two reasons that this justification is not valid in a $\Sigma\Delta$ context:

1. The required scaling may be frequency-dependent. Signal distortion results from the non-unity signal transfer function of the encoder, and from a linearized point of view, the encoder has a signal transfer function between its input and output of

$$H_X(z) = \frac{H(z)}{1 + H(z)}$$

where $H(z)$ is the open-loop transfer function. In general the signal transfer function is not constant over baseband, and thus a simple frequency-independent constant gain is insufficient to equalize it. On the other hand, an equalization filter can be designed to compensate for the ripples

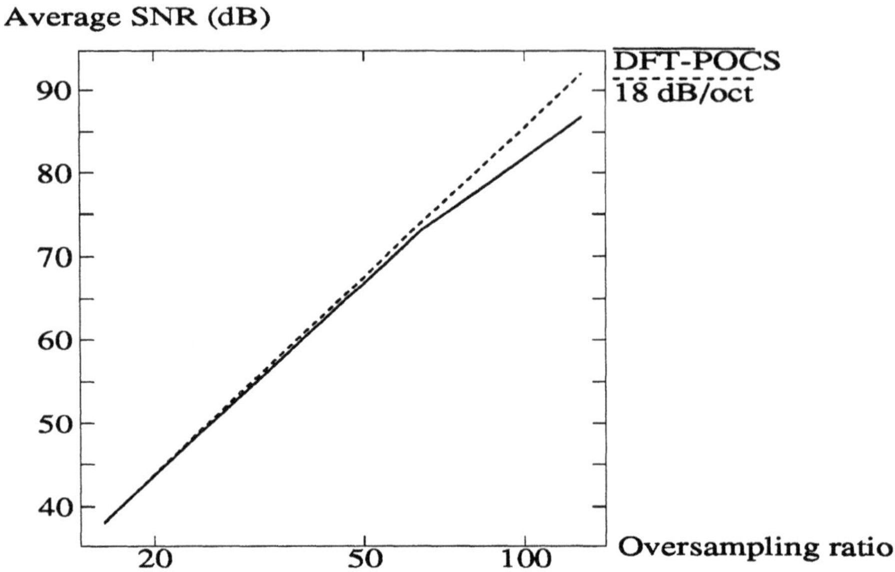


Figure 5.14: *Average SNR as a function of oversampling ratio for the two-stage encoder with single-block decoding and the DFT-POCS decoder. Also shown is line with 15 dB/octave and 18 dB/octave slopes.*

in $H_X(z)$ over baseband, such that the product of its transfer function with $H_X(z)$ approaches unity over baseband.

2. The signal frequency error in general depends nonlinearly on the signal amplitude and phase as well as on the presence of other signals. This is because a $\Sigma\Delta$ encoder is a nonlinear circuit.

To illustrate the importance of signal frequency error, we consider a situation where the signal frequency component at the encoder output equals a constant $K$ times the sinusoidal input signal. We assume for simplicity that $K$ is real. The signal frequency error energy equals $(1-K)^2$, so if the SNR is defined to include signal frequency error, the SNR is limited to $-20\log_{10}|1-K|$. For instance, if $K = 1+10^{-4}$, the SNR cannot exceed 80 dB. For a linear filter, an inaccuracy of $1-K = 10^{-4}$ corresponds to a ripple of 0.0009 dB. This example shows that the passband demands on the equalizing filter are quite stringent. In addition, the filter must act as a good low-pass filter outside of baseband. Even if these requirements are met, the nonlinear signal distortion inherent in the encoder cannot be cancelled.

Although our POCS decoding algorithm does not explicitly take signal frequency errors into account, we have found numerically that the time-domain constraint can act to somewhat reduce these errors. The POCS algorithm

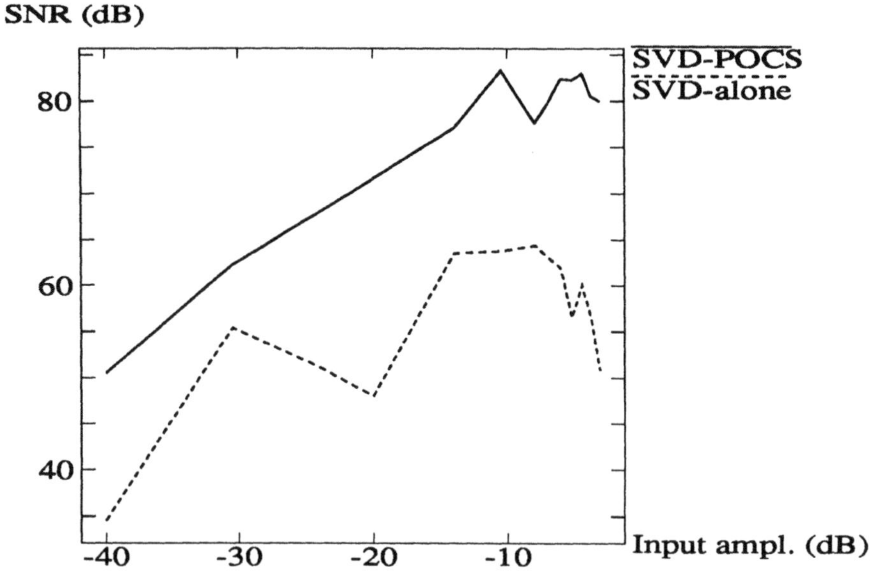


Figure 5.15: *SNR as a function of input amplitude for a specific fourth order interpolative encoder. The two decoders are the SVD-based low-pass decoder and the SVD-based POCS decoder.*

avoids the need to explicitly design a high-precision linear filter, and furthermore the time-domain constraint implies that the algorithm does not neglect the nonlinear signal frequency errors. These observations are further discussed in section 5.3.5.3.

#### 5.3.5.2 Simulation Results

Figure 5.17 is analogous to figure 5.5 and shows SNR results for the single-loop encoder when the SNR is obtained by discarding linear errors. The band limitation is DFT-based. The DFT-alone curves can be taken as upper bounds on the performance of linear filtering. For OSR = 64, the peak SNRs are 63 dB for the DFT-POCS decoder and 58 dB for the DFT-alone decoder. For OSR = 128, the peak SNRs are 84 dB and 66 dB, respectively. Figure 5.17 shows that when excluding linear errors, the POCS method gains about 3-7 dB over the DFT-alone method for OSR = 64, and 3-18 dB for OSR = 128. Not shown are simulation results for OSR = 32 where we find a peak SNR for the DFT-alone method of 49 dB. This result is comparable to a result in [37] for an OSR of approximately 36 where a peak SNR of about 50 dB is shown. Comparing figures 5.5 and 5.17, we find that the gap between the DFT-alone and DFT-POCS curves decreases when the SNR definition is modified to

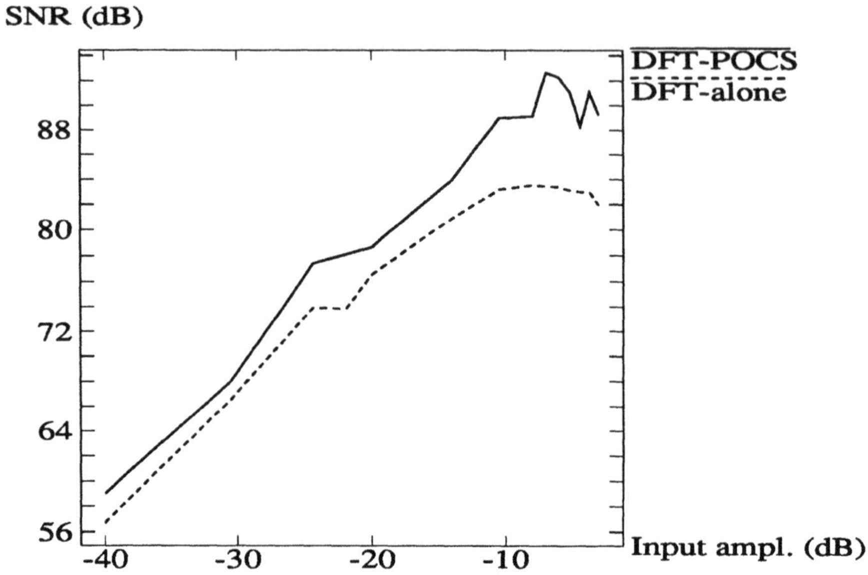


Figure 5.16: *SNR as a function of input amplitude for a specific fourth order interpolative encoder. The two decoders are the DFT-based low-pass decoder and the DFT-based POCS decoder.*

exclude signal frequency errors. This indicates that the DFT-POCS method better suppresses these errors.

Figure 5.18 shows SNR results for the fourth order encoder with a DFT-based band limitation. The results are obtained exactly in the same way as those in figure 5.16, except that the SNR excludes linear errors, and windowing is performed as described in section 5.3.5.1. This enables comparison with [32] in which a peak SNR of about 95 dB is reported. In figure 5.18 the peak SNRs are 96 dB for the DFT-alone decoder and 101 dB for the DFT-POCS decoder. The DFT-POCS decoder is 3-10 dB superior to the DFT-alone decoder.

Other simulation results show that if we use the SVD-based method for band limitation, and if the SNR is defined to exclude linear errors and to include windowing of the DFT, the peak SNRs are 93 dB for the SVD-alone decoder and 98 dB for the SVD-POCS decoder. Under these conditions, the SVD-POCS decoder is 3-15 dB superior to the SVD-alone decoder.

#### 5.3.5.3 Tone Suppression

In this section we consider the problem of spectral tones which is known to plague single-loop and double-loop encoders [42], [44], [123]. Further comments are made in section 8.3. It has been shown theoretically that in single-loop

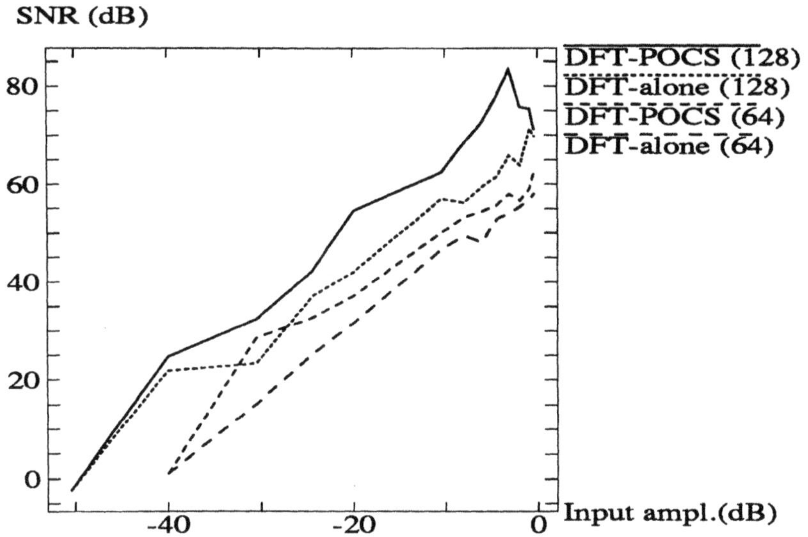


Figure 5.17: *SNR as a function of input amplitude for the single-loop encoder. The SNR calculation excludes linear errors.*

encoders with constant or sinusoidal inputs, the spectrum of the quantization noise is purely discrete [42], [44]. The part of the noise spectrum that falls in signal baseband gives rise to objectionable tones if a linear low-pass decoder is employed. For the double-loop encoder, a similar behavior has been observed empirically [123]. Spectral tones are known to be absent in multi-stage encoders [43], [47], and it is commonly held that tones are also absent in higher-order interpolative encoders. However, there is evidence that this might not be the case [124].

As our proposed POCS algorithm is nonlinear, it is conceivable that the algorithm may be able to exploit time-domain information to suppress baseband tones. On the other hand, it is also conceivable that the algorithm may introduce additional nonlinear distortion. In our simulations, we have found neither of these effects to be significant. To test the performance of our algorithm, we have carried out simulations for a large number of constant inputs, as these are known to give rise to tones. We find that the POCS algorithm quickly converges to an input estimate which contains approximately the same tones as the estimate produced by a linear decoder. As the input estimate resulting from the POCS algorithm satisfies both the time-domain and frequency-domain constraints, there is no deterministic way to distinguish it from the actual constant input. We therefore conclude that the presence of tones in single-loop and double-loop modulators is an inherent property of the encoders which cannot be suppressed by decoders.

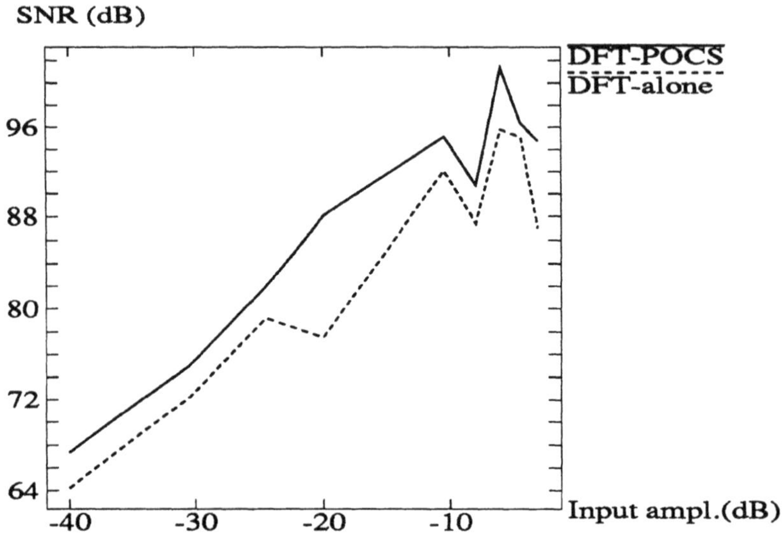


Figure 5.18: *SNR as a function of input amplitude for a specific fourth order interpolative encoder. The decoders are DFT-based as in figure 5.16, but the SNR excludes linear errors, and windowing is performed.*

#### 5.3.5.4 Comparison to Thao and Vetterli's Work

In this section we compare the simulation results of this chapter to those of Thao and Vetterli [117], [118], [119], [120]. We first discuss the asymptotic performance in terms of SNR vs. OSR, and then compare actual SNR levels in dB.

In [118] Thao and Vetterli analytically show the following result: There exists a decoder for an $n$-stage $\Sigma\Delta$ encoder which statistically achieves an asymptotic SNR vs. OSR performance of at least $(2n+2) \cdot 3$ dB/octave, whereas traditional linear decoding is known to be limited to $(2n+1) \cdot 3$ dB/octave. The result is shown under the assumptions that the quantization noise is white, and that the input signal is chosen at random from the set of band-limited signals. They verify numerically that a POCS algorithm obtains $(2n+2) \cdot 3$ dB/octave performance.

In section 5.3.1 we found an SNR vs. OSR performance of approximately 12 dB/octave for the single-loop encoder with the DFT-POCS algorithm. Similarly, we found slopes of approximately 18 dB/octave for the double-loop and two-stage encoders with OSRs less than about 80, whereas the slopes appear to slightly decrease for OSRs exceeding 80. The results of 12 dB/octave and 18 dB/octave are in agreement with the results obtained by Thao and Vetterli. However, the decreasing slopes for large OSRs require an explanation.

One possible explanation is that our space of band-limited sinusoidal encoder inputs has dimension $N/\text{OSR}$ which is on the order of hundreds. On the other hand, Thao and Vetterli implicitly assume the input frequency to be known, so that the dimension of their input space is 3, namely a DC component and the amplitude and phase of a sinusoidal component. It is possible that knowing the input frequency gives their algorithm an asymptotic advantage. As Thao and Vetterli's theorem is derived under the unrealistic assumption of white quantization noise, the theorem does not show that our algorithm should achieve an asymptotic SNR performance of $(2n + 2) \cdot 3$ dB per octave of oversampling.

Other factors in the simulation set-ups may also play a part in the observed differences. In particular, we consider sinusoidal inputs with a range of amplitudes and no DC component, whereas Thao and Vetterli consider sinusoidal inputs with a fixed amplitude of 0.5, but a random DC component between $-0.5$ and 0.5. The untraditional inclusion of a sizable DC component in the input signal as well as in the SNR calculations may conceivably mask small performance changes in the dynamic behavior. This is because encoders with filter poles at DC achieve relatively better SNRs for DC inputs than for dynamic inputs, as the filter magnifies DC errors with larger gain than it magnifies dynamic errors.

One might speculate that the probable cause of the decreasing slopes for large OSRs is that we keep the sample size $N$ constant at 4096, whereas Thao and Vetterli let $N$ be proportional to the OSR. Clearly, if the sample size is finite, the SNR must also be finite even for infinite OSR, that is, constant signals; it therefore appears that we might be beginning to see the effect of finite $N$ at SNRs of about 85-90 dB. However, we find by simulation that no gain in SNR is obtained by doubling the sample size.

We now discuss the absolute performance of the POCS algorithms, and their performances compared to linear filtering. In our results, we found improvements in SNR over linear filtering for all practical input amplitudes and OSRs. Thao and Vetterli do not report simulation results for linear filtering, but rather use a standard linearized formula to approximate its performance. In [119] they report that for the single-loop encoder, the SNR gain over linear filtering ranges from 0 dB at OSR $= 20$ to about 6 dB at OSR $= 128$. For the two-stage encoder, the SNR gains are $-3$ dB at OSR $= 20$ and $+4$ dB at OSR $= 128$, and for the double-loop encoder, the gain ranges from $-8$ dB at OSR $= 20$ to $-2$ dB at OSR $= 128$. The POCS algorithm is reported to be inferior to linear filtering for OSRs less than about 50 for the two-stage encoder, and for OSRs less than about 256 for the double-loop encoder. The authors attribute the negative SNR gains to the fact that they use the zero signal rather than the quantizer output as their initial input estimate; they justify this choice by stating that it avoids artifacts in the evaluation of the alternating projection algorithm. We would find it natural for the authors to justify their claim by simulation, or to use an initial estimate that provides better reconstruction. Another possible explanation for the reported SNR loss may be that the authors do not simulate the performance of linear filtering, and the theoretical expression for the performance of linear filtering that they use appears to exclude linear errors.

| Encoder | OSR | Nom. SNR | Gain | Leak | State | Quant. | Noise |
|---|---|---|---|---|---|---|---|
| Single-loop | 64 | 61 dB | N/A | 1,500 | 2.5% | 2.5% | 0.75% |
| | 128 | 84 dB | N/A | 7,500 | 2.0% | 0.5% | 0.15% |
| Double-loop | 64 | 79 dB | 1% | 5,000 | ≥15% | 10% | 0.2% |
| | 128 | 92 dB | 0.1% | 50,000 | ≥15% | 5% | 0.01% |
| Two-stage | 64 | 86 dB | 1% | 10,000 | ≥15% | ≥15% | 0.1% |
| | 128 | 101 dB | 0.005% | 150,000 | 5% | 2.5% | 0.0075% |
| 4th order | 48 | 93 dB | ≥5% | 750 | 3% | 2% | 0.01% |

Table 5.2: *Sensitivity of the POCS-DFT decoder towards various non-idealities which are unknown to the decoder. The shown non-idealities result in a peak SNR loss of approximately 2 dB. The quantities given in percent are measured relative to the quantizer step size. The quantities given for the leak are the minimum acceptable operational amplifier (op-amp) gains.*

Finally, we compare the actual SNR numbers obtained with our POCS algorithm to those obtained by Thao and Vetterli for the same encoders and input amplitudes. This comparison cannot be made fair, as Thao and Vetterli include a random DC component in their input signal; they also fix the input frequency at the baseband edge, which results in a lower SNR than for input frequencies well within baseband. With these reservations, we find for the single-loop encoder that our SNR is 54 dB, whereas Thao and Vetterli report approximately 46-47 dB. For the two-stage encoder, we find an SNR of 79 dB, whereas Thao and Vetterli report approximately 73 dB. No absolute numbers are reported for the double-loop encoder. We find these numbers to be in acceptable agreement. Without detailed knowledge of both simulations, it does not appear possible to say whether some SNR loss can be attributed to Thao and Vetterli's approximate, but faster time-domain projection.

## 5.4 Decoding in the Presence of Non-Idealities

In this section we present simulation results that describe the effects of a number of non-idealities on the performance of the POCS decoding algorithm. The numerical values of the non-idealities are assumed unknown, but can easily be compensated for in the algorithm if known. The non-idealities we consider are integrator gain and leak, initial state offset, quantizer offset and input noise, all of which are described in appendix B.3. The encoders we consider are the single-loop, double-loop, two-stage and fourth order interpolative encoders. The results are summarized in table 5.2. We first describe the assumptions underlying the results, then present detailed results for the fourth order encoder, and finally comment on the results summarized in table 5.2.

| Coeff. | POCS-FFT | | FFT-alone | |
|---|---|---|---|---|
| | +5% | −5% | +5% | −5% |
| $A_0$ | +0.39 | +1.43 | −0.69 | −0.11 |
| $A_1$ | −0.06 | +3.59 | −0.12 | −0.27 |
| $A_2$ | −1.06 | +3.43 | 0.00 | +0.07 |
| $A_3$ | +1.60 | +3.54 | +0.21 | −0.30 |
| $A_4$ | +1.46 | +1.65 | +0.26 | −0.44 |
| $B_1$ | +1.99 | +3.76 | +0.20 | 0.00 |
| $B_2$ | −1.86 | +0.11 | +0.58 | −0.73 |
| $B_3$ | +2.64 | +3.28 | −0.17 | +0.06 |
| $B_4$ | +1.83 | +3.06 | −1.15 | +0.84 |

Table 5.3: *Peak SNR changes in dB for the fourth order interpolative encoder as a function of perturbations in open-loop filter gains.*

The simulation results in table 5.2 are generated using the DFT as the band limitation and considering a single 4K block of samples. The parameters of the simulations are the same as in section 5.3.1–5.3.4. The SNR is calculated in the same way as in those sections, that is, by including linear errors and not performing windowing. The peak SNR is used as a concise statistic to describe the performance of a modulator. For encoders with more than one integrator or quantizer, the non-idealities are assumed for simplicity to be the same for all elements that are nominally identical. The acceptable levels of non-idealities reported in table 5.2 result in a peak SNR loss of approximately 2 dB.

We now present detailed results for the fourth order encoder showing the sensitivity of the POCS decoder as well as the DFT-based linear decoder towards non-idealities. As the encoder contains nine scaling factors, we have not exhaustively graphed the sensitivities towards perturbations in each gain. However, table 5.3 shows the changes in peak SNR resulting from perturbations of ±5% in each gain from its nominal value. We see that in no case does the perturbation result in more than a 2 dB degradation in peak SNR, and in fact the peak SNR of the POCS decoder often increases as a result of a perturbation. Figure 5.19 shows that the DFT-POCS decoder requires an op-amp gain of approximately 750, whereas the DFT-alone decoder requires a gain of 300. Figure 5.20 shows the sensitivity of the decoders towards initial integrator states. For simplicity, we assume that all state offsets are identical. We see that the DFT-POCS decoder can tolerate an initial state offset of about 3%, whereas the linear decoder is insensitive to initial integrator states. Figure 5.21 shows that the DFT-POCS and DFT-alone decoders require a quantizer offset no larger than 2% and 5%, respectively. Figure 5.22 shows that the noise levels should be no larger than 0.01% and 0.03% for the DFT-POCS and DFT-alone

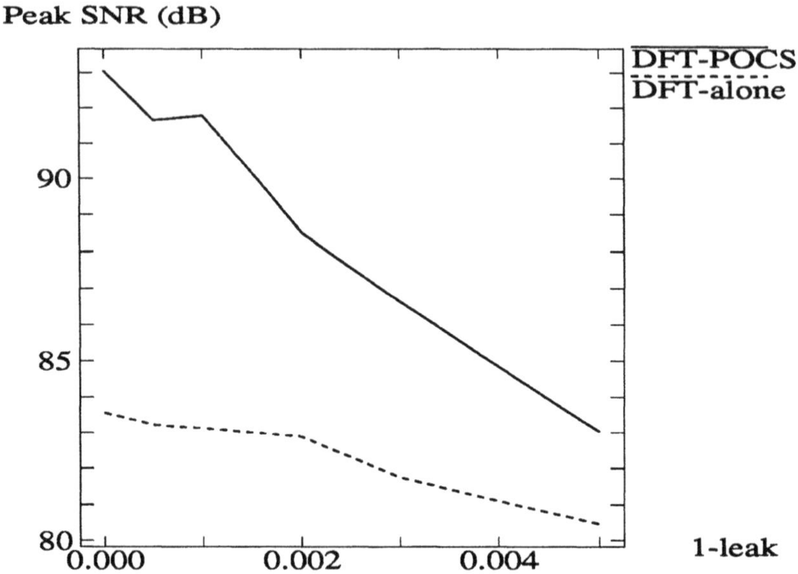


Figure 5.19: *Peak SNR as a function of integrator leak for the fourth order encoder. The two decoders are the DFT-based low-pass decoder and the DFT-based POCS decoder.*

decoder, respectively.

In general, the results in table 5.3 indicate that the sensitivity increases as the OSR and hence the SNR increases. We also see that for comparable SNR performance, the different encoders mostly have comparable sensitivities. The exception to this rule is the fourth order interpolative encoder which exhibits quite low sensitivity towards non-idealities, taking into account its peak SNR of 93 dB. The exact reason for the decreased sensitivity is unclear, but it appears reasonable to conjecture that it is related to the pole locations of the fourth order encoder. Specifically, the single-loop, double-loop and two-stage encoders all have their poles at DC, whereas the fourth order encoder has its poles at non-DC frequencies within signal baseband.

## 5.5 Summary

We have shown by simulations that for the single-loop, double-loop and two-stage encoders and for a specific fourth order interpolative encoder, significant gains in SNR can be achieved using sophisticated decoding techniques rather than linear decoding. The results are valid both for block decoding and, more realistically, for sliding block decoding.

The presented decoding method is computationally complex. However, our

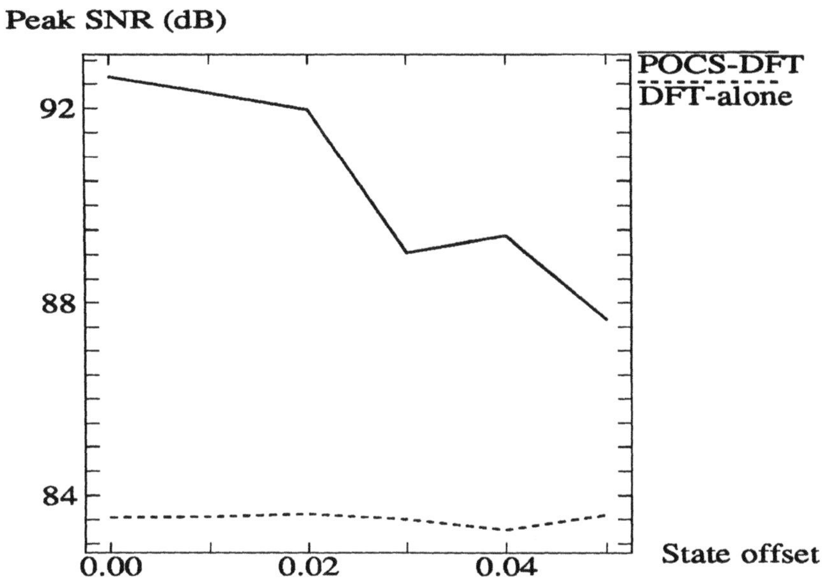


Figure 5.20: *Peak SNR as a function of integrator state offsets for the fourth order encoder. The two decoders are the DFT-based low-pass decoder and the DFT-based POCS decoder.*

results can serve as upper bounds on the achievable SNR, and would be useful for future research in simplified, nonlinear decoding algorithms.

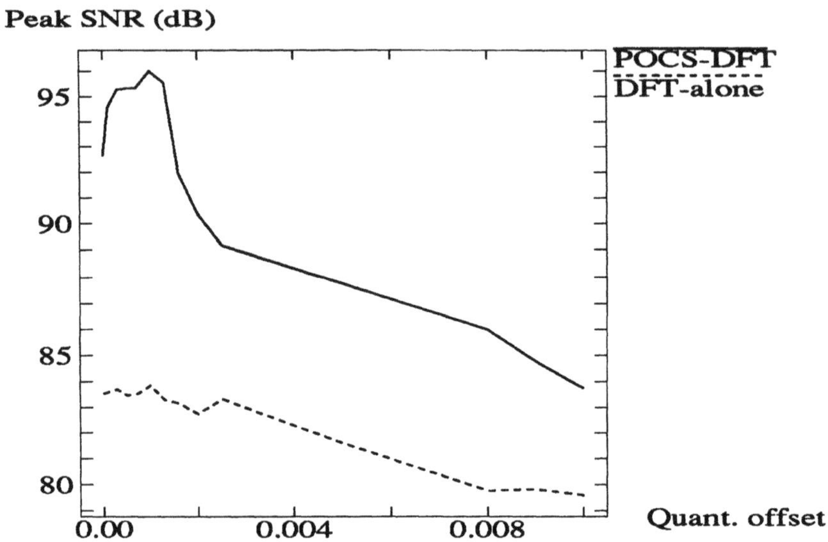


Figure 5.21: *Peak SNR as a function of quantizer offset for the fourth order encoder. The two decoders are the DFT-based low-pass decoder and the DFT-based POCS decoder.*

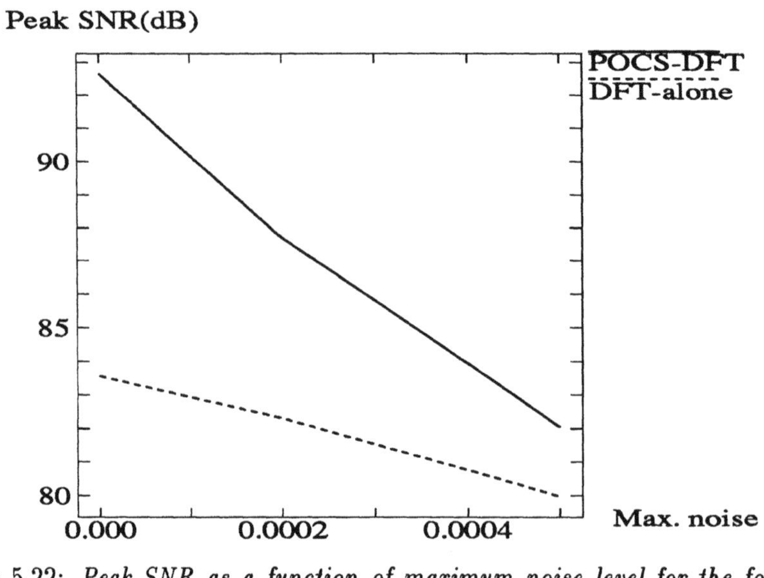


Figure 5.22: *Peak SNR as a function of maximum noise level for the fourth order encoder. The two decoders are the DFT-based low-pass decoder and the DFT-based POCS decoder.*

# Chapter 6

# Halftone to Continuous-Tone Conversion of Error-Diffusion Coded Images

In this chapter[1] we consider the problem of reconstructing a continuous-tone (contone) image from its halftoned version, where the halftoning process is done by error diffusion. As there is a natural connection between error diffusion and $\Sigma\Delta$ modulation, the reconstruction technique developed in chapter 5 for one-dimensional signals readily generalizes to the image reconstruction problem. We present an iterative nonlinear decoding algorithm for halftone-to-contone conversion, and show simulation results that compare the performance of the algorithm to that of conventional linear lowpass filtering. We find that the new technique results in subjectively superior reconstruction.

## 6.1 Introduction

Halftoning is the process of converting continuous-tone (contone) images into a representation suitable for display on binary devices [79], [126], [127], [128]. The importance of halftoning is due to the fact that images such as photographs can have a large dynamic range, whereas display devices such as newspaper printers, laser printers and some computer screens can only display binary images. The idea behind halftoning is to create a binary image whose low-frequency content

[1] Based on [125], ©IEEE.

approximates that of a contone image, and whose high-frequency content is not visibly correlated with the low-frequency content. At a suitable distance, the human visual system acts approximately as a low-pass filter, and so the halftone image creates the illusion of a contone image. If the halftone image is too close to the observer or if the image is sampled too coarsely, however, the illusion does not occur.

The halftoning process has been used in the printing industry for over a hundred years, and a large body of practical and theoretical knowledge has been accumulated [126]. The work reported in this chapter and other research like it [27], [129] is mainly concerned with digital or electronic halftoning rather than manual halftoning which involves contact screens. In the former problem type, an image is manipulated as a two-dimensional array of discrete samples, whereas in the latter problem type, parameters such as the sizes and shapes of ink dots can be varied. For simplicity, we only consider Black-and-White (B/W) contone images.

A number of algorithms for halftoning images are known. The reader is referred to reviews such as [126], [127], [128] for a detailed discussion of the relative merits of these algorithms. In this chapter, we consider the algorithm introduced in [130] called Error Diffusion (ED). Specifically, we address the problem of reconstructing a contone image from its halftoned version. This problem arises in many situations when only the halftone image is available. One situation occurs when the best possible image quality is required for immediate viewing by humans. Another situation occurs when the contone image underlying the halftone image needs to be processed. Such processing could include decimation, interpolation, extrapolation, linear and nonlinear filtering and enhancement, or any other form of manipulation. In these situations, it is natural to first reconstruct the contone image, then perform the desired operations on it, and finally halftone the result. Finally, in many applications such as facsimile and desk-top publishing, it may be desired to convert back and forth between different halftone images, for display on printers and monitors with varying dynamic range and resolution characteristics.

There are many existing approaches to the problem of halftone to contone conversion. For instance, Fan [131] describes a logic filtering approach for halftone images obtained by screening. The most common conversion approach is to apply a low-pass filter [126]. However, closer examination reveals that there is more information present in the halftone image than is extracted by a linear filter; specifically, linear filtering does not exploit knowledge of the detailed mechanism by which the particular halftone image was generated. Our approach is to use this extra information to obtain sharper and more subjectively pleasing contone image reconstructions. We propose a class of iterative, nonlinear algorithms based on the theory of Projections Onto Convex Sets (POCS) to reconstruct contone images from halftone ones, and show that our algorithms result in higher reconstruction quality than linear filtering. The theory of POCS has previously been applied to numerous signal and image reconstruction problems [27], [83], including reconstruction of contone images

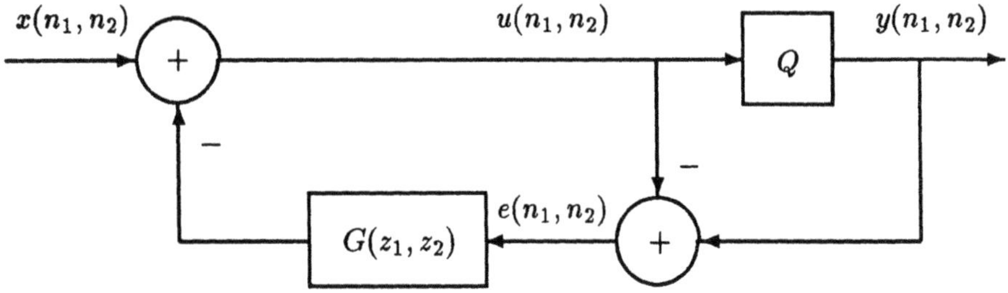


Figure 6.1: *Discrete-time model of an error diffusion encoder.*

from halftones generated using screening [129], [131]. In contrast, our approach is aimed at halftones generated using ED. Furthermore, unlike in [129], [131] where the screening function is estimated, we assume the ED filter to be known.

Our proposed decoding approach is inspired by our work on nonlinear decoding algorithms for $\Sigma\Delta$ modulators in [93], [113], [116]. We thus exploit the relationship between error diffusion and $\Sigma\Delta$ modulation first noted in [79].

The chapter is organized as follows. Section 6.2 presents a brief introduction to error diffusion, and section 6.3 describes our proposed nonlinear decoding algorithm. Section 6.4 shows sample results of our proposed algorithm, and section 6.5 contains a summary.

## 6.2 Error Diffusion Coding

In this section we describe the error diffusion process mathematically. The description is in three parts. In section 6.2.1 we give a brief, general introduction to error diffusion, and in sections 6.2.2 and 6.2.3 we lay the ground for the reconstruction algorithm described in section 6.3. Specifically, in section 6.2.2 we discuss the choice of a computational procedure which is needed to obtain a halftone image from a contone one. While our chosen computational procedure differs slightly from the traditional one, it tremendously simplifies the reconstruction algorithm that we propose in section 6.3. In section 6.2.3 we derive the space-domain description of error diffusion which results from our choice of computational procedure.

### 6.2.1 Introduction

Figure 6.1 shows a diagram of an ED encoder which is a two-dimensional discrete-time system. For comparison, figure 1.4 shows a discrete-time model of an interpolative $\Sigma\Delta$ modulator which is a one-dimensional system used in analog-to-digital and digital-to-analog conversion. Similar to a $\Sigma\Delta$ encoder, an ED encoder contains a 1-bit quantizer and a linear filter in a negative feedback loop, and has a continuous-amplitude input and binary output. Unlike the linear filter in an interpolative $\Sigma\Delta$ encoder, however, the filter in an ED encoder operates on the quantization error and is located in the feedback path. Thus, an ED encoder is closer to being a predictive coder. To be implementable, the two-dimensional filter $G(z_1, z_2)$ must have finite-extent input and output masks, and the system must be recursively computable [132]. The particular input and output masks that we consider for $G$ are shown in figure 6.2, and correspond to the original Floyd-Steinberg ED encoder [130]. Since the introduction of the Floyd-Steinberg ED filter, a number of improvements on the filter have been proposed to overcome problems with artifacts [128]. However, for simplicity and convenience we will use the original filter throughout this chapter.

In our reconstruction algorithm, we will assume that we have perfect knowledge of the particular filter $G(z_1, z_2)$ used in the ED encoder. In applications where this assumption is not justified, an interesting filter estimation problem arises. For an approach to a similar problem, see [129]. Because the encoder is generally implemented in software, we will also assume that there are no non-idealities in the encoding process. In this respect, the reconstruction problem is simpler than the $\Sigma\Delta$ decoding problem in which circuit imperfections must be taken into account [93], [113].

### 6.2.2 Computational Procedure

Two-dimensional discrete-time systems require computational procedures that describe the way in which the output is computed from the input [132]. In this section we define the particular computational procedure that we will use throughout the chapter. As we will see, we slightly modify the standard way of implementing the halftoning process in order to significantly reduce the complexity of our reconstruction algorithm, while minimally affecting the quality of the halftoning. Our computational procedure is inspired by the general procedure described in Lim [132] for linear discrete-time systems, and is outlined below. Using our procedure, we derive a space-domain description of the ED encoder relating the unknown input $\{x(n_1, n_2)\}$, the known output $\{y(n_1, n_2)\}$ and known *space-invariant* impulse responses which are determined by $G$. We have found that if we instead apply the standard computational procedure for ED encoding, the space-domain description involves space-varying impulse responses which would significantly increase the storage requirements and complexity of the decoder that we propose in section 6.3.

We now propose a computational procedure for error diffusion, and show that the procedure results in a significant simplification of the mathematical description. To compute the output of the recursive nonlinear ED encoder, we must specify both boundary conditions and a recursive computational order [132]. We will begin with the boundary conditions. Let us pretend for a moment that the system shown in figure 6.1 is linear and that the quantizer $Q$ is absent. We will denote by $R_x$ the square region

$$R_x = \{(n_1, n_2) \mid 0 \le n_1 \le N-1, 0 \le n_2 \le N-1\}$$

over which the input image $\{x(n_1, n_2)\}$ is non-zero; $N$ denotes the image size.

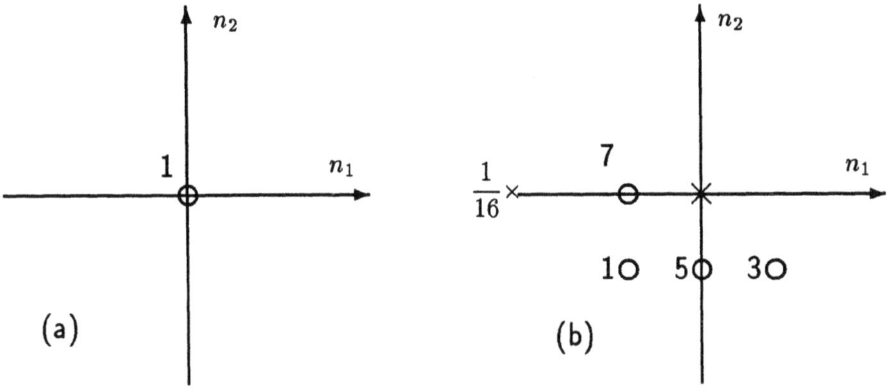


Figure 6.2: *(a) Input and (b) output masks for a particular ED encoder filter* $G(z_1, z_2)$.

Let us consider the operation of the encoder in the absence of the quantizer. The impulse response of the specific filter $G$ is given by the masks in figure 6.2, and has the infinite-extent Region Of Support (ROS) $R_g$ shown in figure 6.3(a). As the input mask is a Kronecker $\delta$-function, the ROS of the output of $G(z_1, z_2)$ is also $R_g$. Therefore the filter output is zero in the boundary condition region $R_{\mathrm{BC}}$ which is the complement of $R_g$.

We now take into account the nonlinear element $Q$, which we assume for specificity to be given by

$$Q(u) = \begin{cases} 0 & \text{for } u \leq L_{\mathrm{mid}} \\ L_{\mathrm{high}} & \text{for } u > L_{\mathrm{mid}} \end{cases} \tag{6.1}$$

where $L_{\mathrm{high}} = 255$ and $L_{\mathrm{mid}} = \frac{1}{2}L_{\mathrm{high}} = 127\frac{1}{2}$. These quantization levels are chosen to match our application to 8-bit images in which input samples take on integer values between 0 and 255. Since we have chosen $Q(0)$ to be 0, inserting the quantizer into the linear circuit within the region $R_{\mathrm{BC}}$ does not alter the functionality of the encoder or violate the difference equations.[2] The boundary condition region $R_{\mathrm{BC}}$ is thus unaffected by the insertion of the quantizer.

Having specified boundary conditions, we now complete our computational procedure for the ED encoder by specifying a recursively computable order [132] in which input samples can be processed. A natural order of computation would be to go through $R_g$ line by line. However, we only need the output of the ED encoder over the square region of support of the input, $R_x$, shown in figure 6.3(b). We can therefore discard any lines above $R_x$ in the $n_2$ direction,

[2]Note that if we choose $Q$ to be the signum function as for $\Sigma\Delta$ encoders, we get $Q(0) = -1$, and thus there would not exist a boundary region $R_{\mathrm{BC}}$ for which we could find a consistent procedure for calculating the encoder output. Strictly speaking, the encoder with such a quantizer would be ill-posed.

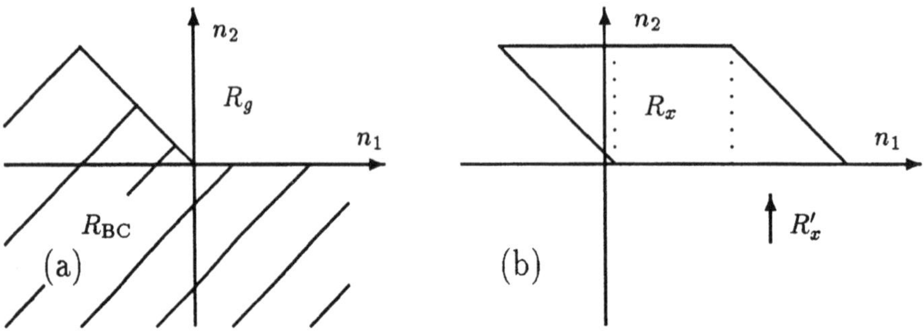


Figure 6.3: *(a) Region of support $R_g$ for the impulse response of $G(z_1, z_2)$. (b) Region of support $R_x$ for the input, shown with dotted lines and wholly contained in the larger region $R'_x$ over which the filter output must be calculated.*

and also any samples with large enough values of $n_1$ that they do not affect the encoder output over $R_x$. To calculate the encoder output over $R_x$, we can restrict computations to the finite region $R'_x$ shown in figure 6.3(b), and the input can be processed line by line over $R'_x$. In contrast, the standard way of implementing ED encoding forces the quantization error to be zero outside of $R_x$, which leads to a violation of the governing difference equations over $R_c$; here we have defined $R_c$ to be the part of $R'_x$ which does not overlap with $R_x$.

While our proposed computational procedure simplifies the decoding algorithm, it has one potential problem: The decoder only has access to the encoder output over $R_x$ and not $R_c$; but due to the feedback, the encoder output over $R_x$ may be affected by encoder outputs over $R_c$. Thus it is conceivable that the discarded encoder outputs over $R_c$ can result in a more complex decoding algorithm. However, as we show in appendix E.1, an encoder output sample is necessarily zero whenever the corresponding encoder input sample is zero, provided that a number of conditions involving the input and the encoder are satisfied; these conditions are all satisfied for the system in question. Therefore the encoder output can only be non-zero over $R_x$, and there is no information about the input in the discarded encoder outputs, as these are all zero.

Although our proposed computational procedure generates the same binary image as the standard procedure over $R_c$, it generates a different image over $R_x$. This is because our procedure requires the encoder to process all of $R'_x$, rather than only $R_x$, which affects the encoder states. Nevertheless, we find that the two halftoning techniques result in subjectively similar quality.

### 6.2.3 Space-Domain Description

We will now derive a space-domain description of the ED encoder based on the above computational procedure. Referring to figure 6.1, our goal is to write the quantizer input $\{u(n_1, n_2)\}$ as a linear combination of input samples $\{x(n_1, n_2)\}$

and output samples $\{y(n_1, n_2)\}$. For given input and output masks of the error diffusion filter $G(z_1, z_2)$, we can use standard methods [132] to calculate the space-invariant impulse responses $\{g_x(n_1, n_2)\}$ and $\{g_y(n_1, n_2)\}$ which describe the relationship between the input and output images, respectively, and the quantizer input. These impulse responses are defined so that

$$u(n_1, n_2) = \sum_{m_1=-\infty}^{\infty} \sum_{m_2=-\infty}^{\infty} [g_x(n_1 - m_1, n_2 - m_2)x(m_1, m_2) + g_y(n_1 - m_1, n_2 - m_2)y(m_1, m_2)] \tag{6.2}$$

Note that even though the quantizer is nonlinear, the relationship between $\{x(n_1, n_2)\}$, $\{y(n_1, n_2)\}$ and $\{u(n_1, n_2)\}$ is linear. By inspection, $g_x(0,0) = 1$, $g_y(0,0) = 0$, and for $(n_1, n_2) \neq (0,0)$, $g_x(n_1, n_2) + g_y(n_1, n_2) = 0$. If our modified computational procedure of section 6.2.2 is not used, the impulse responses $g_x$ and $g_y$ become space-varying, and the storage requirement for the impulse responses increases from $O(N^2)$ to $O(N^4)$. However, the quantizer input would still depend linearly on the encoder input and output.

The summation limits in (6.2) can be reduced by making use of knowledge of the regions of support for the involved sequences. First, we have $x(m_1, m_2) = 0$ for $(m_1, m_2) \notin R_x$. Second, we have the less obvious result that $y(m_1, m_2) = 0$ for $(m_1, m_2) \notin R_x$ as shown in appendix E.1; this is a form of stability result, because it follows from the fact that the quantization error is bounded. Finally, the regions of support of the impulse responses can be calculated for given input and output masks. For instance, $g_x$ and $g_y$ for the masks shown in figure 6.2 are both zero whenever $n_2 < 0$ or $n_1 + n_2 < 0$. Putting these facts together, we find that for our particular ED encoder,

$$u(n_1, n_2) = \sum_{m_2=0}^{\min\{n_2, N-1\}} \sum_{m_1=0}^{\min\{n_1+n_2-m_2, N-1\}} [g_x(n_1 - m_1, n_2 - m_2)x(m_1, m_2) + g_y(n_1 - m_1, n_2 - m_2)y(m_1, m_2)] \tag{6.3}$$

We will use this space-domain characterization in section 6.3 to derive a part of our decoding algorithm.

## 6.3 Decoding Algorithm

In this section we describe our proposed algorithm for the signal reconstruction problem of halftone to contone conversion. The algorithm is based on the theory of POCS [82], [133], and makes use of two operations: A space-domain projection $P_1$ that modifies its input to satisfy the bounds imposed by the particular halftone image, and a frequency-domain operation $P_2$ that band-limits its input. This is entirely analogous to the approach in section 5.2 for $\Sigma\Delta$ modulators. The POCS algorithm alternates between the two operations to find a signal invariant under both, thus extracting all information from the halftone image. In sections 6.3.1 and 6.3.2 we describe the space-domain and frequency-domain projections, respectively.

### 6.3.1 Space-Domain Projection

In this section we present a matrix space-domain description of the error diffusion encoder, based on equation (6.3) of section 6.2. We then show that the projection $P_1$ of an image onto the set of images satisfying the bounds imposed by the halftone image is a Quadratic Programming (QP) problem.

In our reconstruction problem, we only have access to the output image $\{y(n_1,n_2)\}$ and not the quantizer input sequence $\{u(n_1,n_2)\}$. Thus the halftone image provides us with a series of bounds on linear combinations of input samples $\{x(n_1,n_2)\}$ of the form

$$u(n_1,n_2) > L_{\text{mid}} \text{ if } y(n_1,n_2) = L_{\text{high}}, \quad u(n_1,n_2) \le L_{\text{mid}} \text{ if } y(n_1,n_2) = 0 \tag{6.4}$$

Due to our modification of the halftoning algorithm in section 6.2, equation (6.4) can be written more succinctly in matrix notation.[3] To this end, we introduce the $N^2$-dimensional vectors

$$\begin{aligned}\mathbf{x} &= \{x_0, \cdots, x_{N^2-1}\}^\dagger \\ &= \Big\{x(0,0), \cdots, x(0,N-1) \,\vdots\, \cdots \,\vdots\, x(N-1,0), \cdots, x(N-1,N-1)\Big\}^\dagger\end{aligned}$$

$$\begin{aligned}\mathbf{y} &= \{y_0, \cdots, y_{N^2-1}\}^\dagger \\ &= \Big\{y(0,0), \cdots, y(0,N-1) \,\vdots\, \cdots \,\vdots\, y(N-1,0), \cdots, y(N-1,N-1)\Big\}^\dagger\end{aligned}$$

where $\mathbf{x}^\dagger$ denotes the transpose of $\mathbf{x}$. We also define the $N^2 \times N^2$ block-Toeplitz lower-triangular matrix

$$\mathbf{G}_x = \begin{bmatrix} \mathbf{G}_0 & \mathbf{0} & \cdots & \mathbf{0} & \mathbf{0} \\ \mathbf{G}_1 & \mathbf{G}_0 & \cdots & \mathbf{0} & \mathbf{0} \\ \vdots & \vdots & & \vdots & \vdots \\ \mathbf{G}_{N-2} & \mathbf{G}_{N-3} & \cdots & \mathbf{G}_0 & \mathbf{0} \\ \mathbf{G}_{N-1} & \mathbf{G}_{N-2} & \cdots & \mathbf{G}_1 & \mathbf{G}_0 \end{bmatrix} \tag{6.5}$$

where $\mathbf{0}$ denotes the $N \times N$ zero matrix, and the $N \times N$ Toeplitz matrices $\mathbf{G}_n$

[3] With the conventional computational procedure for error diffusion, the halftone image still induces bounds on linear combinations of input samples. However, the structure of the bounds is simplified by our modification.

are given by

$$\mathbf{G}_n = \begin{bmatrix} g_x(n,0) & g_x(n,-1) & \cdots & g_x(n,-N+2) & g_x(n,-N+1) \\ g_x(n,1) & g_x(n,0) & \cdots & g_x(n,-N+1) & g_x(n,-N) \\ \vdots & \vdots & & \vdots & \vdots \\ g_x(n,N-2) & g_x(n,N-3) & \cdots & g_x(n,0) & g_x(n,-1) \\ g_x(n,N-1) & g_x(n,N-2) & \cdots & g_x(n,1) & g_x(n,0) \end{bmatrix} \tag{6.6}$$

Finally, we define $\mathbf{I}$ to be the $N^2 \times N^2$ identity matrix, and $\mathbf{Q}$ to be the diagonal matrix whose $n$th diagonal element is $-1$ if $y_n = L_{\text{high}}$ and $+1$ if $y_n = 0$. We can then write equation (6.4) as

$$\mathbf{Q}\mathbf{G}_x\mathbf{x} \leq \mathbf{Q}\left[L_{\text{mid}}\mathbf{I} + (\mathbf{G}_x - \mathbf{I})\mathbf{y}\right] \tag{6.7}$$

where the inequality sign is to be taken coordinate-wise. The space-domain projection problem can now be stated as follows: Given an arbitrary image with vector representation $\mathbf{x}$, find a vector $\hat{\mathbf{x}}$ which satisfies (6.7) and which minimizes the 2-norm distance

$$d^2 = \|\hat{\mathbf{x}} - \mathbf{x}\|_2^2 \tag{6.8}$$

Equations (6.7) and (6.8) constitute a linearly constrained QP problem which can be solved with techniques available in the literature [121]. The QP problem is of the exact same form as the one stated in section 5.2.2 for $\Sigma\Delta$ encoding. Therefore the methods described in that section for approximately solving the QP problem while drastically reducing computational complexity, are directly applicable to the present problem. Specifically, we can solve a number of QP sub-problems of size $L_{\text{QP}}$ rather than one QP problem of size $N^2$. We can also employ the idea described in section 5.2.2 of overlapping QP sub-blocks in order to more closely approximate the solution to the actual $N^2$-dimensional QP problem. There is no need, however, for the analogy of sliding block decoding, because the entire image is treated as one block.

### 6.3.2 Frequency-Domain Projection

In this section we discuss a number of candidate methods for performing the frequency-domain projection $P_2$ in our reconstruction algorithm. The methods include the Discrete Fourier Transform (DFT), the Discrete Cosine Transform (DCT), a Singular Value Decomposition (SVD) based transform, linear filtering and a projection used in sub-band coding schemes. Because the problem of finding a good projection is more difficult for images than for one-dimensional signals, this section is of a more speculative and empirical nature than section 6.3.1.

Our overall goal is to find an operator on an arbitrary $N \times N$ image that satisfies two conditions:

1. It efficiently discriminates between images and "non-images". To make this description objective, it would be ideal to have a mathematical distinction between images and non-images. However, no such distinction is presently known, and we must therefore appeal to an intuitive description. For instance, a heavily distorted version of an image, with ringing or graininess, would be a non-image. Thus, the visible effect of applying the sought operator on an actual image should be minimal, whereas the operator should significantly alter the appearance of non-images.

2. It is an orthogonal projection in the 2-norm, so that the theory of POCS guarantees the convergence of the algorithm.

Unfortunately, finding an operator that satisfies the first condition is extremely difficult, and simultaneously satisfying both conditions is yet more difficult [134], [135]. There is no inherent requirement that the desired operator be a frequency-domain based one [136]. However, we find a frequency-domain approach most attractive due to the approximate low-pass nature of many naturally occurring images [136]. We will discuss a number of band limitation methods in section 6.3.2.1 below, and comment on their applicability.

#### 6.3.2.1 Candidate Projections

An image can be band-limited by taking its two-dimensional (2-D) DFT, setting the frequency components outside of a region around DC to zero, and taking the inverse DFT. One problem with this approach is the periodicity assumption inherent in the DFT, which creates visible ringing artifacts mainly around the edges of an image. To alleviate ringing, the DCT can be used [132]. Both the DFT and the DCT are 2-norm projections, and they can both be applied to a whole image or, to improve computational efficiency, to blocks of an image at a time. The blocks may be overlapped to reduce blocking effects. We have implemented both forms of band limitations in a number of variants, and we have found that when band-limiting our test image, keeping less than 25-30% of the transform coefficients results in poor quality. On the other hand, keeping that many transform coefficients results in a not very restrictive frequency-domain projection, which leaves too many "non-images" unaltered, thus reducing reconstruction quality.

The third transform method for band limitation that we have considered is the 2-D generalization of the signal-independent SVD method presented in chapter 4. This transform method is also a 2-norm projection, and due to its maximal energy concentration, we would expect it to perform acceptable band limitations with fewer coefficients than the DFT or the DCT. Indeed, we find that when keeping approximately 18% of the SVD coefficients, the resulting band limitation is superior to the DFT or the DCT with more than 25% of their coefficients. Because a smaller number of coefficients are necessary in the SVD method, we conclude that the method captures more of the essence of images, and we prefer it over the two other transform methods. The 2-D SVD method is described in appendix E.2.

We have also considered using one low-pass branch of a sub-band coding system [137], [138] as the band limitation. This approach has the advantage that the result is approximately a projection.[4] However, the orthogonality constraints on analysis and synthesis filters which are required to obtain a projection, also restrict the achievable performance compared to traditional linear filtering with the same mask sizes. In our particular example, we found this restriction to be significant, and we therefore prefer linear filtering over a sub-band based scheme.

Finally, we consider a band limitation method that does not satisfy the projection requirement. The disadvantage of relaxing this requirement is that the theory of POCS no longer guarantees convergence. On the other hand, 2-norm measures such as Signal-to-Noise Ratio (SNR) and Mean Squared Error (MSE) are often not adequate indicators of image quality, and thus it is possible that an algorithm with stagnating convergence properties can perform subjectively better than an algorithm with guaranteed convergence. We therefore consider the use of a simple low-pass filter such as a Gaussian filter [139]. The particular linear filter that we use is a Gaussian $9 \times 9$ filter with varying width $\sigma$ [132]. Its taps are given by

$$h(n_1, n_2) = \frac{1}{K} \exp\left(\frac{n_1^2 + n_2^2}{2\sigma^2}\right), \quad -4 \le n_1, n_2 \le +4 \tag{6.9}$$

where the normalization constant is given by

$$K = \sum_{n_1=-4}^{4} \sum_{n_2=-4}^{4} \exp\left(\frac{n_1^2 + n_2^2}{2\sigma^2}\right)$$

## 6.4 Results

In this section we show a number of images that illustrate the reconstruction theory presented in section 6.3. Throughout the section, we use the $512 \times 512$ *Lenna* image in figure 6.4 as our test case. For all POCS results, the QP block size for the space-domain projection is set to $L_{\mathrm{QP}} = 64$ with an overlap of 50% between blocks. Whenever the frequency-domain projection is SVD-based, it is calculated as described in appendix E.1. The number of POCS iterations is ten. The Peak SNR (PSNR) used is defined as

$$\mathrm{PSNR} \triangleq 10 \log_{10} \frac{255^2}{\frac{1}{N^2} \sum_{i=0}^{N-1} \sum_{j=0}^{N-1} (x_{ij} - \hat{x}_{ij})^2}$$

where $\mathbf{x}$ and $\hat{\mathbf{x}}$ are two $N \times N$ images.

[4] It is a projection for infinite-extent signals.

Figure 6.4: *The original* 512 × 512 *8-bit Lenna image.*

Figure 6.5 shows the effect of band-limiting the Lenna image using the SVD projection. It is the halftone version of this image that we use in the reconstructions shown in this section. We find that the subjective quality of figure 6.5 is close to that of figure 6.4, although some detail is lost in the straws on the hat and in the hair falling down on the right. In addition, some ringing is introduced by the upper right edge of the hat brim.

Figure 6.6 shows the halftoned version of figure 6.5 from which the reconstruction is done. Note that we halftone the SVD-limited image rather than the original image in order to ensure that at least one image is invariant under both the space-domain and frequency-domain projections, so that the theory of POCS guarantees convergence. This approach may not be realistic, but is made to facilitate our investigation.

Figure 6.7 shows the effect of low-pass filtering the halftone image in figure

Figure 6.5: *The SVD-limited Lenna image. The PSNR with respect to figure 6.4 is 32.3 dB.*

6.6 using a Gaussian $9 \times 9$ filter with $\sigma = 2.0$, where the expression for the filter is given in (6.9). We find empirically that $\sigma = 2.0$ results in the subjectively best low-pass filtering; choosing a larger $\sigma$ results in a more blurred image, whereas choosing a smaller $\sigma$ results in a more grainy appearance. Comparing figures 6.4 and 6.7, we see that linear filtering results in a visually pleasing but quite blurred image. Note for instance the blurring of the straws on the hat, the eyebrows, the eyes and the upper brim of the hat. Note also the graininess of the leftmost part of the cheek, and the contouring in the background to the right of the cheek. The image in figure 6.7 is the benchmark against which we will compare the results of our reconstruction algorithm.

Figure 6.8 shows the result of applying the POCS algorithm to the halftone image in figure 6.6 using the SVD as the band limitation. The reconstruction is

Figure 6.6: *The halftoned Lenna image.*

quite sharp and visually pleasing compared to figure 6.7. In large magnification, however, figure 6.8 exhibits some visually objectionable artifacts in the form of a grainy texture in the lower right part of the face. In addition, ringing is present along the upper right edge of the hat brim as in figure 6.5. This is to be expected as figure 6.5 is the image that we are reconstructing.

Figures 6.9 and 6.10 show the results of applying the POCS algorithm with two different frequency-domain projections, both based on a $9 \times 9$ linear Gaussian filter whose $\sigma$ depends on the iteration number.[5] The idea of varying $\sigma$, or equivalently the filter bandwidth, is inspired by the approach in [129]. By

[5]Even though the SVD is not used in the frequency-domain operation, we still SVD-limit the original image before halftoning. This is to enable comparison with the remaining reconstructed images of this section. However, we find that virtually no visual change in the reconstructions occur if the original image is not SVD-limited before halftoning.

Figure 6.7: *The halftoned Lenna image filtered with a 9×9 Gaussian filter. The PSNR with respect to figure 6.5 is 29.1 dB.*

starting with a large $\sigma$, the initial iterates become quite low-pass and blurred, thus filtering out the halftone artifacts. Subsequently, decreasing $\sigma$ reduces the effect of the filter. In figure 6.9, $\sigma$ varies in the following way with the iteration number $n$:

$$\sigma(n) = \begin{cases} 2.0\left(1 - \frac{n}{5}\right) & \text{for } n < 5 \\ 0.4 & \text{for } n \geq 5 \end{cases} \tag{6.10}$$

whereas in figure 6.10, $\sigma$ is given by

$$\sigma(n) = \begin{cases} 2.0\left(1 - \frac{3n}{20}\right) & \text{for } n < 5 \\ 0.8 & \text{for } n \geq 5 \end{cases} \tag{6.11}$$

Figure 6.8: *The result of ten POCS iterations using the SVD as the band limitation. The PSNR with respect to figure 6.5 is 34.8 dB.*

As expected, the image in figure 6.10 is somewhat more blurred than the one in figure 6.9, because $\sigma(n)$ in (6.11) is larger than $\sigma(n)$ in (6.10). On the other hand, figure 6.9 has a "pearly" appearance, with white dots appearing along the edges of the hat among other places. We ascribe this to the fact that as $\sigma$ decreases, the effect of the filter also decreases, and so the space-domain projection dominates the algorithm. As the space-domain projection is not restricted to make smooth modifications to the iterates, the local white dots arise. The pearliness is absent in figure 6.10 due to the smaller bandwidth of the filter employed. In choosing the filter bandwidth, there is thus a trade-off between sharpness and pearliness. Both images are less sharp than figure 6.8. However, figures 6.9 and 6.10 do not exhibit the graininess of figure 6.8 in large magnification.

Figure 6.9: *The result of ten POCS iterations using linear filtering as the band limitation. The PSNR with respect to figure 6.5 is 32.1 dB.*

It is interesting to compare figures 6.9 and 6.10 to figure 6.7, even though the bandwidths of the Gaussian filters used are not the same. We see that both figures 6.9 and 6.10 are sharper than figure 6.7 and do not suffer from the graininess of figure 6.7. Using the POCS algorithm rather than linear filtering thus improves the reconstruction quality on both these counts. In addition, the filter used in the POCS iteration can be chosen to avoid pearly artifacts as discussed above.

Finally, figure 6.11 shows the result of applying the POCS algorithm with a mixed band limitation as the frequency-domain projection. Specifically, the band limitation is chosen to consist of the linear filter characterized by (6.10), followed by the SVD. The purpose of this is to combine the best of the SVD and linear filtering for band limitation, namely the convergence and sharpness

| Figure | Sharpness | Artifacts | Convergence |
|---|---|---|---|
| 6.7 | low | graininess | N/A |
| 6.8 | high | ringing | yes |
| 6.9 | high | pearliness | no |
| 6.10 | medium | low | no |
| 6.11 | high | low | stagnating |

Table 6.1: *Summary of image reconstruction quality.*

properties of the SVD and the smoothing effect of linear filtering. We find the image in figure 6.11 to be subjectively pleasing.[6] The main artifacts of the image are in the white dots on the chin and the dark shade to the right of the nose. In addition, there is some graininess in the facial skin. On the other hand, figure 6.11 is sharper than figure 6.10. We find that figure 6.11 is the best reconstruction presented in this section, and that it is superior to figure 6.7.

To give some quantitative measure of the convergence rates involved in the above reconstructions, figure 6.12 shows the energy modification due to the frequency-domain projection as a function of iteration number for four variations on the POCS algorithms. The four variations in the frequency-domain projection are the pure SVD, linear filtering based on (6.10) and on (6.11), and the mixed band limitation. As expected, the pure SVD has the best convergence curve. The linear filtering curve corresponding to (6.11), that is, the smaller bandwidth, has the worst convergence. Neither of the two linear filtering curves correspond to converging algorithms, as linear filtering is not a projection. Finally, the mixed band limitation approaches the convergence performance of the pure SVD as the bandwidth of the linear filter increases. We emphasize that although it is mathematically desirable for an reconstruction algorithm to converge, an image processing algorithm should be judged primarily on its visual qualities and secondarily on strict convergence properties.

The subjective reconstruction quality of the presented images is tabulated in table 6.1. In summary, low-pass filtering can be used on its own as in figure 6.7, but this results in either graininess or blurriness depending on the filter bandwidth. In contrast, figures 6.9 and 6.10 show the effect of linear filtering in a POCS iteration; the blurriness is clearly reduced, but pearliness results if the filter bandwidth is too large. Using SVD in a POCS iteration also reduces blurriness and results in fast convergence, but introduces some ringing artifacts. Finally, in our opinion, a combination of SVD and linear filtering in a POCS iteration results in a reasonable compromise between sharpness and artifact-free

[6] The reconstruction quality is slightly decreased if the original image is not SVD-limited before halftoning.

images, as shown in figure 6.11.

## 6.5 Summary

In this chapter we described a POCS-based iterative algorithm for reconstructing a contone image from the halftone image obtained by error diffusion coding. We showed simulation results which demonstrate that the proposed algorithm provides subjectively superior reconstruction to conventional linear filtering. The algorithm is a direct generalization of the nonlinear decoding algorithm presented in chapter 5 in the context of $\Sigma\Delta$ encoders.

While this chapter presents the mathematical framework for our algorithm and demonstrates its feasibility, the chapter does not provide definitive answers to the contone reconstruction problem. Our conclusions are thus tentative, and require psychovisual testing for verification. In addition, the behavior of the algorithm on a number of different images, or with a number of different error diffusion filters, could fruitfully be examined. Other questions include the optimal design of error diffusion filters in the context of the presented algorithm, and the problem of reconstruction when the error diffusion filter is unknown and must be estimated jointly with the contone image. Reducing the computational complexity of the algorithm, particularly the space-domain projection, is also an interesting problem. Finally, the absence of a completely satisfactory frequency-domain projection is a problem whose solution would have implications not only for our algorithm, but also for other applications.

Figure 6.10: *The result of ten POCS iterations using a more narrow-band form of linear filtering as the band limitation. The PSNR with respect to figure 6.5 is 31.7 dB.*

Figure 6.11: *The result of ten POCS iterations using both linear filtering and the SVD as the band limitation. The PSNR with respect to figure 6.5 is 34.0 dB.*

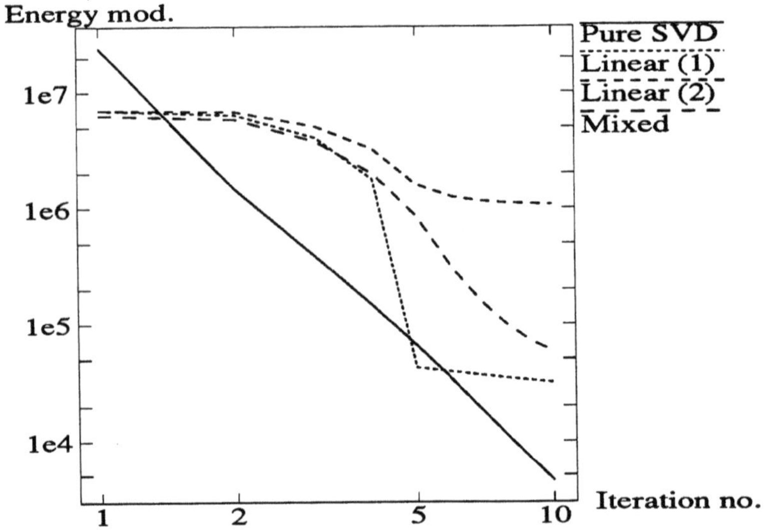


Figure 6.12: *Convergence curves for four band limitations as a function of POCS iteration number. Linear (1) and linear (2) curves refer to equations (6.10) and (6.11), respectively.*

# Chapter 7

# Stability of ONS Encoders

In this chapter[1] we propose a framework for stability analysis of $\Sigma\Delta$ encoders, and argue that limit cycles for constant inputs are natural objects to investigate in this context. We present a number of analytical and approximate techniques to aid the stability analysis of the double-loop and interpolative encoders, and use these techniques to propose ways of improved design that explicitly take stability into account.

## 7.1 Introduction

The purpose of this chapter is to discuss stability as an integral part of analysis and design of $\Sigma\Delta$ modulators. It may be argued that the design effort should first concentrate on SNR performance, and that stability problems can be solved by subsequently scaling circuit coefficients. We argue instead that scaling by itself does not entirely solve the instability problem, and that some sacrifice of SNR may be necessary in addition to scaling to stabilize higher order encoders. Specifically, we demonstrate a trade-off between SNR and stability for interpolative modulators, and show that stability concepts can be applied even to the double-loop encoder which is usually labeled as stable. We do not claim to provide definitive answers to all questions of optimal design, but intend to give a frame of reference for stability considerations, as well as to present a number of analytical techniques that aid the design process.

Our approach to the stability problem throughout the chapter is to examine the large-amplitude limit cycle behavior of the double-loop and interpolative encoders for constant inputs. This approach is justified in more detail in section 7.2, but the main motivation is that constant inputs are a special case of more general inputs, and that limit cycles characterize the long-term behavior of the encoders under constant inputs. Therefore stability under constant inputs is a

[1] Based on [140], ©1991 IEEE, [141], ©1992 IEEE, and [142], ©IEEE.

necessary condition for stability under more general inputs. Furthermore, we show that results from limit cycle analysis can be used in the design process.

The chapter is organized as follows. Section 7.2 contains a discussion of stability issues, and proposes an operational definition of stability. Section 7.3 addresses stability issues for the double-loop encoder; this encoder is in itself of interest, and is also important as a building block in cascaded encoders. Section 7.4 considers the class of interpolative encoders, and section 7.5 contains a summary. Some variations of architectures proposed in the literature [65], [143] may be accommodated by correspondingly minor modifications in the analysis, while others may require more substantive changes.

## 7.2 Framework for Stability Analysis

In this section we first briefly consider the effect of integrator clipping. In section 7.2.1 we suggest a definition of stability which appears to be more operational and better suited to $\Sigma\Delta$ encoders than the definitions of traditional nonlinear systems theory. In section 7.2.2 we argue that the study of limit cycles for constant inputs provides insight into stability issues.

$\Sigma\Delta$ encoders are generally built around a number of integrators; an example is the single-loop encoder shown in figure 7.2 which contains one integrator. Integrator limiting or clipping is an important practical effect that occurs because of voltage saturation in the operational amplifiers of internal integrators. A simple clipping model employs a saturation characteristic of the form

$$\mathrm{sat}(x) = \begin{cases} x & \text{for } |x| \leq L \\ L\ \mathrm{sign}(x) & \text{otherwise} \end{cases} \tag{7.1}$$

where $L$ is the clipping level, and $\mathrm{sign}(\cdot)$ is the signum function. Encoders are typically designed so that the clipping level is not much larger than the feedback voltage; for instance, the single-loop encoder in figure 7.2 would have $L$ close to $b$. Clipping entails loss of state variable information, and hence performance degradation: All other things being equal, it should be avoided. We will refer to practical encoders suffering from clipping as clipped encoders, and encoders with ideal integrators, that is, encoders with no clipping, as unclipped encoders. Comparing the clipped and unclipped encoders, it is clear that the added nonlinearity shown in (7.1) in each integrator further complicates analytical attacks on the already nonlinear system. Such problems necessitate a high degree of reliance on computer simulations to assess performance, and emphasize the need for accurate behavioral models of circuit non-idealities [144]. To circumvent these complications, we suggest a way in section 7.2.1 to address the stability problem analytically which avoids introducing the nonlinearity (7.1).

### 7.2.1 Scaling and Stability

For a given $\Sigma\Delta$ encoder, the only way to avoid clipping is to scale integrator gains, feedback coefficients and other circuit parameters to keep the signal levels throughout the encoder below saturation most of the time. This description of scaling, however, is vague in two respects:

1. The maximum values of signal levels depend not only upon the encoder, but also upon the class of its input signals.

2. There is an important distinction between scaling that preserves the functionality of the encoder, and one that modifies it. More explicitly, by the functionality of an encoder we mean the transformation it applies to its input to produce the output bit stream. For instance, the functionality of the double-loop encoder in figure 7.3 is not affected if we multiply $G$ and $b$ by the same number; we refer to this case as equivalent scaling . On the other hand, changing only $G$ or $b$ affects the functionality; we refer to this case as functional scaling. In section 7.3 we discuss the difference in more detail.

These two points are treated separately in this and the following section. Equivalent scaling is straightforward and may sufficiently reduce signal levels that clipping occurs infrequently under normal operation. There is a simple connection between signal levels at the nodes of the unscaled and scaled, unclipped encoders, since care is taken that the performed scaling at each node only affects the signal level at that node. The main problem with the limited approach of equivalent scaling is that the required scale factors may be excessive, and in practice, very large or small filter coefficients are not easily implementable as capacitor ratios. A more attractive option is to first use functional scaling to improve stability, typically at the cost of some SNR performance, and subsequently use equivalent scaling to further reduce the signal levels so that clipping rarely occurs. We take the latter approach in this chapter. However, functional scaling also has its problems. First, the effects of functional scaling on signal levels are more unpredictable than those of equivalent scaling, since functional scaling at a node is not restricted to only have local consequences for that node. Second, the effects of functional scaling on such performance parameters as SNR are not easily predictable, because linearized analyses may be misleading. To use functional scaling appropriately, it is necessary to examine the trade-off involved between SNR performance and stability. This trade-off is a central theme of sections 7.3 and 7.4.

In order to better address the problems of functional scaling, we propose to define the stability of $\Sigma\Delta$ encoders in terms of the maximum signal levels occurring throughout the unclipped system, that is, we consider stability to be a matter of degrees. We will call a system $K$-stable if the signal levels are bounded in absolute value by $K$ for a given class of input signals. We will also call a system very stable or very unstable according to whether maximum signal levels in the corresponding unclipped system are very much smaller or larger than the clipping level, for a given class of inputs. This definition is reasonable because the signal levels dictate whether clipping occurs. The definition is in contrast to traditional stability definitions such as Bounded Input Bounded

Output (BIBO) and Bounded Input Bounded State (BIBS) which are only concerned with categorizing systems as either stable or unstable [84]. Although the input dependence is present in traditional stability theorems [84], these typically assume bounded, square integrable or summable inputs, and allow for no way of distinguishing between, say, two different constant inputs. In contrast we consider the dependence of stability upon DC level or sinusoidal amplitude to be extremely important.

The described viewpoint on stability avoids dealing with the nonlinear clipping operation (7.1) which would require a more complicated nonlinear analysis, and suggests that stability can be assessed by judging the ratio between the maximum signal levels in unclipped systems and the clipping level $L$. This viewpoint appears to be more useful in the present context than traditional definitions of stability; for instance, real $\Sigma\Delta$ encoders are always stable in the BIBO and BIBS sense, in that clipping keeps all voltages bounded. The viewpoint means that stability can be seen as an integral part of the design process for any $\Sigma\Delta$ encoder, even a second-order one. We thus argue that clipping should be precluded by design, and our goal is to guide designs with that objective. This approach is in contrast with the existing one [55] where specialized circuitry is used to detect the saturation of integrators and reset these.

In passing we mention that in addition to the stability concept, the concepts of complete observability and controllability take interesting turns in the context of $\Sigma\Delta$ encoders. Specifically, no encoder is completely observable, or it would not be a quantizer. In addition, it is undesirable that an encoder be completely controllable with inputs in a given class, since it would imply that arbitrarily large integrator outputs were achievable with inputs in that class. The objective of $\Sigma\Delta$ encoder design may therefore be phrased as the problem of making the system as observable and uncontrollable as possible.

### 7.2.2 Limit Cycles

This section discusses the applicability of limit cycle analysis to stability investigations. As described above, the purpose of scaling is to keep the maximum values of integrator outputs below the clipping level. Of course, too much downscaling will decrease the ratio between signal and circuit noise levels excessively, and will thus adversely affect performance. In practice a trade-off must therefore be found between the frequency and effect of clipping on one side, and noise sensitivity on the other. This trade-off may be viewed as designing for the right amount of stability.

For a given encoder, the signal levels depend on the input signal class. To eliminate transient phenomena and to focus on long-term behavior, we will consider limit cycles or periodicities for constant inputs. The following arguments justify this:

1. Limit cycles are essential to the operation of $\Sigma\Delta$ encoders, as evidenced by their prominent position in several papers including [31], [66] and [145]. In [87], chapter 2 and other places is shown that for the single-loop and

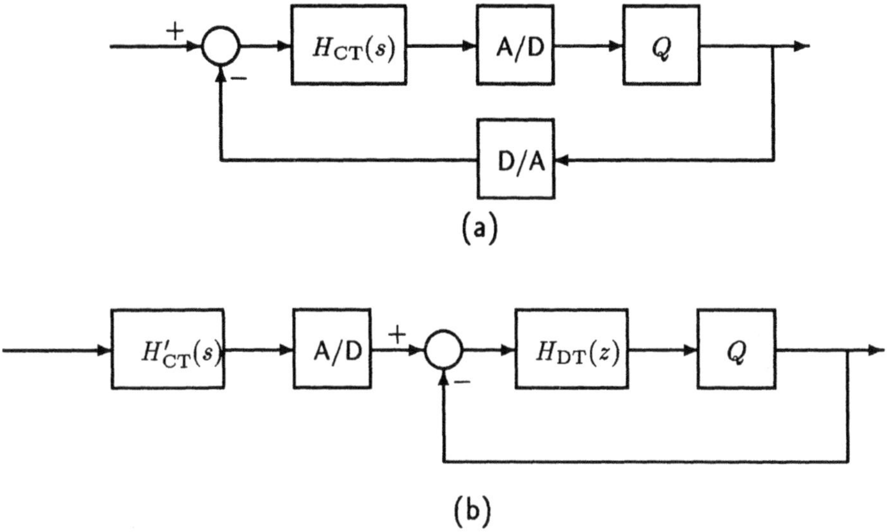


Figure 7.1: *(a) Continuous-time interpolative encoder with CT filter $H_{CT}(s)$, sample-and-hold A/D converter, 1-bit digital quantizer and ideal D/A converter. (b) Equivalent discrete-time encoder with DT filter $H_{DT}(z)$ and CT pre-filter $H'_{CT}(s)$, both depending only upon $H_{CT}(s)$.*

double-loop encoders, limit cycles occur only when the constant input is a rational fraction of the quantizer step size. Limit cycles can thus be seen as a natural result of approximating constant inputs using $\Sigma\Delta$ encoders. Furthermore, a recent paper [48] shows that for a single-loop encoder whose integrator has its pole inside the unit circle, almost all constant inputs generate limit cycles.

2. The oversampling of the input in practical situations implies that it appears approximately constant to the encoder.

3. Any encoder designed for dynamic low-pass inputs must be able to handle constant inputs as a special case; therefore stability under constant inputs is a necessary condition for stability under more general conditions.

4. The assumption of constant input allows us to make statements which hold for both Continuous-Time (CT) and Discrete-Time (DT) $\Sigma\Delta$ encoders. This is because it can be shown that any CT encoder can be converted into a DT encoder [67]; to make the two encoders equivalent, the CT input must be converted into a DT sequence by prefiltering it with a CT filter and then sampling it, as shown in figure 7.1. The required CT prefilter depends on the specific CT encoder. However, the prefilter can simply be represented by a constant gain for the analysis of DC inputs.

5. Constant inputs simplify analytical attacks; indeed, a number of illuminative results have been based on this simplifying assumption [41], [87].

Based on the above discussion, our approach in the following sections is to use limit cycle analysis for constant inputs to find maximum signal levels in $\Sigma\Delta$ encoders. We also examine functional scaling as a way of trading off SNR performance for stability.

### 7.2.3 Comparison with Existing Work

In this section we compare our stability approach to previous results on Delta modulation.

Gersho [146] considers single-integration Delta modulation with stochastic stationary input processes, and either perfect or leaky integration. The stability concept underlying his approach is consistent with the approach of this paper. For the modulator, he derives upper bounds on the error signal. The corresponding stability result for single-loop $\Sigma\Delta$ modulators with ideal integration is well known, and is stated in section 7.3 below. Gersho's method does not appear to generalize to higher-order modulators.

Nielsen [147] considers a special form of double-integration Delta modulation with zero input, and examines the specific limit cycle type consisting of a number of positive output bits followed by the same number of negative bits. He states that the limit cycle length is a measure of the stability, and he numerically optimizes a particular modulator parameter for stability. The approach is less general than the one presented in this paper, because we consider general constant inputs and arbitrary limit cycles, as well as more general modulators.

Finally, Steele [148] considers double-integration Delta modulation with ideal integration. For the case of zero input, he derives the peak-to-peak value of the feedback signal. He then introduces prediction around the second integrator to reduce feedback oscillations. Due to the special form of the double-loop $\Sigma\Delta$ modulator with two feedback paths for the quantizer output, Steele's results are not applicable to our problem.

## 7.3 Double-Loop Encoder

In this section we discuss stability issues for the double-loop $\Sigma\Delta$ encoder within the framework of section 7.2. In section 7.3.1 we present a number of exact analytical methods for investigating the limit cycle behavior of the encoder, and in section 7.3.2 we derive exact upper bounds on the largest integrator outputs occurring on limit cycles. We use these results to arrive at a design and scaling technique for double-loop encoders that provides a more favorable SNR performance than the standard one.

Some of the results of this section may be viewed as extensions of results for the single-loop encoder given in [41] and other papers. For completeness we provide a concise overview of these results in the language of the present

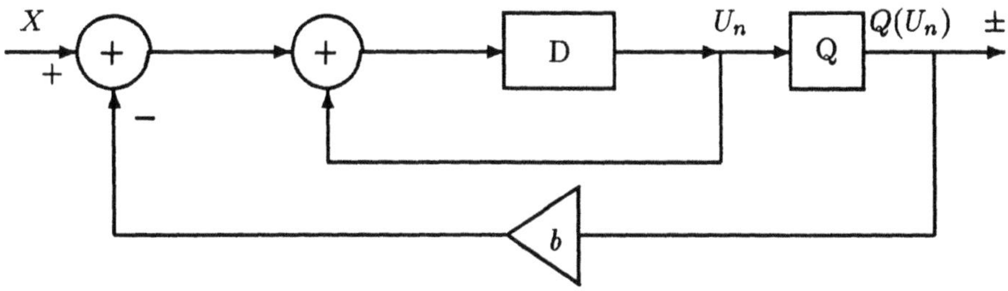


Figure 7.2: *Discrete-time model of the single-loop ΣΔ encoder.*

chapter, drawing also on the results in [87]. Figure 7.2 shows the block diagram of the single-loop encoder with constant discrete-time input $X$. Note that the quantizer step size has been factored out as a variable scaling factor $b$, so that the 1-bit quantizer is given by

$$Q(U) = \begin{cases} +1 & \text{for } U > 0 \\ -1 & \text{for } U \leq 0 \end{cases}$$

It is shown in [41] that for any constant input $X \in (-b, +b)$, the state variable $U_n$ at any time $n > 0$ lies in the interval $(X - b, X + b)$ provided only that the initial state at time 0 lies in that interval. This result essentially resolves the stability issue for the single-loop encoder: $b$ must be chosen such that $2b$ is less than the clipping level of the integrator. If the constant input $X$ equals some irreducible integer fraction $p/q$ of the DAC feedback voltage $b$, there exists exactly one limit cycle; it has period $2q$ if either $p$ or $q$ is even, and period $q$ otherwise; the average of the quantizer outputs $Q(U_n)$ over one period equals the normalized input $X/b$ [87]. The limit cycles or periodicities show up as spikes in the spectrum of the quantization noise sequence [42].

Figure 7.3 shows a block diagram of the general double-loop encoder. The encoder contains four scaling factors, namely two for the integrators and two for the feedback paths. The scaling factors corresponding to the outer integrator are denoted by the upper-case letters $B$ and $G$, and the scaling factors corresponding to the inner integrator are denoted by the lower-case letters $b$ and $g$. The double-loop encoder is of interest in itself, and its analysis and implementation have been described in a number of papers including [122] and [149]. However, the encoder is also important as a building block in higher-order cascaded encoders, as evidenced in [143], [150] and other papers. Using linearization, the encoder can be viewed as a two-pole digital filter in a feedback loop, and considering measures such as phase margin, the encoder may loosely be characterized as barely stable.

### 7.3.1 Detection of Specific Limit Cycles

In this section we consider in detail the limit cycles of the canonical unclipped double-loop encoder, that is, the encoder shown in figure 7.3 with $b = B = g = G = 1$ and ideal integrators. The methods readily translate to the more general structure. In sections 7.3.1.1 and 7.3.1.2 we address the following problem: Given a $P$-bit sequence $\mathbf{Y} = \{Y_0, \cdots, Y_{P-1}\}$, does there exist a constant input $X$ and a limit cycle with period $P$ such that $\mathbf{Y}$ corresponds to the encoder output sequence $\{Q(U_0), \cdots, Q(U_{P-1})\}$? If so, what are the largest values of the state variables occurring as the encoder goes through one period of the limit cycle? Our technique for solving these problems makes use of the standard Tsypkin method of opening the loop and thus circumventing the nonlinearity [151]. A similar approach was used in [152] for the specific case of zero-input symmetric limit cycles with a number of positive bits followed by the same number of negative bits. In section 7.3.1.3 we present a numerical technique to assess the regions in state space that are parts of limit cycles.

#### 7.3.1.1 Existence of Specific Limit Cycles I

From figure 7.3 with $b = B = g = G = 1$, we obtain the following state equations for the variables $U_n$ and $V_n$:

$$\begin{aligned} U_n &= U_{n-1} + V_{n-1} - Q(U_{n-1}) \\ V_n &= V_{n-1} + X - Q(U_n) \end{aligned} \tag{7.2}$$

Inserting $Y_n = Q(U_n)$ and $Y_{n-1} = Q(U_{n-1})$ in (7.2), we arrive at a closed-form solution derived in chapter 2:

$$\begin{aligned} U_n &= U_0 + nV_0 + \frac{1}{2}n(n-1)X - Y_0 - \sum_{i=1}^{n-1}(n+1-i)Y_i \\ V_n &= V_0 + nX - \sum_{i=1}^{n} Y_i \end{aligned} \tag{7.3}$$

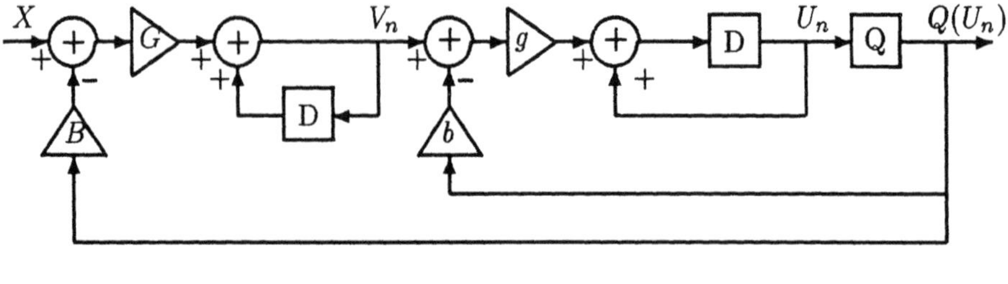


Figure 7.3: *Discrete-time model of the double-loop ΣΔ encoder.*

The double-loop encoder is defined to have $\mathbf{Y}$ as the output sequence on a limit cycle if and only if two conditions are satisfied [151]:

**C1** The state variable sequences $\{U_n\}$, $\{V_n\}$ obtained by using $\mathbf{Y} = \{Y_0, \cdots, Y_{P-1}\}$ in (7.3) are periodic with period $P$, that is, $U_{P+n} = U_n$, $V_{P+n} = V_n$ for all $n$. It follows from (7.2) that this condition holds if and only if $U_P = U_0$ and $V_P = V_0$.

**C2** For each $0 \leq n \leq P-1$, the sign of the quantizer input matches the corresponding bit of the sequence $\mathbf{Y}$, that is, $Y_n = Q(U_n)$. This condition corresponds to closing the loop and checking the consistency of the resulting system.

To satisfy $V_P = V_0$ in condition C1, (7.3) implies

$$\frac{1}{P}\sum_{n=0}^{P-1} Y_n = X \tag{7.4}$$

There is thus only one constant input that might give rise to the limit cycle under investigation, and this constant input is rational and equals the average output. To satisfy $U_P = U_0$ in condition C1, we use the first equation in (7.3) to obtain

$$V_0 = -\frac{1}{2}(P-1)X + \frac{1}{P}\left(Y_0 + \sum_{n=1}^{P-1}(P+1-n)Y_n\right) \tag{7.5}$$

To check condition C2 we may proceed in the following way: Let $R_0$ be the set of all values of $U_0$ such that the consistency relation at time $n = 0$ is satisfied. For $n \geq 1$, use equation (7.3) to recursively compute $U_n$ in terms of $U_{n-1}$. Determine the set $R_n$ of all values of $U_0$ such that the consistency relation $Q(U_n) = Y_n$ is satisfied; this amounts to a linear inequality in $U_0$.[2] At time $P$, determine the intersection $R$ of all the sets $R_0, \cdots, R_{P-1}$. If $R$ is empty, the combined constraints on $U_0$ are impossible to satisfy simultaneously, and the sequence $\mathbf{Y}$ is not a limit cycle for the double-loop encoder. If $R$ is non-empty, the sequence $\mathbf{Y}$ does exist as a limit cycle for any initial state in $R$. In this case we can step through (7.3) to determine the largest values of the state variables over the limit cycle $\mathbf{Y}$. For brevity, we shall refer to these maxima as the amplitudes of the limit cycles in the state variables $U_n$ and $V_n$. Without loss of generality we only consider positive constant inputs, and search for maximum absolute values of $U_n$ and $V_n$.

#### 7.3.1.2 Existence of Specific Limit Cycles II

The method presented in section 7.3.1.1 makes use of the fact that a solution (7.3) to the difference equation (7.2) is available. A more general method which

[2] In the more general case of a multi-bit quantizer $Q$, the consistency relation $Q(U_n) = Y_n$ is unchanged, and also gives rise to linear inequalities in $U_0$.

does not make use of (7.3) and which is easily generalized to higher-order systems can however also be devised. In this section we show the application of this technique to the double-loop encoder, and in section 7.4.1.1 we will use it on the general interpolative encoder.

We can rewrite the two first-order state equations shown in (7.2) as a single second-order state equation,

$$U_{n+2} - 2U_{n+1} + U_n = -2Y_{n-1} + Y_n + X \tag{7.6}$$

We will assume that the state variable sequence $\{U_n\}$ is periodic with period $P$ and enforce the consistency requirement C2. Assuming the periodicity condition C1 holds, equation (7.6) can be written as a linear vector equation

$$\begin{bmatrix} -2 & 1 & 0 & \cdots & 0 & 0 & 1 \\ 1 & 0 & 0 & \cdots & 0 & 1 & -2 \\ \vdots & \vdots & & & \vdots & \vdots & \vdots \\ 0 & 1 & -2 & \cdots & 0 & 0 & 0 \\ 1 & -2 & 1 & \cdots & 0 & 0 & 0 \end{bmatrix} \begin{bmatrix} U_0 \\ U_1 \\ \vdots \\ \vdots \\ U_{P-1} \end{bmatrix} =$$

$$\begin{bmatrix} -2 & 0 & \cdots & 0 & 1 \\ 0 & 0 & \cdots & 1 & -2 \\ \vdots & \vdots & & \vdots & \vdots \\ 0 & 1 & \cdots & 0 & 0 \\ 1 & -2 & \cdots & 0 & 0 \end{bmatrix} \begin{bmatrix} Y_0 \\ Y_1 \\ \vdots \\ \vdots \\ Y_{P-1} \end{bmatrix} + X \begin{bmatrix} 1 \\ 1 \\ \vdots \\ \vdots \\ 1 \end{bmatrix} \tag{7.7}$$

where the two $P \times P$ matrices are circulant, and the equations for $n = 0$ and $n = P - 1$ are at the bottom and top, respectively. The $P \times P$ matrix on the left-hand side is singular and has rank $P - 1$. By adding together the $P$ scalar equations, we again arrive at the requirement (7.4) for the constant input $X$. The right-hand side of (7.7) is thus known for a given $\mathbf{Y}$. Equation (7.7) can be solved by simple forward substitution with, say, $U_0$ and $U_1$ as independent variables. As this procedure reaches $n = P - 1$, it produces a linear constraint involving only $U_0$ and $U_1$, which implies that one of $U_0$ and $U_1$ can be used as the sole independent variable, and so the entire sequence $\{U_n\}$ can be specified in terms of only $U_0$. Enforcing the consistency requirement C2 proceeds in the same manner as before: $U_0$ is chosen, if possible, such that for $0 \leq n \leq P - 1$, $Q(U_n) = Y_n$.

The described method can be interpreted geometrically in the following way: Equation (7.7) specifies a line in the $P$-dimensional space $\mathcal{R}^P$ of $P$-element real sequences to which $\mathbf{U} = \{U_0, \cdots, U_{P-1}\}$ belongs. On the other hand, the consistency requirement C2 limits the allowable region of the space $\mathcal{R}^P$ to a single octant, because all elements $U_n$ of $\mathbf{U}$ must have signs specified by the corresponding element $Y_n$ of the sequence $\mathbf{Y}$. Combining the two constraints, the limit cycle $\mathbf{Y}$ exists if and only if the intersection between the line and the octant is non-empty. The points in the one-dimensional intersection each represent a possible limit cycle in the quantizer input $\{U_n\}$.

We can also explain directly why there is some latitude in choosing $U_0$: The left-hand side matrix of (7.7) has rank $P-1$, and the eigenvectors corresponding to the eigenvalue 0 are of the form $\mathbf{K} = \{k, k, \cdots, k\}$. Therefore, if $\mathbf{U} = \{U_0, \cdots, U_{P-1}\}$ is a solution to (7.7), $\mathbf{U} - \mathbf{K} = \{U_0 - k, \cdots, U_{P-1} - k\}$ is another solution for any $k$. As long as the consistency requirement C2 is not violated, both $\mathbf{U}$ and $\mathbf{U} - \mathbf{K}$ are limit cycles in the state variable $U_n$. More precisely, if $\mathbf{U}$ is a limit cycle, then for all $k$ in the following range of width denoted by $\Delta U$, the shifted sequence $\mathbf{U} - \mathbf{K}$ is another limit cycle:

$$\max_{U_n \le 0} U_n \le k < \min_{U_n > 0} U_n \tag{7.8}$$

Using the described technique or that of section 7.3.1.1, we can obtain information about limit cycles with small periods $P$ by exhaustively searching through binary sequences with average value $X$. Table 7.1 summarizes such information for periods up to 24 for the cases $X = 0$ and $X = 0.5$. For each period $P$ we list the pair $(U_n, V_n)$ that achieves the largest quantizer input $U_n$ while belonging to a limit cycle with period $P$. We also list the quantity $\Delta U$ corresponding to the limit cycle that maximizes $U_n$. We observe that as $X$ increases, the maximum value of $U_n$ increases comparatively more than that of $V_n$, so in a practical clipped encoder, $U_n$ will clip before $V_n$. At both inputs 0 and 0.5, several limit cycles exist at each period for all but the smallest periods.

Table 7.1 suggests that as the period $P$ increases, the maximum integrator outputs exhibit an increasing trend and approach limits.[3]. In fact, we conjecture that for $X = 0$, the limits equal $\frac{8}{3}$ for $U_n$ and $\frac{5}{3}$ for $V_n$; for $X = 0.5$, the limit is about 4.16 for $U_n$. If our observations are valid for all constant inputs, the results suggest that relatively short limit cycles are good indicators of the maximum signal levels encountered also on longer limit cycles.

#### 7.3.1.3 Graphical State Space Method

The main drawback of the methods presented in sections 7.3.1.1 and 7.3.1.2 is the requirement to examine all binary sequences with average value $X$ in order to detect limit cycles for the constant input $X$. We now present a more graphical approach to obtain an overview of the limit cycle behavior. The approach is based on a state space representation of the double-loop encoder where pairs of state variable values $(U_n, V_n)$ are points in a plane with $U_n$ and $V_n$ along the horizontal and vertical axes, respectively. For a given constant input $X$, each point in state space completely specifies a trajectory that can be found by stepping through the difference equations (7.2). It is possible in principle to determine for each point whether or not the corresponding trajectory is a

[3]This observation is substantiated by considering limit cycles with periods up to several hundred and limit cycles for other constant inputs. These limit cycles were not generated exhaustively, but found with the method described in section 7.3.1.3.

| $X$ | Period | # Limit cycles | $(U_n, V_n)$ w/max. $U_n$ | $\Delta U$ |
|---|---|---|---|---|
| 0 | 2 | 1 | (1.500, 0.500) | 1.500 |
| | 4 | 1 | (2.000, 1.000) | 1.000 |
| | 6 | 3 | (2.500, 1.500) | 0.500 |
| | 8 | 2 | (2.250, 1.250) | 0.500 |
| | 10 | 4 | (2.300, 1.300) | 0.400 |
| | 12 | 2 | (2.333, 1.333) | 0.333 |
| | 14 | 8 | (2.643, 1.643) | 0.071 |
| | 16 | 6 | (2.625, 1.625) | 0.125 |
| | 18 | 8 | (2.611, 1.611) | 0.167 |
| | 20 | 4 | (2.600, 1.600) | 0.200 |
| | 22 | 10 | (2.545, 1.591) | 0.182 |
| | 24 | 4 | (2.500, 1.583) | 0.167 |
| 0.5 | 4 | 1 | (2.250, 1.250) | 1.250 |
| | 8 | 3 | (3.500, 2.000) | 0.500 |
| | 12 | 3 | (3.333, 1.917) | 0.250 |
| | 16 | 4 | (3.750, 2.125) | 0.125 |
| | 20 | 7 | (3.950, 2.150) | 0.200 |
| | 24 | 3 | (4.000, 2.167) | 0.167 |

Table 7.1: *Number of limit cycles in the double-loop encoder with constant inputs $X = 0$ and $X = 0.5$. Also shown are the pairs $(U_n, V_n)$ maximizing the quantizer input, and the amount by which the limit cycles in the state plane can be shifted down in the $U_n$ direction.*

limit cycle.[4] The set of states that belong to limit cycles, or equivalently the collection of limit sets [153], can then be shown in a plot such as figure 7.4. In practice the process can be implemented numerically by discretizing state space. The number of grid points per unit, referred to as the grid density $\mu$, should be chosen such that $X$ is a grid point, that is, $\mu X$ is an integer. This is because in (7.5), the quantity $Y_0 + \sum(P+1-n)Y_n$ is an integer, so if $\mu V_0$ is to be an integer, $\mu X$ must in general also be an integer.

Figures 7.4 and 7.5 show state space plots obtained for constant inputs $X = 0$ and $X = 0.5$ with grid densities 80 and 50 respectively. The limit cycles that make up the plots have periods varying from two to many hundred. The maximum state variable values reported for short limit cycles in table 7.1 match the plots well. As the constant input increases, the limit cycles result in larger values of the state variables. It turns out that for a fixed constant input, the state space plots for different, but sufficiently large grid densities look very similar. On the other hand the periods of the limit cycles that make up the plots

[4] This also holds for the more general case of multi-bit quantization.

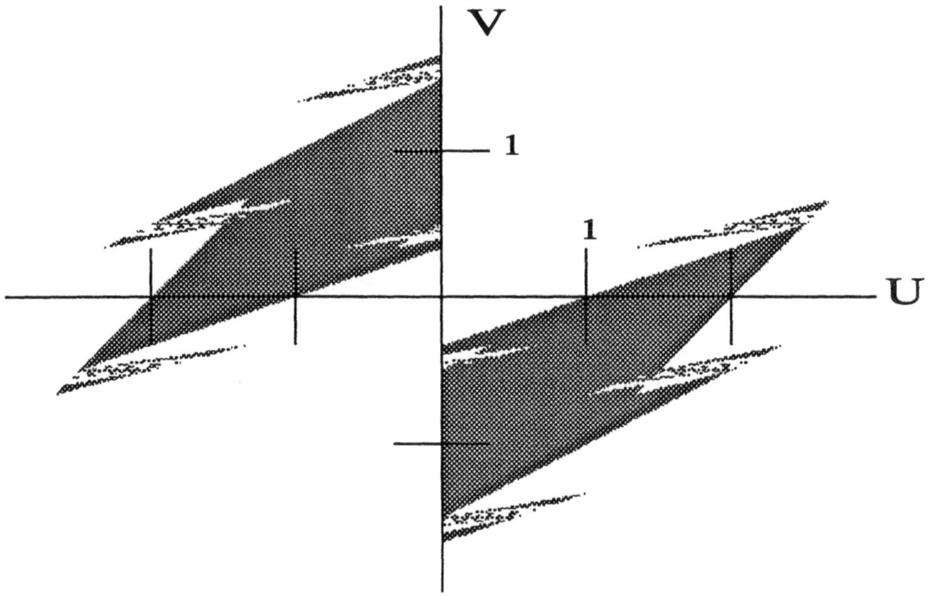


Figure 7.4: *Plot of the points in state space that lie on a limit cycle for a constant input of* $X = 0$. *The grid density is 80, and each axis tick represents one unit.*

can be quite different, and not all limit cycles materialize for an arbitrary grid density. This partly explains that the collection of limit sets appears ragged and irregular. The observed periods of limit cycles tend to share a number of prime factors with the grid density, although we found no general rule.

To verify that the results based on limit cycles are indicative of the behavior of the double-loop encoder, figure 7.6 shows simulation results for the maximum value of the quantizer input $U_n$ as a function of the constant input $X$. The results are obtained by using a large number of random starting points in state space, simulating the encoder for a large number of time steps, and registering the largest quantizer input value whether or not it occurs on a limit cycle. To get over transients, maxima occurring on the first 1000 time steps are discarded; this is not to deny the importance of transients. For inputs $X = 0$ and 0.5 there

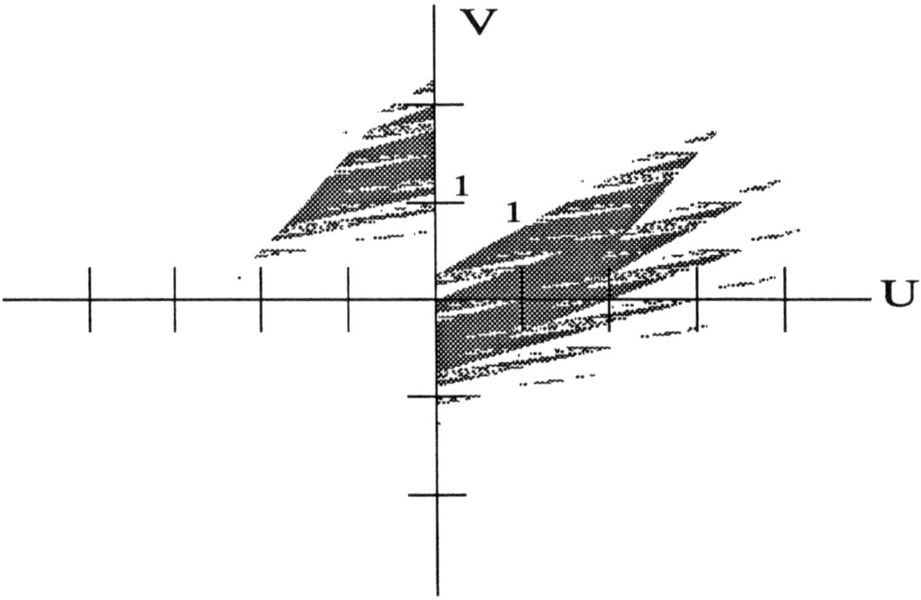


Figure 7.5: *Plot of the points in state space that lie on a limit cycle for a constant input of* $X = 0.5$*. The grid density is 50, and each axis tick represents one unit.*

is good agreement with the limit cycle based results in table 7.1 and figures 7.4 and 7.5, and the agreement is confirmed for other constant inputs. Figure 7.6 shows that as $X$ approaches unity, the maximum value of $U_n$ increases rapidly, that is, the encoder becomes less stable and more prone to clipping.

### 7.3.2 Bounds on Limit Cycles

In section 7.3.2.1 we present analytical upper bounds on limit cycle amplitudes for the general, unclipped double-loop encoder with scaling factors $B, G, b$ and $g$ shown in figure 7.3. In section 7.3.2.2 we use the results to propose an encoder with scaling factors that are optimized with respect to both stability and an approximate measure of SNR, and in section 7.3.2.3 we compare the performance of the clipped encoder to previously suggested scaled encoders.

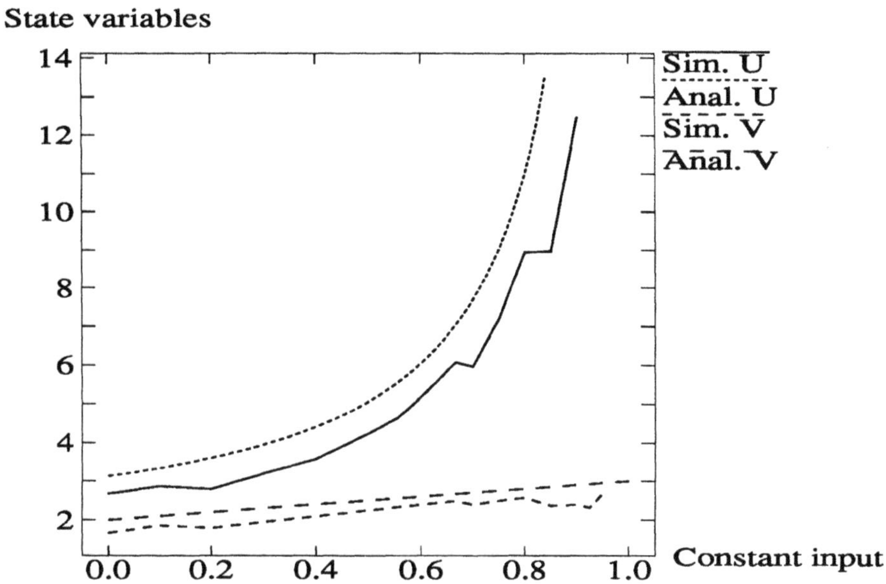


Figure 7.6: *Maximum absolute values of state variables $U_n$ and $V_n$ as a function of the constant input $X$. The analytical curves show the bounds (7.10) and (7.11), while the other curves show simulated maximum absolute values.*

#### 7.3.2.1 Derivation of Bounds

Our upper bounds on limit cycle amplitudes are derived in appendix F.1. To concisely express them, it is convenient to first transform the double-loop encoder in figure 7.3 into an equivalent encoder, shown in figure 7.7 and specified in terms of normalized quantities. The equivalence can be confirmed by a series of block diagram manipulations and the observation that the gain $g$ only has the effect of scaling $U_n$. This is because $Q(U_n)$ only depends on the sign of $U_n$. To summarize the results in appendix F.1, we introduce the normalized

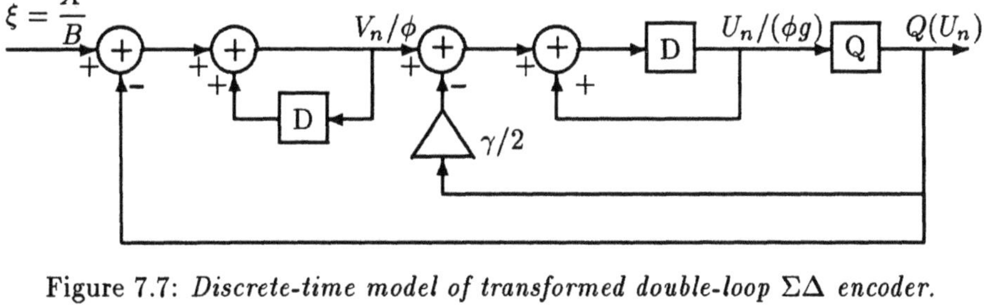


Figure 7.7: *Discrete-time model of transformed double-loop ΣΔ encoder.*

quantities

$$\xi = \frac{X}{B}, \quad \phi = GB, \quad \gamma = \frac{2b}{GB} = \frac{2b}{\phi} \tag{7.9}$$

where $X$ is the constant input, and $b$, $B$ and $G$ are defined in figure 7.3. These three normalized quantities replace $X$, $G$ and $b$ as independent variables, so we change our set of independent variables from $(X; G, B, g, b)$ to $(\xi; \phi, B, g, \gamma)$. Some immediate observations on the effects of changing the values of the original variables can be made based on the new set of variables. Multiplying $b$ and $G$ by the same factor only has the effect of increasing $\phi$, so the signal levels of $U_n$ and $V_n$ are scaled proportionally. This is equivalent scaling. However, changing only $G$ has the effects of scaling $\phi$ proportionally and scaling $\gamma$ in inverse proportion. The change in $\gamma$ affects the signal levels nonlinearly, since the functionality of the encoder is modified. Similarly, changing only $B$ has effects on $\xi$, $\phi$ and $\gamma$, and the combined effect on signal levels is nonlinear. These are examples of functional scaling. We show in appendix F.1 that assuming

$$0 \le \xi < 1 \quad \text{and} \quad \gamma > 1$$

the limit cycle amplitudes of the state variables $U_n$, $V_n$ are upper bounded by

$$\frac{8(1-\xi)|U_n|}{\phi g} \le \begin{cases} \left[(\gamma-1)(1-\xi) + \dfrac{2\gamma}{\gamma-1}\right]^2 & \text{for } 1 < \gamma \le f_1(\xi) \\ 16 + \gamma^2(1-\xi)^2 & \text{for } f_1(\xi) < \gamma \le f_2(\xi) \\ [\gamma(1+\xi)+2]^2 & \text{for } f_2(\xi) < \gamma \end{cases} \tag{7.10}$$

and

$$\frac{2|V_n|}{\phi} \le \begin{cases} \gamma\left(\xi + \dfrac{2}{\gamma-1}\right) & \text{for } 1 < \gamma \le f_1(\xi) \\ (\gamma+1)\xi + 3 & \text{for } f_1(\xi) < \gamma \end{cases} \tag{7.11}$$

where

$$f_1(\xi) = 1 + \frac{2}{1+\xi}, \quad f_2(\xi) = \frac{-1 + \sqrt{1 + 2\xi\left(\frac{4}{1-\xi} - \frac{1}{1+\xi}\right)}}{\xi}$$

We denote the bounds on $|U_n|$ and $|V_n|$ in (7.10) and (7.11) by $U_{\max}$ and $V_{\max}$, respectively. Similar bounds hold for $-1 < \xi \le 0$. For the standard double-loop encoder with $b = B = g = G = 1$ or equivalently $\phi = 1$ and $\gamma = 2$, we find in particular that for $0 \le X < 1$,

$$U_{\max} = \frac{(5-X)^2}{8(1-X)}, \quad V_{\max} = X + 2 \tag{7.12}$$

Figure 7.6 shows the bounds in (7.12) as well the maximum signal levels actually observed in a simulation of the standard double loop encoder. The simulated results are obtained by choosing 500 random pairs of initial states, running the modulator for 1000 samples to get over transients, and observing the largest absolute state variables on the following 1000 samples. The bounds correspond to $\gamma = 2$, $\phi = 1$, and are seen to be relatively tight.

We find in general that for $1 < \gamma < 2$, the derived bounds on state variables are valid, but not extremely tight. For $\gamma > 2$ the bounds are tighter, especially for moderate and large constant inputs $\xi$, and are thus suitable for design. Interestingly, the analytical bounds that are valid for $1 < \gamma \leq f_1(\xi)$ are very good general approximations to simulated maximum state variable values, even when $\gamma > f_1(\xi)$. When $\gamma < 1$, the technique in appendix F.1 does not yield upper bounds on limit cycle amplitudes. This of course does not imply that the encoder is unstable. However, it is interesting to compare with [67] in which it is claimed that under a number of approximations, the double-loop encoder with $b = B = 1$ is stable provided $G < 2$, or equivalently, $\gamma > 1$.

#### 7.3.2.2 Design Implications

In this section we use the results of section 7.3.2.1 to design a double-loop encoder that takes into account both stability and SNR performance. The design is based on the constant input assumption, but we show simulation results for sinusoidal inputs to verify the design. We assume that our circuit technology dictates a given clipping level $L$ defined in (7.1).

The design problem has five degrees of freedom, namely $(X_{\max}; G, B, g, b)$ where $X_{\max}$ is the largest constant input for which the design guarantees absence of clipping. Equivalently, we can use the parameters $(\xi_{\max}; \phi, B, g, \gamma)$ where $\xi_{\max} = X_{\max}/B$. We reduce the number of degrees of freedom to two by the following three equality constraints:

$$U_{\max} = L, \quad V_{\max} = L, \quad B(1 + \xi_{\max}) = L \tag{7.13}$$

The first two constraints are stability constraints to avoid saturation, and the last constraint states that the maximum signal level at the output of the input summer, $B + X_{\max} = B(1 + \xi_{\max})$, should also equal the clipping level. The constraints ensure that we take maximum advantage of the dynamic ranges of the circuit elements while maintaining stability.

We use the remaining two degrees of freedom, $\xi_{\max}$ and $\gamma$, to find a trade-off between two goals, namely maximizing an approximate measure of SNR performance and maximizing the largest constant input $X_{\max}$ which does not saturate the integrators. The SNR measure that we use is the product $gG$ which is shown in appendix F.2.1 to approximately control the baseband noise suppression. We show in appendix F.2.2 that for a given $\xi_{\max} \in (\sqrt{5} - 2 = 0.2361, 1)$, the product $gG$ is maximized by choosing $\gamma = 1 + 2/(1 + \xi_{\max})$. We are thus left with a single degree of freedom, $\xi_{\max}$, on which both the product $gG$ and the largest allowed constant input given by

$$X_{\max} = L \cdot \xi_{\max}/(1 + \xi_{\max})$$

depend. Figure 7.8 shows $gG$ and $X_{\max}/L$ as functions of the normalized maximum input $\xi_{\max}$. We see that the design goals of maximizing both quantities

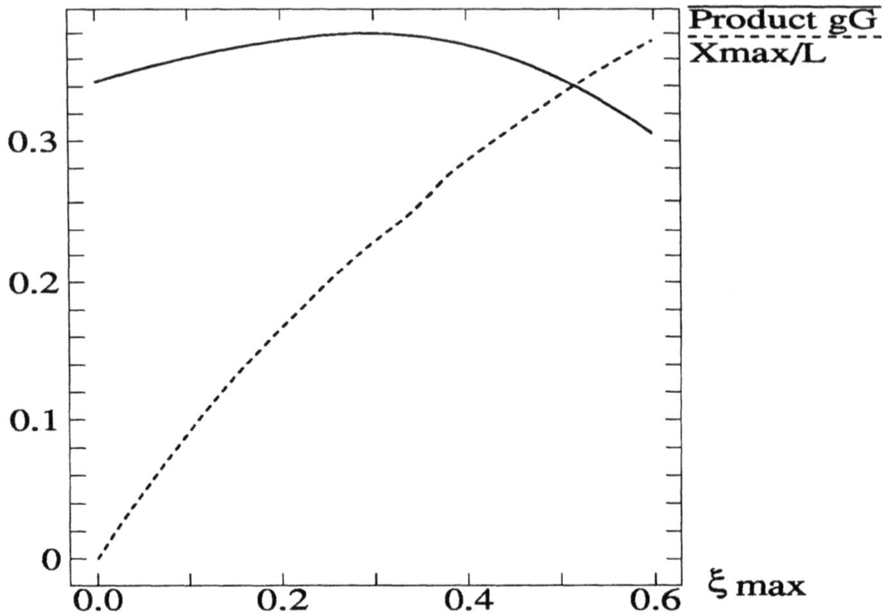


Figure 7.8: *The product gG is an approximate measure of the SNR performance. The dynamic range is the range of constant inputs permissible according to the design. These are shown as a function of the constant normalized input* $\xi_{\max}$.

are conflicting. As a compromise, we choose $\xi_{\max} = 0.5$; this choice is commented on in section 7.3.2.3. Using (7.10), (7.11), and (7.13), we then find $\gamma = 2.33$, $\phi = 0.429L$, and the scaling factors

$$B = \frac{L}{1+\xi_{\max}} = 0.667L, \quad g = \frac{L}{\phi} \cdot \frac{8(1-\xi_{\max})}{16+\gamma^2(1-\xi_{\max})^2} = 0.538,$$
$$G = \frac{\phi}{B} = 0.643, \quad b = \frac{\gamma\phi}{2} = 0.500L \qquad (7.14)$$

which results in a maximum permissible constant input of $X_{\max} = B\xi_{\max} = 0.333L$, and a performance product of $gG = 0.346$. For a given clipping level $L$, equation (7.14) presents the designer with a choice of scaling factors that take into account both stability and SNR performance.

#### 7.3.2.3 Design Comparisons

For comparison we mention the scaling schemes that have appeared in the literature. In [122] and [143] scaling factors of $G = g = \frac{1}{2}$, $B = b = 1$ are found

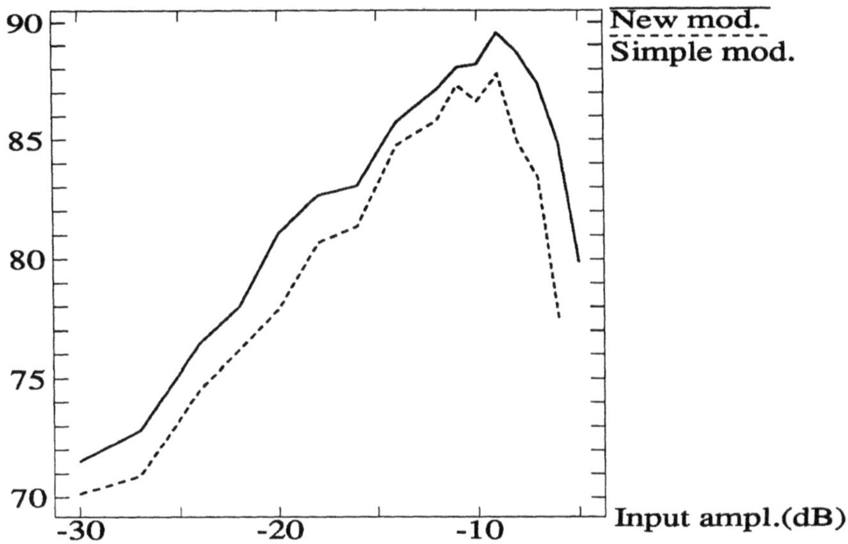


Figure 7.9: *Comparison of two differently scaled double-loop encoders with sinusoidal input. The curve labeled "Simple mod." represents scaling factors* $(G, B, g, b) = (0.5,\ 1,\ 0.5,\ 1)$, *while the other curve represents scaling factors* $(G, B, g, b) = (0.64,\ 1.12,\ 0.54,\ 0.85)$.

to make the double-loop encoder sufficiently stable with a clipping level of $L = 1.7$. In [150] scaling factors of $G = \frac{1}{4}$, $g = 1$, $B = 1$, $b = \frac{1}{2}$ are chosen, although $G$ is split into two gains of $\frac{1}{2}$ each: One before and one after the outer integrator. The clipping level is not reported. These two designs both correspond to $\gamma = 4$ and $gG = 0.25$. In the set-up of the present chapter, the value of $\gamma$ would appear to be somewhat large, and the product $gG$ correspondingly small; however, considerations such as ease and regularity of implementation may also have played a part in the described choices.

Figure 7.9 shows the simulated SNR performance of an encoder designed with our technique, and an encoder with $B = b = 1$, $G = g = 0.5$, for a clipping level of $L = 1.7$ [122], [143]. The vertical axis shows the SNR, while the horizontal axis represents the amplitude of a sinusoidal input with a fixed frequency of 1020 Hz relative to a sampling frequency of 1.024 MHz. The amplitude is measured in dB relative to the level 1. A $sinc^3$ decimation filter is used for both encoders, and the oversampling ratio is 128. The plot shows that the peak SNR for our modulator is 1.5-2 dB above that of the simpler modulator, and the dynamic range is 2-3 dB larger. The increase in dynamic range is due to our modulator's ability to operate on larger inputs, and indicates a more stable design. The increase in peak SNR may reflect the explicit design with respect to an SNR performance measure, even though the measure is approximate.

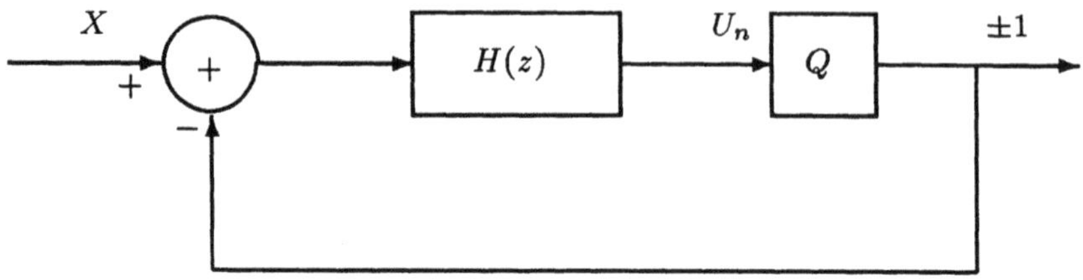


Figure 7.10: *Discrete-time model of the interpolative ΣΔ encoder.*

Figure 7.9 demonstrates that although our modulator design is based on a limit cycle analysis for constant inputs, the results are also useful for dynamic inputs. For instance, reducing $\xi_{\max}$ to 0.3 results in a reduction in dynamic range. However, we should bear in mind that the analysis is a worst-case one, and that it is not strictly valid for time-varying inputs. For instance, we find numerically that choosing $\xi_{\max}$ between 0.5 and 0.7 has little effect on the dynamic range.

## 7.4 Interpolative Encoder

This section discusses stability issues for the interpolative ΣΔ encoder. Section 7.4.1 presents an exact analytical method to determine the existence and amplitude of given limit cycles. Section 7.4.2 addresses the problem of finding the maximum amplitudes of limit cycles without requiring knowledge of their precise form. Specifically, in section 7.4.2.1 we derive an approximate result based on the describing-function approach and aimed at systems with open-loop poles close to the unit circle, and in section 7.4.2.2 we present a numerical method for finding upper bounds on limit cycle amplitudes. Throughout results are exemplified using the fourth order interpolative encoder introduced in [32], often referred to below as just "the fourth order encoder". The present section is similar in form to section 7.3, but significant differences in the methods and results will be pointed out.

Figure 7.10 shows a discrete-time model of the interpolative ΣΔ encoder. As mentioned in chapter 1, high-order interpolative designs appear to be inherently plagued with the potential for large-amplitude low-frequency oscillations. These oscillations are detrimental to performance because they can drive the encoders into sustained modes of integrator saturation. The occurrence of large oscillations is not predictable from simple linearized analysis which indicates that an interpolative encoder is stable if and only if the zeros of $1 + H(z)$ are inside the unit circle, and that stability is independent of the initial states of the integrators or the level of DC inputs. In contrast, we find empirically that both these factors profoundly affect the behavior of the encoder. In addition, we show in section 7.4.2.1 that the proximity of the poles of the open-loop filter $H(z)$ to the unit circle can be important to stability. This is despite the fact that the poles of $H(z)$ do not manifest themselves in the linearized signal transfer function $H_X(z)$, and only appear as zeros in the linearized error transfer function $H_E(z)$ (1.1). An immediate, but new observation demonstrating the importance of the poles of $H(z)$ is that an encoder is guaranteed to be BIBS stable if only the poles of $H(z)$ are inside the unit circle. This is because both the quantizer output and the encoder inputs are bounded, implying that the open-loop filter input is also bounded. Another such result emphasizing the natural role of the poles of $H(z)$ is presented in section 7.4.1. The shift of focus from the zeros of $1 + H(z)$ to the poles of $H(z)$ is important, especially in view of the fact that interpolative encoders are frequently designed with open-loop poles on the unit circle itself [32], [55].

The limitations of linearized analysis suggest that the phenomenon of large oscillations should be considered from a state space point of view, and that the limit cycle framework set forth in section 7.2 may provide insight. As the set-up of the present section is more general than that of section 7.3, we find it convenient to focus on the quantizer input $U_n$ as representative of limit cycle amplitudes, and to not consider amplitudes of oscillations in other internal state variables. This is because the filter $H(z)$ may be realized in various ways leading to different natural choices of state variables, and we wish to separate the problems of realization and transfer function design.

The flow of the proposed design process is as follows. We assume that we are given an unclipped encoder with satisfactory SNR performance, and that there exists a realization of the encoder filter such that its filter coefficients are all of the same order of magnitude. The encoder could therefore potentially be implemented in switched-capacitor technology. We assume further that the encoder has stability problems at one or more internal nodes. Finally we assume that equivalent scaling is insufficient to solve the stability problem, because the required scaling factors would result in capacitor ratios that are too large to be implemented in practice. Our goal is to perform functional scaling to make the encoder more stable, so that subsequent equivalent scaling will not result in excessive scaling factors.

### 7.4.1 Detection of Specific Limit Cycles

By way of motivation, consider the fourth order interpolative encoder with transfer function [32]

$$H(z) = \frac{0.8653 - 2.2692z^{-1} + 2.0064z^{-2} - 0.59714z^{-3} + 0.000035z^{-4}}{1 - 3.99646z^{-1} + 5.992922z^{-2} - 3.996460866z^{-3} + 1.000000433z^{-4}} \cdot z^{-1} \tag{7.15}$$

This open-loop filter $H(z)$ has both its two pole pairs on the unit circle in signal baseband, and the filter is realized with a cascade of integrators from which outputs are fed forward and backward. For constant inputs and zero initial integrator outputs, the encoder is reported in [32] to be unstable when $|X| > 0.65 - 0.7$, in the sense that the SNR decreases dramatically. Whether or not the oscillations are in fact bounded, this behavior is undesirable as it limits the dynamic range. Expanding on this, we find that if the initial integrator states are all chosen to be 1, large oscillations occur when the constant input exceeds approximately 0.2803. This result and others like it underscore the influence of the initial states on the system trajectory, and shows that even small inputs may excite large oscillations. It appears difficult to exactly describe the relationship between initial states and open-loop filter that gives rise to large oscillations for various inputs. Therefore, rather than avoiding initial states that might result in large oscillations, it is desirable to design encoders that do not exhibit large oscillations for *any* initial states. As seen in section 7.4.2 such designs effectively sacrifice SNR to improve stability.

In this section we describe in more detail the instability problems of interpolative encoders. In section 7.4.1.1 we present a method that can be used to answer the following questions: Given a $P$-bit pattern $\mathbf{Y} = \{Y_0, \cdots, Y_{P-1}\}$, does there exist a constant input $X$ and an initial encoder state such that the corresponding output sequence is a periodically repeated version of $\{Y_0, \cdots, Y_{P-1}\}$? If so, what is the amplitude of the limit cycle seen at the quantizer input? As in section 7.3, our method is based on the standard technique of opening the loop frequently associated with Tsypkin's name [154]. In section 7.4.1.2 we present results of applying the presented method to the fourth order encoder. In section 7.4.1.3 we discuss the conditions under which the limit cycles of an interpolative encoder are attracting.

#### 7.4.1.1 Existence of Specific Limit Cycles

Consider the general interpolative encoder of order $N$. We may write its open-loop transfer function as

$$H(z) = \frac{\sum_{n=0}^{N} A_n(z-1)^{N-n}}{(z-1)^N - \sum_{n=1}^{N} B_n(z-1)^{N-n}} \cdot z^{-1} \qquad (7.16)$$

where $A_0, \cdots, A_N$ and $B_1, \cdots, B_N$ are filter coefficients.[5] Let us define an $N$-dimensional state vector $\mathbf{S}_n$ for the open-loop filter, where $n$ is the time step. Further, let us define an $N \times N$ matrix $\mathbf{B}$ in terms of the filter coefficients, and an $N$-dimensional input vector $\mathbf{L}_n$ depending on the constant input $X$ as well as quantizer outputs, such that the state space representation of the system is of the form

$$\mathbf{S}_{n+1} = \mathbf{B}\mathbf{S}_n + \mathbf{L}_n \qquad (7.17)$$

[5] For the fourth order encoder with transfer function (7.15), $(A_0, \cdots, A_4) = (0.8653, 1.1920, 0.3906, 0.06926, 0.005395)$ and $(B_1, \cdots, B_4) = (-3.540 \cdot 10^{-3}, -3.542 \cdot 10^{-3}, -3.134 \cdot 10^{-6}, -1.567 \cdot 10^{-6})$.

It is well known from systems theory that there exist many system representations of the form (7.17). For the fourth order encoder, one such representation is obtained by defining the four-dimensional state vector $\mathbf{S}_n$ such that

$$\mathbf{B} = \begin{bmatrix} 1 & 1 & 0 & 0 \\ 0 & 1 & 1 & 0 \\ 0 & 0 & 1 & 1 \\ B_4 & B_3 & B_2 & 1+B_1 \end{bmatrix}, \quad \mathbf{L}_n = \begin{bmatrix} 0 \\ 0 \\ 0 \\ X - Q(U_n) \end{bmatrix}$$

In general, $\mathbf{L}_n$ depends linearly on the constant input $X$. Finally, let us define two $N$-dimensional vectors $\mathbf{C}$ and $\mathbf{D}$, depending on the filter coefficients, such that the filter output satisfies

$$\mathbf{U}_{n+1} = \mathbf{C}^\dagger \mathbf{S}_n + \mathbf{D}^\dagger \mathbf{L}_n \tag{7.18}$$

where $\dagger$ denotes transposition. For instance, the fourth order encoder with transfer function (7.16) has

$$\mathbf{C} = \begin{bmatrix} A_4 + A_0 B_4 \\ A_3 + A_0 B_3 \\ A_2 + A_0 B_2 \\ A_1 + A_0 B_1 \end{bmatrix}, \quad \mathbf{D} = \begin{bmatrix} 0 \\ 0 \\ 0 \\ A_0 \end{bmatrix}$$

We now address the question stated above, that is, does there exist a limit cycle of period $P$ corresponding to the binary output sequence $\mathbf{Y} = \{Y_0, \cdots, Y_{P-1}\}$? It is convenient to associate with each sequence $\mathbf{Y}$ a vector $\mathbf{L}'_n$ which is obtained from $\mathbf{L}_n$ by replacing $Q(U_n)$ with $Y_n$. Specifically, $\mathbf{L}'_n = \{0, 0, 0, X - Y_n\}^\dagger$ for the fourth order encoder. To answer the question we must determine whether or not there exists an $X$ such that two conditions hold:

**C1** Periodicity, that is, $\mathbf{S}_{P+n} = \mathbf{S}_n$ for all $n$. From (7.17), this condition is equivalent to $\mathbf{S}_P = \mathbf{S}_0$.

**C2** Consistency, that is, as we step through the difference equation (7.17) using $\{\mathbf{L}'_n\}$ as the input vectors, we have $Y_n = Q(U_n)$ for all $0 \le n \le P-1$.

Condition C1 can be used to find the initial state vector $\mathbf{S}_0$ as a linear function of $X$ in the following way: From (7.17) we find

$$\mathbf{S}_P = \mathbf{B}^P \mathbf{S}_0 + \sum_{n=0}^{P-1} \mathbf{B}^{P-1-n} \mathbf{L}'_n$$

We set $\mathbf{S}_P = \mathbf{S}_0$ to enforce condition C1:

$$\mathbf{S}_0 = (\mathbf{I} - \mathbf{B}^P)^{-1} \sum_{i=0}^{P-1} \mathbf{B}^{P-1-n} \mathbf{L}'_n \tag{7.19}$$

where $\mathbf{I}$ is the $N \times N$ identity matrix, and $\mathbf{I}-\mathbf{B}^P$ is assumed invertible. If $\mathbf{I}-\mathbf{B}^P$ is singular, a generalization of the technique in section 7.3.1.1 must be used instead.[6] The right-hand side of (7.19) depends linearly on $X$. Using $\mathbf{S}_0$ from (7.19), we can step through the difference equation (7.17) of the system, and at each time step use (7.18) to find $U_n$ as a linear function of $X$. The sequence $\mathbf{Y}$ is then a limit cycle output sequence if and only if condition C2 holds, that is, if there exists an $X$ such that all the linear inequalities $Q(U_n) = Y_n$ in $X$ can be satisfied simultaneously.[7] The amplitude of the limit cycle, if it exists, is also found by stepping through equations (7.17) and (7.18).

Our procedure is similar in spirit to the derivation in section 7.3.1.2 for the double-loop encoder. If we were to apply the above technique to this encoder,[8] however, we find that with the state vector $\mathbf{S}_n = (U_n, V_n)^\dagger$,

$$\mathbf{B} = \begin{bmatrix} 1 & 1 \\ 0 & 1 \end{bmatrix} ; \quad \mathbf{B}^n = \begin{bmatrix} 1 & n \\ 0 & 1 \end{bmatrix} , \quad n \geq 1$$

Therefore $\mathbf{I} - \mathbf{B}^n$ is singular. This is an indication that for a given $X$ there either exist infinitely many initial states satisfying $\mathbf{S}_P = \mathbf{S}_0$ or none at all, confirming the result of section 7.3.1.

Another difference between the double-loop and the general interpolative encoder is that in the former there is a range of initial states supporting a given limit cycle, but there is only one specific constant input supporting it. In the latter there is also a range of initial states supporting a given limit cycle, but there is only one possible initial state $\mathbf{S}_0$, specified by (7.19), for each constant input, and there is a range of possible constant inputs. This difference is due to the finite DC gain $H(1)$ of the open-loop filter. Specifically, consider the average input to the open-loop filter over one period,

$$Z = X - \frac{1}{P} \sum_{n=0}^{P-1} Y_n \tag{7.20}$$

If the open-loop filter has infinite DC gain, $Z$ must be zero to maintain the limit cycle, and the situation is analogous to that of the double-loop encoder. However, if $H(1)$ is finite, the DC level of the quantizer input sequence is $ZH(1)$. From a time-domain point of view the constant input $X$ can be varied around a nominal value without affecting the output sequence as long as no quantizer

[6] Assuming that the realization of $H(z)$ is minimal, the characteristic polynomial of the matrix $\mathbf{B}$ is the denominator of the open-loop filter $H(z)$, so $\mathbf{B}$ has the eigenvalue 1 if and only if $H(z)$ has a pole at DC. Since $\det[\mathbf{I}-\mathbf{B}] = 0 \implies \det[\mathbf{I}-\mathbf{B}^P] = 0$, a DC pole of $H(z)$ implies non-invertibility. The complete result is that $\det[\mathbf{I}-\mathbf{B}^P] = 0 \iff H(e^{2\pi n/P}) = \infty$ for some integer $n$. Therefore the outlined technique will not work if and only if the open-loop filter has a pole on the unit circle with an argument which is a multiple of $2\pi/P$, where $P$ is the period in question.

[7] The method generalizes easily to multi-bit quantization.

[8] The argument is somewhat imprecise because the structure of the double-loop encoder is inherently different from that of an interpolative encoder, but the argument can be made rigorous.

input $U_n$ is shifted so much that it changes sign. All other things being equal, it is undesirable that several values of $X$ give rise to indistinguishable output sequences, because it implies that any decoder is inherently limited in resolution when the encoder is on such limit cycles.

#### 7.4.1.2 Numerical Results on Specific Limit Cycles

To illustrate the method of section 7.4.1.1, we consider the fourth order encoder with transfer function (7.15) [32], as well as a variation of this encoder in which the poles of the open-loop transfer function are scaled by a factor of 0.98 to move them inside the unit circle.[9] We find empirically that the limit cycles with the largest amplitudes are the ones with relatively large periods, that is, low frequencies, and that the output sequences on these limit cycles tend to take on the special form of a number of positive bits $q$ followed by a number of negative bits $r$. We focus on these limit cycles, and characterize them for brevity by pairs of the form $(q, r)$.

For the fourth order encoder with transfer function (7.15), we find that many limit cycles with fairly short periods exist, but limit cycles of the form $(q, r)$ with periods around 100-140 fail to materialize. As we will see in section 7.4.2.1, this is the range in which we would expect to find large-amplitude limit cycles. We attribute the absence of these limit cycles to the fact that the poles of $H(z)$ are on the unit circle, so if the input to the open-loop filter contains a sinusoidal component at a pole frequency, the filter output will contain an unbounded oscillation at the pole frequency with linearly increasing amplitude. Although unbounded oscillations do not necessarily occur when $H(z)$ has poles on the unit circle, they are a possibility which manifests itself in the present case.

We next turn to the encoder whose open-loop pole moduli have been reduced by two percent. For this encoder we find a large number of limit cycles of the form $(q, r)$ with period $P = q + r = 117$. Table 7.2 summarizes the characteristics of all the limit cycles with at least 50% positive bits, that is, $P/2 \leq q \leq P$. We find similar results for other periods close to 117. The first two columns of the table show the center and width of the $X$-interval supporting the limit cycle, while the next two columns show the maximum and minimum values of the quantizer input $U_n$ on the limit cycle. The table shows that as $q$ moves from its smallest to its largest value, the width of the input interval supporting the limit cycle first increases from close to zero, then reaches a maximum and finally returns to zero. The maximum value of the quantizer input $U_n$ follows the same pattern, while the most negative value of $U_n$ is an increasing function of $q$. As shown in section 7.4.2.1 the limit cycles in question are close to sinusoidal, so the average of the extremes of $U_n$ is a good estimate of the DC level

[9] In terms of the filter coefficients of (7.16), $(A_0, \cdots, A_4)$ are unchanged and $(B_1, \cdots, B_4) = -8.347 \cdot 10^{-2}, -6.010 \cdot 10^{-3}, -1.752 \cdot 10^{-4}, -3.053 \cdot 10^{6})$.

| $q$ | Average $X$ | $10^3 \times$ width | max. $U_n$ | min. $U_n$ | $\Delta$ (%) |
|---|---|---|---|---|---|
| 73 | 0.4645 | 1.68 | 1400 | –647 | –1.5 |
| 74 | 0.4932 | 3.88 | 1410 | –618 | –1.8 |
| 75 | 0.5212 | 6.10 | 1419 | –588 | –1.6 |
| 76 | 0.5486 | 8.33 | 1426 | –560 | –1.8 |
| 77 | 0.5753 | 10.6 | 1431 | –532 | –1.7 |
| 78 | 0.6012 | 12.7 | 1434 | –504 | –1.8 |
| 79 | 0.6265 | 14.9 | 1435 | –477 | –1.8 |
| 80 | 0.6509 | 17.0 | 1434 | –451 | –1.8 |
| 81 | 0.6745 | 19.0 | 1431 | –425 | –1.8 |
| 82 | 0.6973 | 20.8 | 1425 | –400 | –1.8 |
| 83 | 0.7187 | 21.3 | 1417 | –376 | –1.7 |
| 84 | 0.7393 | 19.1 | 1401 | –350 | –1.9 |
| 85 | 0.7596 | 15.9 | 1385 | –323 | –2.0 |
| 86 | 0.7791 | 12.9 | 1367 | –296 | –1.8 |
| 87 | 0.7977 | 10.0 | 1347 | –271 | –2.0 |
| 88 | 0.8154 | 7.31 | 1325 | –248 | –2.0 |
| 89 | 0.8323 | 4.81 | 1301 | –225 | –2.1 |
| 90 | 0.8482 | 2.52 | 1275 | –203 | –2.1 |
| 91 | 0.8633 | 0.45 | 1248 | –183 | –2.0 |

Table 7.2: *Characteristics of limit cycles of the form $(q,r)$ with period $P = q+r = 117$ for a fourth order interpolative encoder. We show all such limit cycles with at least 50% positive bits, that is, $P/2 \leq q \leq P$. The first two columns show the center and width of the $X$-interval supporting the limit cycle, while the next two columns show the maximum and minimum values of the quantizer input $U_n$ on the limit cycle. The last column measures the difference in percent between the actual DC input to the quantizer and an approximation to this.*

of the quantizer input. We therefore expect the following quantity to be small:

$$\Delta = \frac{\frac{1}{2}(U_{\max} + U_{\min}) - ZH(1)}{ZH(1)}$$

where $Z$ is defined in (7.20). The last column of table 7.2 confirms that the quantity $\Delta$ is indeed small, namely on the order of 2%. The amplitudes of limit cycles of the form $(q,r)$ with period 117 are upper bounded by approximately 1435, which is a rather large number.

#### 7.4.1.3 Attracting Limit Cycles

In this section we show that if the open-loop filter $H(z)$ has all its poles inside the unit circle, almost all limit cycles are attracting. More precisely, if we take almost any limit cycle in state space and consider a sufficiently small region around any point on the limit cycle, then for all initial states in the region, the system trajectory will converge to the limit cycle. This result follows from the fact that if all points on a limit cycle satisfy $U_n \neq 0$, the collection of Lyapunov exponents for that limit cycle equals the set of eigenvalues of the matrix $\mathbf{B}$, or equivalently the poles of $H(z)$ [153]. Therefore, if the poles of $H(z)$ are all inside the unit circle, the limit cycle is attracting or stable. Note that the concept of stability of limit cycles is different from the concept of stability of $\Sigma\Delta$ encoders [153].

The described result calls for further comparison between the double-loop and the interpolative encoder. The double-loop encoder only has limit cycles for rational constant inputs, and since its open-loop transfer function has poles on the unit circle, its limit cycles are not attracting. An interpolative encoder with a stable open-loop transfer function, on the other hand, has limit cycles in many intervals of constant inputs, and its limit cycles are attracting. This may imply that limit cycles play an even greater role for interpolative encoders than for single-loop and double-loop encoders.

### 7.4.2 Amplitudes of Limit Cycles

In this section we take a different view on stability issues, and relax the requirement of known limit cycles. In section 7.4.2.1 we apply the describing function approach to obtain approximate relationships between characteristics of an open-loop filter $H(z)$ and the corresponding large-amplitude limit cycles. We will see that the approximation is useful for filters with poles inside and close to the unit circle, as is the case for many practical interpolative encoders. In section 7.4.2.2 we present a numerical method for deriving upper bounds on limit cycle amplitudes.

#### 7.4.2.1 Describing Function Approximation

In this section we use the describing function method to obtain approximate relationships between the open-loop filter $H(z)$ and the corresponding large-amplitude limit cycles. The approximation is valid for filters with their poles close to the unit circle, and improves in accuracy as the poles move closer to the unit circle. We will use the analysis to demonstrate design trade-offs between SNR and stability.

Our approximate describing function approach used in this section bears some similarity to the work of Ardalan and Paulos [67] who use a frequency-domain approach and model the quantizer as two linearized gains, namely one for either a DC or a sinusoidal component and one for the residual which is assumed Gaussian. Unlike that work, however, we are specifically interested in situations where instability can occur, and we can thus use a relatively simple approach without a Gaussian assumption. In addition, we do not assume that the open-loop transfer function has a pole at DC.

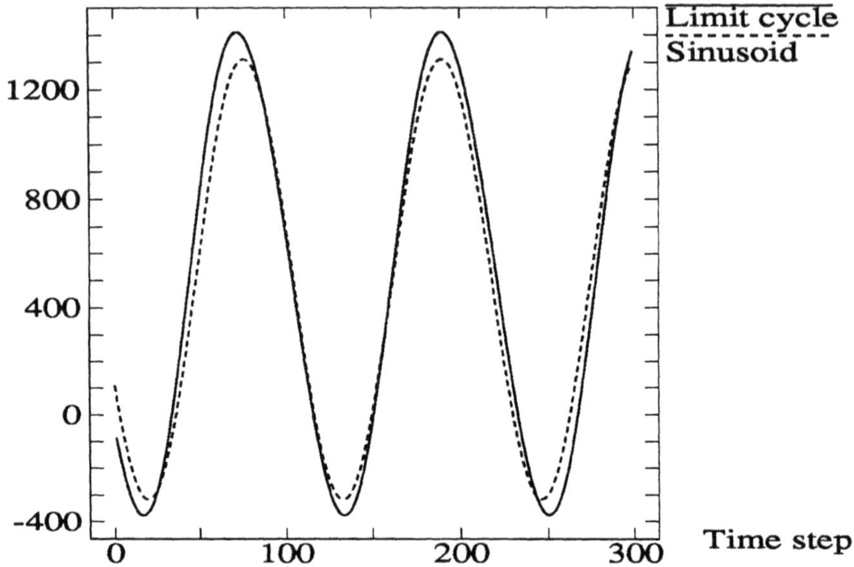


Figure 7.11: *Limit cycle in the quantizer input sequence $U_n$ for the fourth order encoder with pole moduli scaled by 0.98. The constant input is $X = 0.7$. Also shown is the describing function approximation.*

Our motivation for using the describing function method is the observation that large-amplitude limit cycles are often close to sinusoidal. For example, figure 7.11 shows a limit cycle in the fourth order encoder (7.15) with its pole moduli scaled by 0.98. The figure also shows the result of using the describing function approximation derived below. We see that the maximum value on the limit cycle is predicted to within about 10%.

In applying the describing function method, we assume that the quantizer input sequence is of the form

$$U_n = C - A \sin \omega_0 n, \quad 0 \leq C < A \tag{7.21}$$

where $\omega_0$ is the frequency of oscillation, and $A$, $C$ and $\omega_0$ are unknowns which are to be determined for a given open-loop filter $H(z)$ and a given constant input $X \geq 0$. Equation (7.21) is clearly an approximation, although bounds on its quality may be derived using the methods of [155]; in general terms the approximation is best if higher harmonics of the fundamental frequency $\omega_0$ are highly suppressed by $H(z)$. As a further approximation we consider the problem in continuous time rather than discrete time,

$$U(t) = C - A \sin \omega_0 t, \quad 0 \leq C < A$$

| $r$ | $X$ | LC period | | $k_{DC}$ | $k_{GM}$ | LC extremes | |
|---|---|---|---|---|---|---|---|
| | | sim. | theor. | | | sim. | theor. |
| 0.99 | 0.7004 | 127 | 125 | 2845 | 2194 | +3700,–1400 | +3600,–1400 |
| 0.98 | 0.7839 | 117 | 113 | 1767 | 808 | +1370,–300 | +1250,–220 |
| 0.97 | 0.8292 | 101 | 96 | 1004 | 293 | +350,–25 | +450,–50 |

Table 7.3: *Simulated and analytical results for limit cycle periods and extreme values for the fourth order encoder with its pole moduli scaled by $r$. The constant input is chosen to be the smallest input resulting in a large-amplitude limit cycle if the initial states of $H(z)$ are all zero. Also shown are the DC gain and gain margin of $H(z)$.*

In appendix F.3 we show that the unknown constants $A$, $C$ and $\omega_0$ can be found as follows: The frequency $\omega_0$ is given by $\angle H(e^{j\omega_0}) = -180^{\circ}$, where the phase of $H$ is reduced to $[-180^{\circ}, +180^{\circ})$; therefore the expected limit cycle period is $P = 2\pi/\omega_0$. The constants $A$ and $C$ are found by solving a set of nonlinear equations depending only on $X$, the DC gain $k_{DC} = H(1)$ and the gain margin $k_{GM} = |H(e^{j\omega_0})|$, namely

$$\begin{aligned} C &= k_{DC}\left(X - \frac{2}{\pi}\arcsin\frac{C}{A}\right) \\ A &= k_{GM}\cdot\frac{4}{\pi}\sqrt{1-\left(\frac{C}{A}\right)^2} \end{aligned} \tag{7.22}$$

We show in appendix F.3 that as a first order approximation, the nonlinear equations (7.22) are also valid for determining $A$ and $C$ when the quantizer is 2-bit or infinite-bit rather than one-bit;[10] these quantizers are defined in the appendix. This result indicates that for interpolative encoders with large-amplitude limit cycles, the quantizer resolution does not affect stability; the stability problem is intrinsically linked to the open-loop transfer function.

Returning to 1-bit quantization, tables 7.3 and 7.4 illustrate the results for variations on the fourth order encoder (7.15) [32]. Specifically, in table 7.3 we scale the pole moduli by a factor $r$, and in table 7.4 we instead scale the pole angles by a factor $a$. The tables show simulated and analytical results for limit cycle periods and extreme values; the constant input is in each case chosen to be the smallest input resulting in a large-amplitude limit cycle if the initial states of $H(z)$ are all zero. The predictions of the describing function approximation are generally close to the observed amplitudes, and the approximation is better for poles close to the unit circle.

[10] It is found by simulation that using a two-bit or infinite-bit quantizer does not change the amplitude or the period of the resulting large-amplitude limit cycles.

| $a$ | $X$ | LC period | | $k_{DC}$ | $k_{GM}$ | LC extremes | |
|---|---|---|---|---|---|---|---|
| | | sim. | theor. | | | sim. | theor. |
| 1.2 | 0.7253 | 104 | 102 | 1459 | 1245 | +2100,–800 | +2050,–800 |
| 1.0 | 0.7004 | 127 | 125 | 2845 | 2194 | +3700,–1400 | +3600,–1400 |
| 0.8 | 0.6928 | 161 | 159 | 6248 | 4324 | +7400,–2600 | +7100,–2500 |

Table 7.4: *Simulated and analytical results for limit cycle periods and extreme values for the fourth order encoder with its pole arguments scaled by a factor a and its pole moduli fixed at* $r = 0.99$. *The constant input is chosen to be the smallest input resulting in a large-amplitude limit cycle if the initial states of* $H(z)$ *are all zero. Also shown are the DC gain and gain margin of* $H(z)$.

We also show in appendix F.3 that for a given open-loop filter, the largest value of $U(t)$ for any constant input, that is, the largest limit cycle amplitude equals

$$U_{\max} = \max_X(A + C) = \frac{3\sqrt{3}}{\pi} k_{GM} \approx 1.65 k_{GM}$$

This quantity is seen to only depend on the gain margin and not on the DC gain. Although the analysis is approximate, it does suggest the existence of a design conflict: On one hand the gain margin should be kept small to minimize limit cycle amplitudes and thus maximize system stability. On the other hand, a linearized system model suggests that the magnitude of the open-loop transfer function $H(z)$ should be large over all of baseband, including the frequency $\omega_0$, so that the baseband noise suppression and thus the SNR are maximized [32]. In fact, we can use the DC gain as a rough indicator of the SNR, because the DC gain sets the level of the transfer function magnitude in baseband for interpolative encoders such as the one in [32]. In this set-up, the trade-off is between maximizing the DC gain and minimizing the gain margin.

The trade-off between stability and SNR performance can be explored in various ways. As an illustration we consider the following problem: For the fourth order encoder with transfer function (7.15), how should the pole moduli of the open-loop transfer function be modified to maximize stability, given that only a certain degradation of SNR is acceptable? We consider $k_{DC}$ and $k_{GM}$ to be indicators of the SNR and maximum limit cycle amplitudes, because the SNR depends approximately linearly on $20 \log_{10} k_{DC}$, and $U_{\max} \approx 1.65 k_{GM}$. Let us denote the modulus scaling factors for the lower frequency (LF) and higher frequency (HF) poles by $r_{LF}$ and $r_{HF}$, respectively. For a given DC gain we carry out the optimization numerically by choosing the pair $(r_{LF}, r_{HF})$ resulting in the smallest $k_{GM}$ subject to $H(1) = k_{DC}$ and $r_{LF} < 1$.

Figure 7.12 shows the optimum scaling factors as functions of the DC gain. The main observation is that $r_{LF}$ and $r_{HF}$ remain close to 1 for moderate de-

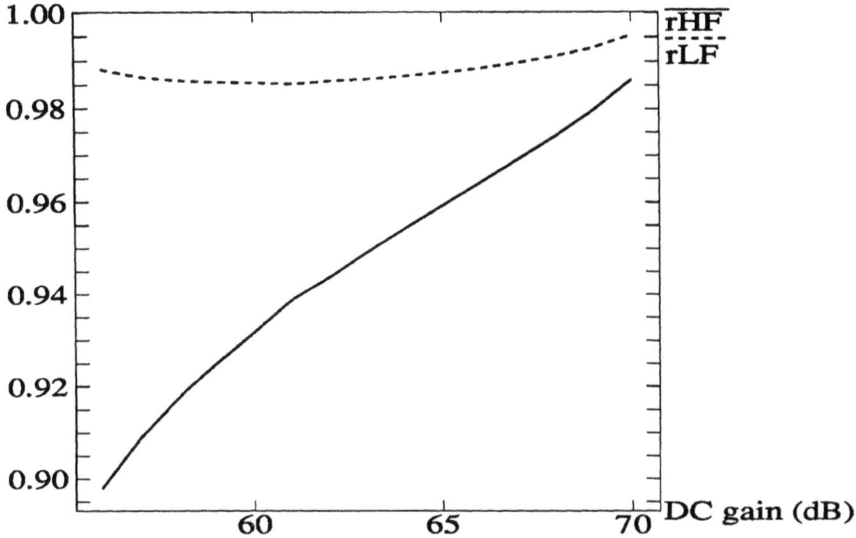


Figure 7.12: *Optimum pole modulus scaling factors for the fourth order interpolative encoder. For a given DC gain, these scaling factors achieve the smallest gain margin.*

creases in DC gain from the nominal 70.7 dB. In general, the HF pole pair is scaled more than the LF one for a given DC gain. This is intuitively understandable, because the oscillation frequency is close to the frequency of the HF pole. Therefore the most efficient way of reducing $k_{GM}$ is to increase the distance between the HF pole pair and $e^{\pm j\omega_0}$, where $\omega_0$ is the oscillation frequency. Figure 7.13 shows the trade-off between $k_{DC}$ and $U_{max} = 1.65k_{GM}$ obtained by using the optimum scaling factors $r_{LF}$ and $r_{HF}$ from figure 7.12. As an example, if the designer is willing to sacrifice 6 dB of DC gain, the limit cycle amplitudes can be reduced from essentially infinite to approximately 800. This is still a large number, but it can be remedied by equivalent scaling. This way the encoder will not be as heavily clipped as the original encoder, and due to the more stable design, it will be able to leave such saturation modes more gracefully.

#### 7.4.2.2 Bounding the Amplitude

In appendix F.1 we derived analytical upper bounds on limit cycle amplitudes for the double-loop encoder. The idea was to divide trajectories in state space into two types of segments in time: Positive and negative ones as specified by the sign of the quantizer input. For each segment type we defined a potential

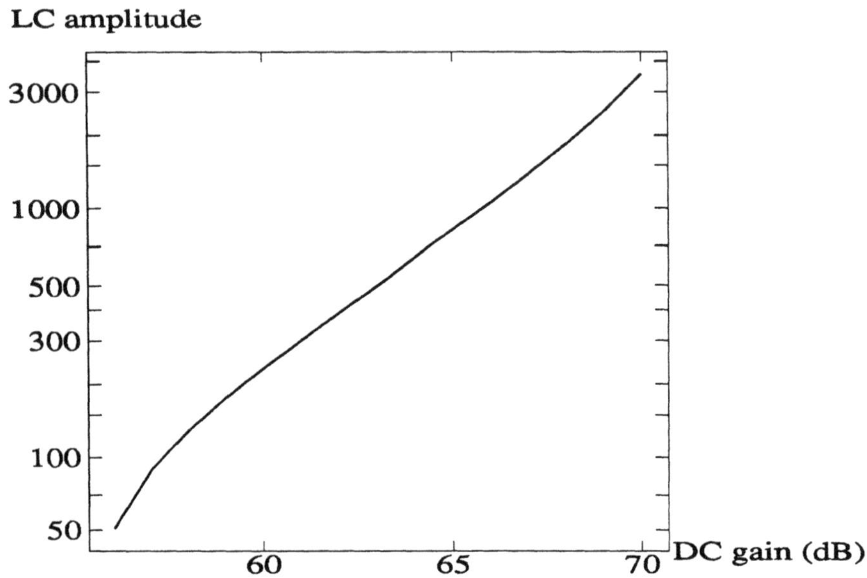


Figure 7.13: *Best achievable trade-off between DC gain and limit cycle amplitudes. Note the logarithmic axes.*

related to the maximum values that the state variables can assume on the segment. We derived an upper bound on the potential obtainable on a positive segment as a function of the potential on the previous positive segment; the bound was a monotonically increasing function for all but very small values of the potential. We argued that an upper bound on the positive potential on any limit cycle could be found by considering all initial potentials for which the next potential can be as least as large: The largest such initial potential is the desired upper bound, because potentials exceeding the bound cannot recur and so cannot correspond to limit cycles.

The outlined technique does not immediately generalize to interpolative encoders. The main reason is that solutions to the difference equations are more involved for higher-order systems. The idea of positive and negative segments can however still be put to use. In this section we describe a numerical method for finding upper bounds on limit cycle amplitudes, and present numerical results in agreement with the limit cycle results in section 7.4.2.1. We first describe our method conceptually, then describe its numerical implementation and show resulting bounds on limit cycle amplitudes.

Our approach is to focus directly on the quantizer input $U_n$, rather than a potential, for a given constant input $X$. Figure 7.14 shows some possible trajectories $\{U_n\}$ as functions of time. We divide all trajectories into positive and negative segments characterized by the sign of $U_n$ as before, and consider

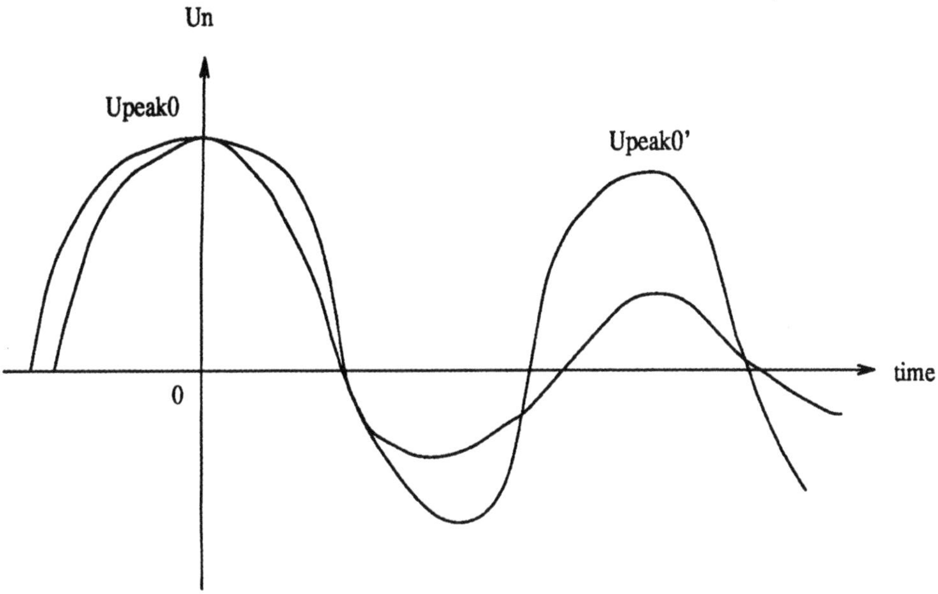


Figure 7.14: *Some time sequences of the quantizer input* $\{U_n\}$ *for a given constant input* $X$. *Both peak at the value* $U_{\text{peak}}$ *within the positive segment around time 0.* $U'_{\text{peak}}$ *is the maximum peak value over all trajectories on the next positive segment.*

the peak values assumed by the quantizer input on each positive segment.[11] The basic idea is to derive an upper bound on the positive peak value as a function of the previous positive peak value. The fact that this function is not in general monotonic is a complicating factor, but as detailed below, it is still possible to use the function to find an upper bound on the positive peak value on any limit cycle. To be more specific, we define a parameter $U_{\text{peak}}$ which physically denotes a peak value in the positive segment around time 0. In figure 7.14 a particular value, $U_{\text{peak0}}$ is used. For each value of $U_{\text{peak}}$, we consider all trajectories that have the following two properties:

**P1** At some arbitrary time $n = 0$, $U_0 = U_{\text{peak}} > 0$.

**P2** Within the positive segment to which $n = 0$ belongs, the trajectory peaks at $n = 0$.

Both trajectories in figure 7.14 have these properties with the particular value $U_{\text{peak}} = U_{\text{peak0}}$. Let us follow all such trajectories forward in time until they

[11] The method can be generalized to multi-bit quantization if a positive segment is understood to be a segment for which all $U_n \geq M$, where $M$ is the smallest number for which $Q(M) = 1$. Negative segments are defined similarly.

reach their next positive segment, and register the largest value of $U_n$, denoted by $U'_{\text{peak}}$, attained by any such trajectory. In figure 7.14, a particular value $U'_{\text{peak}} = U'_{\text{peak0}}$ is shown, assuming that none of the other trajectories with properties P1 and P2 have a value of $U'_{\text{peak}}$ exceeding $U'_{\text{peak0}}$. We can plot $U'_{\text{peak}}$ as a function of $U_{\text{peak}}$, and an example of this is shown schematically in figure 7.15. As follows shortly, it is not necessary for the plot to be monotonically increasing. Our approach only allows us to draw conclusions if there exists some value $U_{\text{max}}$ such that for all $U_{\text{peak}}$ greater than $U_{\text{max}}$, $U'_{\text{peak}}$ is less than $U_{\text{peak}}$, and for all $U_{\text{peak}}$ less than $U_{\text{max}}$, $U'_{\text{peak}}$ is less than $U_{\text{max}}$; if in particular the plot in figure 7.15 is monotonically increasing, $U_{\text{max}}$ is always the intersection of the plot with the $45^o$ line. Geometrically, the two requirements on the plot in figure 7.15 are that there exists some $U_{\text{max}}$ such that

**R1** For all $U_{\text{peak}} > U_{\text{max}}$, the plot is below the $45^o$ line.

**R2** For $0 \leq U_{\text{peak}} \leq U_{\text{max}}$, the plot is completely contained in the square $(U_{\text{peak}}, U'_{\text{peak}}) \in [0, U_{\text{max}}]^2$.

If both requirements on the plot hold, we can conclude that $U_{\text{max}}$ is an upper bound on the amplitude of any limit cycle. This follows by contradiction: Assume that some limit cycle has a peak value $U_{\text{peak}} > U_{\text{max}}$. Then the series of peak values in the subsequent segments must be strictly decreasing as long as the peak values are above $U_{\text{max}}$ because of requirement R1, and once the peak value is below $U_{\text{max}}$, it can never again exceed $U_{\text{max}}$ because of requirement R2. In particular, the peak value can never again reach $U_{\text{peak}}$. Therefore $U_{\text{peak}}$ could not correspond to a limit cycle. The process is illustrated in figure 7.16.

The presented argument can be further refined. Consider the same approach as before, except that instead of observing the sequence $\{U_n\}$ on the next positive segment, we wait a fixed number $s$ of positive segments. Specifically, $U'_{\text{peak}}$ now denotes the largest positive peak on the $s$th positive segment, and the $s-1$ intermediate positive peak values are ignored. The exact same graphical technique of plotting $U'_{\text{peak}}$ against $U_{\text{peak}}$ as in figure 7.15 will then yield an upper bound on $U_{\text{peak}}$ on any limit cycle, as shown in appendix F.4.1. Using a fixed number $s > 1$ of positive segments produces better bounds than using $s = 1$, as shown rigorously in appendix F.4.2. Figure 7.17 shows some actual upper bound curves for different values of $s$ for a specific fourth order encoder, as described in more detail below. We see that the curves tend to become flatter as $s$ increases.

We now outline a numerical implementation of the method for bounding limit cycle amplitudes. For a given constant input and each value of $U_{\text{peak}}$, we need to determine $U'_{\text{peak}}$ so that we can make a plot similar to figure 7.15. We must therefore find all trajectories with properties P1 and P2. For convenience, we express the system equation (7.17) as a scalar difference equation of the form

$$\sum_{m=0}^{N} a_m U_{n-m} = cX + \sum_{m=0}^{N} b_m Q(U_{n-1-m}) \tag{7.23}$$

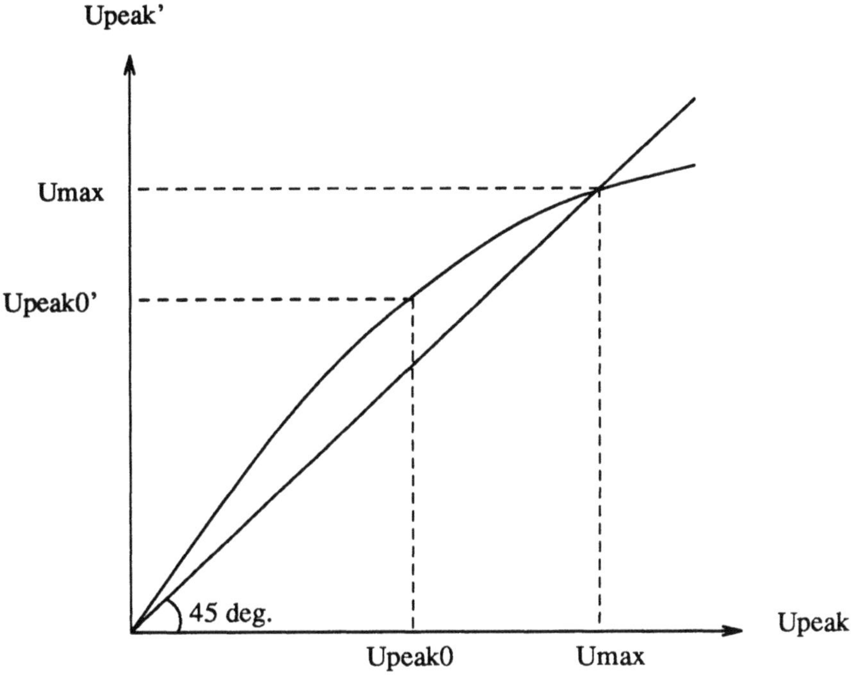


Figure 7.15: *Stylized plot of the largest possible peak value $U'_{\text{peak}}$ as a function of the peak value $U_{\text{peak}}$ on the previous positive segment. Also shown is the 45° line. $U_{\max}$ is an upper bound on limit cycle amplitudes.*

The coefficients $\{a_m\}$ and $\{b_m\}$ can be obtained directly from the denominator and numerator, respectively, of the open-loop transfer function $H(z)$. Restating our goal, we want to find all trajectories of (7.23) subject to P1 and P2.

Equation (7.23) is nonlinear and of order $N+1$. If we consider the special case where all the $N+1$ output bits used on the right-hand side of (7.23) are identical over some period of time, the equation however reduces to an $N$th order linear difference equation,

$$\sum_{m=0}^{N} a_m U_{n-m} \;=\; K \tag{7.24}$$

where the constant $K$ depends on whether the segment is positive or negative.[12] We will limit our numerical technique to trajectories with at least $N+1$ positive bits directly before each value of $U_{\text{peak}}$ considered.[13] This limitation appears

[12] Incidentally, on such segments the system behavior is determined almost exclusively by the $\{a_m\}$ coefficients, that is, by the poles of $H(z)$. The only effect of the $\{b_m\}$ coefficients, that is, the zeros of $H(z)$, is to set the constant value $K$.

[13] The same argument can be made for $N+1$ negative bits.

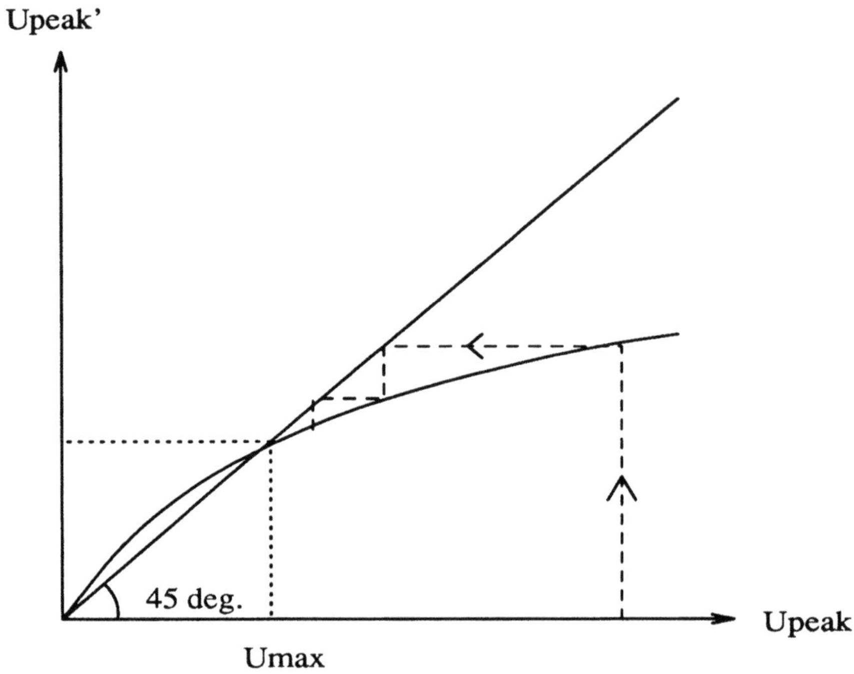


Figure 7.16: *Demonstration that $U_{\text{max}}$ is an upper bound on limit cycle amplitudes. The trajectory marked with arrows is a worst-case scenario where the peak values decrease as slowly as possible. Each point in the plot of $U'_{\text{peak}}$ vs. $U_{\text{peak}}$ is an upper bound, obtained only by one or a few trajectories.*

reasonable for the encoders in question: For example, the fourth-order encoders investigated in section 7.4.2.1 have on the order of 30 positive bits before the positive peaks on their large-amplitude limit cycles.

We show in appendix F.4.3 that an explicit solution can be found to the simplified equation (7.24). The solution is valid over the positive segment to which $U_{\text{peak}}$ belongs, and depends linearly on $N$ arbitrary constants. Our goal is to select those trajectories that have properties P1 and P2 for a given constant input and a given peak value $U_{\text{peak}}$. We also show in appendix F.4.3 that enforcing constraints P1 and P2 reduces the order or the number of degrees of freedom from $N$ to $N-2$. The two constraints manifest themselves as linear equations in the arbitrary constants. To express the constraints conveniently, trajectories are approximated by continuous-time curves in the appendix.

Conceptually, we can now generate a figure similar to figure 7.15 by maximizing $U'_{\text{peak}}$ over a sufficiently fine grid in the $(N-2)$-dimensional space of arbitrary constants, for each value of $U_{\text{peak}}$ and given constant input $X$. For $N = 3$ this task is relatively simple, because the maximization is over a one-

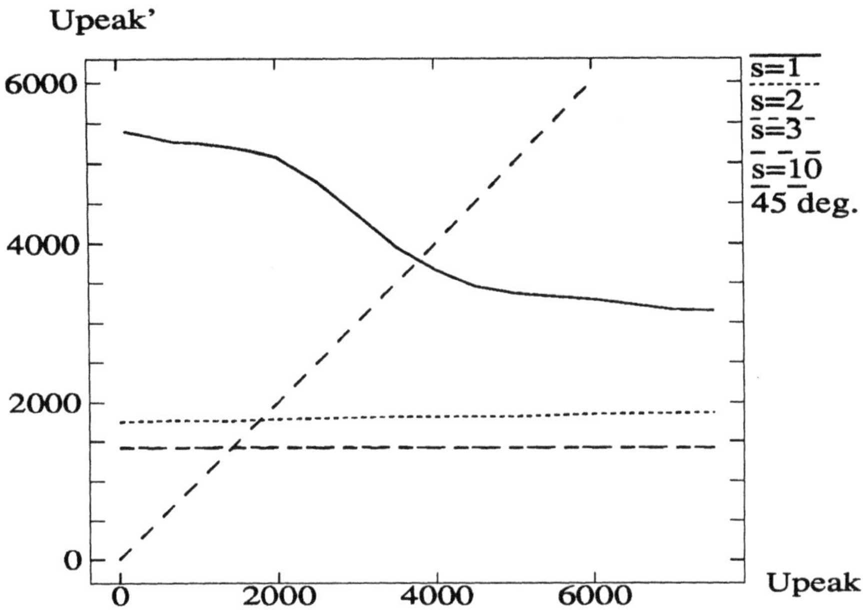


Figure 7.17: *Actual plot of* $U'_{\text{peak}}$ *as a function of* $U_{\text{peak}}$ *for a specific fourth order encoder. Also shown is the* $45^o$ *degree line.*

dimensional space. For $N = 4$ an example is given below. For larger $N$, more sophisticated search methods in the $(N-2)$-dimensional space may be necessary.

Since the figure analogous to figure 7.15 is generated entirely numerically, we have no information about the behavior of the curve for values of $U_{\text{peak}}$ which are not explicitly investigated. However, if the curve satisfies requirements R1 and R2 up to some large value $U_{\text{big}} > U_{\text{max}}$, we can at least make the following statement: No limit cycles with peak values in the range between $U_{\text{max}}$ and $U_{\text{big}}$ can exist, that is, if limit cycles with amplitudes above $U_{\text{big}}$ exist, they cannot be excited from trajectories with any peak value $U_{\text{peak}} \leq U_{\text{big}}$. As $U_{\text{big}}$ tends to infinity, $U_{\text{max}}$ becomes a guaranteed upper bound on amplitudes.

To illustrate our technique, we consider the fourth order encoder (7.15) with its pole moduli reduced by 2%. The encoder has pole arguments $\omega_1 = 0.02277$, $\omega_2 = 0.05498$, and pole moduli $r_1 = r_2 = 0.98$. We consider a constant input of $X = 0.70$, so by table 7.2 we know that a limit cycle with amplitude 1425 exists. If we choose the number of positive segments $s$ in our method to be 10, we obtain the results shown as one of the curves in figure 7.17. We can satisfy requirements R1 and R2, at least up to $U_{\text{big}}$ which was $10^6$ in our case, by choosing $U'_{\text{peak}} = 1413$. We conclude that no limit cycles with peak values between 1413 and $10^6$ can exist; the upper bound 1413 is about 1% below the observed amplitude 1425 due to the continuous-time approximation made in appendix F.4.3, and is thus is good agreement with the observed limit cycle amplitude. The flatness of the plot in figure 7.17 indicates that the encoder tends to a limit cycle with amplitude 1413 regardless of the initial peak value. The results suggest that the limit cycles shown in table 7.2 are the ones with largest amplitudes, and are thus in agreement with the derivations in section 7.4.2.1.

## 7.5 Summary

In this chapter we proposed a framework for stability analysis of $\Sigma\Delta$ encoders, and we argued that limit cycles for constant inputs are natural objects to investigate in this context. We presented the following analytical and approximate techniques to aid the analysis and design of $\Sigma\Delta$ encoders:

- For the double-loop encoder, we presented a variation on a standard technique to determine the existence and amplitude of specific limit cycles. We introduced a graphical state space approach which provides intuition in itself, and used it to derive analytical upper bounds on limit cycle amplitudes in a general, scaled double-loop encoder with constant input. This led to a suggestion for improved design of scaling factors so as to maximize stability as well as an approximate measure of SNR performance. Simulations indicated that the peak SNR of the resulting modulator is about 1.5-2 dB above that of another often used double-loop modulator, and the dynamic range is about 2-3 dB greater.

- For the interpolative encoder, we presented another variation on a Tsypkin-type method to determine existence and amplitudes of specific limit cycles. We used the describing function method to approximately quantify an inherent trade-off between SNR performance and stability, and used the results to suggest a way of functional scaling which leads to as stable encoders as possible for a given acceptable degradation in SNR. We argued that remaining stability problems could be remedied with subsequent equivalent scaling. We also presented a numerical counterpart to the analytical derivation of upper bounds on limit cycle amplitudes for the double-loop encoder.

In concluding this chapter, we emphasize that there is more to be discovered about the important stability aspect. Although our limitations of focus appear to produce useful results, further research and experiments are needed to strictly verify their validity; transient as well as possibly chaotic behavior may require attention.

# Chapter 8

# Conclusions

In this chapter we summarize and discuss the main results of the thesis, drawing more freely on the terminology used throughout than in the introductory section 1.4. Section 8.1 recapitulates our contributions, and section 8.2 describes a number of possible generalizations and extensions of the presented work. In section 8.3 we discuss future directions for research, in particular the tone problem.

## 8.1 Summary

In sections 8.1.1 and 8.1.2 we summarize the decoding results of chapters 2 through 6 and the stability results of chapter 7, respectively.

### 8.1.1 Nonlinear Decoding

In chapters 2 and 3 we presented a time-domain representation of $\Sigma\Delta$ encoders with constant inputs to complement the conventional frequency-domain representation. We used our representation to analyze the fundamental limitations on SNR that any encoder imposes on its modulator. We then derived a conceptually simple decoding algorithm, dubbed the zoomer, whose performance reaches the upper limit on SNR, and we demonstrated that the proposed algorithm achieves large gains in SNR over conventional linear filtering. Alternatively, the zoomer can be used to decrease the acquisition time to obtain a specified accuracy. Such a speed-up over linear filtering may be useful under multiplexed operating conditions, where the converter cyclically scans a series of input channels, because the required time per channel is reduced. The zoomer has the additional advantage that it is tone-free, as it only produces a single number in each conversion cycle.

In chapter 3 we showed numerical results to quantify the performance of the zoomer under ideal circumstances as well as in the presence of circuit nonidealities. We also proposed a novel cascade $\Sigma\Delta$ encoder specifically suited for constant inputs, and we showed that this encoder performs substantially better than the MASH multi-stage encoder which is designed for band-limited inputs.

In chapter 5 we generalized the optimal nonlinear decoding algorithm for constant inputs to the more general case of band-limited inputs. We showed that an iterative algorithm based on the theory of Projections Onto Convex Sets (POCS) extracts all available information from the binary encoder output. The POCS-based algorithm alternates between a time-domain and a frequency-domain projection to find a signal invariant under both. The time-domain projection results in a quadratic programming problem. The frequency-domain projection of a finite block of samples poses a general signal processing problem for which we proposed a Singular Value Decomposition (SVD) based solution in chapter 4. The SVD-based method does not suffer from the same spectral leakage problems as the Discrete Fourier Transform (DFT), and is optimal in an energy-concentration sense. We showed numerical results to quantify the performance of the proposed decoding algorithm under ideal circumstances and in the presence of non-idealities. The results indicated large gains in SNR over linear decoding.

In chapter 6 we generalized the decoding algorithm presented in chapter 5 to two-dimensional signals. Specifically, we showed that the POCS-based algorithm is directly applicable to the reconstruction problem of finding an estimate of a continuous-tone image, given only a halftoned version of that image obtained using error diffusion. We showed simulation results to illustrate the performance of the proposed algorithm as well as linear low-pass filtering. We found that the POCS-based algorithm provides subjectively superior reconstruction performance.

### 8.1.2 Stability

In chapter 7 we proposed a framework for stability analysis of $\Sigma\Delta$ encoders which takes into account the fact that practical circuit voltages are limited. We argued that it is desirable to analyze the true nonlinear behavior of an encoder rather than a linearized approximation, and that limit cycles for constant inputs are natural indicators of stability properties. We derived numerical methods for determining the amplitudes of such limit cycles for the general double-loop and interpolative encoders, and exemplified the methods for the standard double-loop and a specific fourth order interpolative encoder. We used the results to design a fourth order interpolative encoder which trades SNR performance for stability in an approximately optimal way. We also derived analytical upper bounds on the limit cycle amplitudes for the general double-loop encoder, and used those bounds to design an encoder which explicitly takes both SNR performance and stability into account. This design resulted in a gain in peak SNR and dynamic range of about 2 dB.

## 8.2 Extensions

In this section we discuss a number of possible generalizations and extensions of the presented work. Sections 8.2.1 and 8.2.2 address the areas of decoding and stability, respectively.

### 8.2.1 Nonlinear Decoding

Our decoding results are directly applicable to all current $\Sigma\Delta$ encoder architectures, including single-loop, double-loop, multi-stage and interpolative encoders. Indeed, we presented simulation results for all these architectures. Although we only considered encoders with 1-bit quantizers, the decoding algorithms generalize to multi-bit quantizers. The only difference in the algorithms is the following: In the former case, a binary quantizer output supplies either a lower or an upper bound on the quantizer input. In the latter case, however, a quantizer output supplies either a lower bound, an upper bound or both types of bound on the quantizer input. It can be shown that the set of signals generating a given encoder multi-bit output sequence is convex, so that the POCS algorithm is still guaranteed to converge. The detailed analytical time-domain results of chapter 2 do not immediately generalize to multi-bit quantizers, but the concepts of transition points and quantization intervals remain equally valid.

The POCS-based decoding algorithm for band-limited inputs is quite computationally intensive, the bottleneck being the time-domain projection. Part of the complexity derives from the fact that the algorithm estimates all the input samples, rather than a decimated version of the high-rate input. However, it does not appear possible to modify the time-domain projection to operate on decimated inputs.

The POCS algorithm is block-oriented, and therefore uses a block-oriented transform method for its band limitation. It does not appear straightforward to instead use linear filtering in some form, as linear filtering is not block-oriented. For the high SNRs that we consider, the edge effects of linear filtering would appear to limit the possible performance.

The conclusions based on the simulation results of chapters 3 and 5 must be considered tentative, and experimental results are needed to demonstrate the practical applicability of the results in the presence of various circuit imperfections. We mention that the assumption of known initial integrator states can easily be enforced in practical encoders with reset capability, such as encoders fabricated by Analog Devices, Crystal Semiconductor and Harris.

The decoding work reported in this thesis was partly inspired by research on implicit sampling and signal reconstruction from zero crossings [27]. It is plausible that further connections between $\Sigma\Delta$ modulation and zero-crossing based reconstruction can be found and exploited, either for low-pass or band-pass $\Sigma\Delta$ modulators.

The possibility of combining linear filtering with $\Sigma\Delta$ A/D conversion has recently been explored, both for FIR [156] and IIR [157] filters. Another recent paper [158] discusses the possibility of performing nonlinear arithmetic operations directly on a $\Sigma\Delta$ modulated representation of a signal, rather than on a more conventional signal representation. While these applications do not at present appear practical, it would be of interest to apply the nonlinear approach taken in this thesis to such applications.

In applications where the $\Sigma\Delta$ encoder filter is digital, as in D/A converters, it would be computationally advantageous if the filter coefficients were simple rational multiples of one another. In the extreme case, the coefficients could be limited to 0 and $\pm 1$. There exist papers on this type of filter design [159], [160], [161] which might fruitfully be applied to the encoder design problem for these applications.

The connection between $\Sigma\Delta$ and error diffusion encoders was exploited in one particular way in chapter 6. There are no doubt other areas in which the design and analysis problems for the two encoder types overlap. One example is in the choice of encoder filters.

The analytical richness of $\Sigma\Delta$ encoders does not yet appear to be depleted. Examples of approaches that have recently been applied to the encoders include symbolic dynamics, chaos theory, number theory, continued fraction theory, Euclid's algorithm [58], [162], [163], [164] and Sturmian analysis [165]. In general, these approaches appear to work most easily for the single-loop encoder, and to be at best difficult to generalize to other encoders. Finally, we mention the possibility of taking inspiration from the field of sliding-mode control [166] which however addresses continuous-time systems.

### 8.2.2 Stability

The numerical methods for determining limit cycle amplitudes presented in chapter 7 generalize to multi-bit quantization, as indicated in that chapter. However, the analytical method for upper-bounding limit cycle amplitudes in the general double-loop encoder appears difficult to generalize even to 2-bit quantization.

We believe that the stability framework presented in chapter 7 will prove useful in further work on the important stability problem. However, there is still considerable room for new, rigorous results. The ultimate goal of such work could be the derivation of a simple, sufficient and not overly restrictive test for the stability of a general interpolative $\Sigma\Delta$ encoder.

## 8.3 Future Directions

A significant potential problem with $\Sigma\Delta$ encoders is the presence of tones in their output streams. Tones typically appear for small constant or sinusoidal inputs, and are particularly objectionable in audio applications where they may be clearly audible. For the single-loop encoder, it has been shown theoretically that the spectrum of the quantization noise is purely discrete for both constant and sinusoidal inputs [42], [44]. There is empirical evidence for the presence of spectral tones in the double-loop encoder output as well [123]. For ideal multi-stage encoders, spectral tones are known to be absent [43], [47]. However, gain mismatch and other differences between stages can give rise to output tones [124]. Tones may also be present in higher-order encoder outputs [124]. The problem is exacerbated by the fact that tones may be present and yet not be distinguishable in a spectral plot [124]; therefore, even a mathematical method for reliable tone detection appears wanting.

In [124] is listed a number of techniques for breaking up periodicities in $\Sigma\Delta$ encoder outputs. These center on dithering, that is, the deliberate addition of a deterministic or random signal to the encoder input signal in order to randomize the quantization noise [45], [167]. The penalty for dithering is the introduction of additional baseband noise which degrades the SNR performance. The SNR loss is reported to be on the order of 2-4 dB in [124]. Alternatively, if the analog dither can be considered known, it can be compensated for in the digital decoder at the expense of additional circuitry.

Because limit cycles correspond to periodic behavior, it appears plausible that limit cycle analysis can shed light on the tone problem in $\Sigma\Delta$ encoders. For instance, we showed in section 7.4.1.3 that if the open-loop filter of an interpolative encoder has all its poles within the unit circle, almost all limit cycles are attracting or stable. Such limit cycles will tend to persist, and possibly manifest themselves as baseband tones. Conversely, if the open-loop filter has all its poles outside the unit circle, almost all limit cycles are unstable, and will thus not be present for extended periods of time in the quantizer output of a practical circuit. In fact, we conjecture that an interpolative encoder whose filter has poles outside the unit circle will exhibit chaotic behavior [153]. This has been demonstrated rigorously in the simple case of the single-loop encoder in [168]. Such chaotic behavior may be desirable due to its lack of discernible pattern or long-term periodicity.

While the introduction of chaos may alleviate the tone problem, it still leaves important questions unanswered. One such question is the trade-off between filter pole locations and tone levels. We conjecture that as the poles move further outside the unit circle, the persistence of limit cycles will decrease, and so in a practical encoder, tones will have less likelihood of manifesting themselves for enough samples to be subjectively perceived. Another question is the effect on SNR performance of placing filter poles outside the unit circle. From a linearized point of view, the filter poles dictate the noise transfer function zeros, which can be used to control the baseband noise level [32]. Therefore, moving the filter poles away from the unit circle reduces the available noise suppression capability. A final question concerns the stability of encoders with filter poles outside the unit circle. Combining the above conjectures and observations, it appears that there is a complicated trade-off between SNR performance, tone persistence and encoder stability which can fruitfully be explored.

# Appendix A

# Appendix to Chapter 2

## A.1 Derivations for Single-Loop Encoder

Appendices A.1.1 and A.1.2 contain derivations to justify the assertions of sections 2.2.1.1 and 2.2.2.1, respectively.

### A.1.1 Transition Points

For any given number of negative bits $j$, there are in general several consecutive time steps $n$ satisfying (2.4). The largest of these time steps is the position of the $(j+1)$st negative bit, denoted by $n_{j+1}$. Thus, $n_{j+1}$ is determined by the following inequalities:

$$n_{j+1}(b - X) \geq U_0 + 2jb$$

$$(n_{j+1} - 1)(b - X) < U_0 + 2jb$$

Taken together, these two inequalities imply that the position of the $(j+1)$st negative bit is

$$n_{j+1} = \left\lceil \frac{U_0 + 2jb}{b - X} \right\rceil, \qquad 0 \leq j \leq N-1 \tag{A.1}$$

where $\lceil a \rceil$ is the smallest integer greater than or equal to $a$. To illustrate the implications of (A.1), let us assume that the initial state is $U_0 = 0$; the first bit of any codeword is then $Q(U_0) = -b$ regardless of the input. We can make the following detailed observations: For $b - \frac{2b}{N-1} < X < b$, the corresponding $N$-bit codeword consists of all positive bits except for the first one. For $X = b - \frac{2b}{N-1}$, a second negative bit appears at time step $N-1$, and as $X$ decreases, this bit advances to earlier time steps $N-2$, $N-3, \cdots$. At $X = b - \frac{4b}{N-1}$, the second negative bit reaches time step $\lceil (N-1)/2 \rceil$, and a third negative bit appears at time step $N-1$. As $X$ decreases from $b - \frac{4b}{N-1}$ to $b - \frac{6b}{N-1}$, the second bit moves from $\lceil (N-1)/2 \rceil$ to $\lceil (N-1)/3 \rceil$, and the third negative bit moves from $N-1$

to $\lceil 2(N-1)/3 \rceil$. More generally, as $X$ crosses $b - \frac{2bq}{N-1}$, the $(j+1)$st negative bit reaches time step $\lceil j(N-1)/q \rceil$ for $0 \leq j < q$, and the $(q+1)$st negative bit appears at time step $N-1$. To help visualize the process, table A.1 shows the codewords of the single-loop encoder for $N = 12$, as the input $X$ is swept from $+b$ to $-b$. Note the correspondence with Table 2.1 in section 2.2.1.1.

We can further exploit equation (A.1) to find an upper bound on the number of codewords of the single-loop encoder. From (A.1) it is clear that $n_{j+1}$ changes from one time step to the next when $X$ passes the constant input value given by

$$\frac{2jb + U_0}{b - X} = p \quad \Rightarrow \quad X = \frac{pb - (2jb + U_0)}{p}, \qquad 1 \leq p \leq N-1 \tag{A.2}$$

This input value is referred to as a transition point, since it marks a change in the $N$-bit output sequence of the encoder. For any $p$ in the permitted range of integers in (A.2), we can only allow integers $j$ that result in values of $X$ such that $X \in (-b, +b)$ and $U_0 \in (b - X, b + X)$. For instance, we see that for $U_0 = 0$,

$$-b < \frac{pb - 2jb}{p} < +b \;\Rightarrow\; 1 \leq j < p$$

More generally, the requirements $X \in (-b, +b)$, $U_0 \in (X - b, X + b)$ imply that

$$\begin{cases} 0 \leq j < p, & 1 \leq p \leq N-1 \qquad U_0 > 0 \\ 1 \leq j < p, & 1 \leq p \leq N-1 \qquad U_0 = 0 \\ 1 \leq j \leq p, & 1 \leq p \leq N-1 \qquad U_0 < 0 \end{cases} \tag{A.3}$$

The number of transition points is at most the number of permitted parameter sets $(j, p)$ in (A.3); there may be fewer transition points than that, since some parameter sets may correspond to the same transition point in (A.2). The number of codewords is the number of transition points plus one. For $U_0 = 0$ we therefore have a maximum of

$$1 + \sum_{p=1}^{N-1} (p - 1) = \frac{1}{2}(N-1)(N-2) + 1$$

codewords. More generally, the maximum number of codewords the single-loop $\Sigma\Delta$ encoder can generate is

$$C_{\max} = \begin{cases} \frac{1}{2}N(N-1) + 1 & U_0 \neq 0 \\ \frac{1}{2}(N-1)(N-2) + 1 & U_0 = 0 \end{cases}$$

Let us further consider the degenerate case where two parameter sets $(j_1, p_1)$ and $(j_2, p_2)$ lead to the same transition point. From (A.2), this occurs if and only if

$$\frac{p_1 b - (2j_1 b + U_0)}{p_1} = \frac{p_2 b - (2j_2 b + U_0)}{p_2} \quad \Rightarrow \quad \frac{U_0}{b} = 2\,\frac{j_2 p_1 - j_1 p_2}{p_2 - p_1}$$

The right hand side of this expression is rational, so degeneration can only occur when $U_0/b$ is rational. If that is not the case, all transition points are distinct, and the number of codewords is exactly $\frac{1}{2}N(N-1) + 1$.

| Transition point | 0 | 1 | 2 | 3 | 4 | 5 | 6 | 7 | 8 | 9 | 10 | 11 |
|---|---|---|---|---|---|---|---|---|---|---|---|---|
| 1 | − | + | + | + | + | + | + | + | + | + | + | + |
| 9/11 | − | + | + | + | + | + | + | + | + | + | + | − |
| 4/5 | − | + | + | + | + | + | + | + | + | + | − | + |
| 7/9 | − | + | + | + | + | + | + | + | + | − | + | + |
| 3/4 | − | + | + | + | + | + | + | + | − | + | + | + |
| 5/7 | − | + | + | + | + | + | + | − | + | + | + | + |
| 2/3 | − | + | + | + | + | + | − | + | + | + | + | + |
| 7/11 | − | + | + | + | + | + | − | + | + | + | + | − |
| 3/5 | − | + | + | + | + | − | + | + | + | + | − | + |
| 5/9 | − | + | + | + | + | − | + | + | + | − | + | + |
| 1/2 | − | + | + | + | − | + | + | + | − | + | + | + |
| 5/11 | − | + | + | + | − | + | + | + | − | + | + | − |
| 3/7 | − | + | + | + | − | + | + | − | + | + | + | − |
| 2/5 | − | + | + | + | − | + | + | − | + | + | − | + |
| 1/3 | − | + | + | − | + | + | − | + | + | − | + | + |
| 3/11 | − | + | + | − | + | + | − | + | + | − | + | − |
| 1/4 | − | + | + | − | + | + | − | + | − | + | + | − |
| 1/5 | − | + | + | − | + | − | + | + | − | + | − | + |
| 1/7 | − | + | + | − | + | − | + | − | + | + | − | + |
| 1/9 | − | + | + | − | + | − | + | − | + | − | + | + |
| 1/11 | − | + | + | − | + | − | + | − | + | − | + | − |
| 0 | − | + | − | + | − | + | − | + | − | + | − | + |
| −1/11 | − | + | − | + | − | + | − | + | − | + | − | − |
| −1/9 | − | + | − | + | − | + | − | + | − | − | + | − |
| −1/7 | − | + | − | + | − | + | − | − | + | − | + | − |
| −1/5 | − | + | − | + | − | − | + | − | + | − | − | + |
| −1/4 | − | + | − | + | − | − | + | − | − | + | − | + |
| −3/11 | − | + | − | + | − | − | + | − | − | + | − | − |
| −1/3 | − | + | − | − | + | − | − | + | − | − | + | − |
| −2/5 | − | + | − | − | + | − | − | + | − | − | − | + |
| −3/7 | − | + | − | − | + | − | − | − | + | − | − | + |
| −5/11 | − | + | − | − | + | − | − | − | + | − | − | − |
| −1/2 | − | + | − | − | − | + | − | − | − | + | − | − |
| −5/9 | − | + | − | − | − | + | − | − | − | − | + | − |
| −3/5 | − | + | − | − | − | − | + | − | − | − | − | + |
| −7/11 | − | + | − | − | − | − | + | − | − | − | − | − |
| −2/3 | − | + | − | − | − | − | − | + | − | − | − | − |
| −5/7 | − | + | − | − | − | − | − | − | + | − | − | − |
| −3/4 | − | + | − | − | − | − | − | − | − | + | − | − |
| −7/9 | − | + | − | − | − | − | − | − | − | − | + | − |
| −4/5 | − | + | − | − | − | − | − | − | − | − | − | + |
| −9/11 | − | + | − | − | − | − | − | − | − | − | − | − |

Table A.1: *Single-loop encoder: Codewords and transition points for* $N = 12$ *and* $U_0 = 0$. *As* $X$ *crosses a transition point from above, the output sequence changes from the codeword on the line above to the one on the line containing the transition point. Transition points are normalized by* $b$, *and bits are shown as* $+$ *and* $-$.

### A.1.2 Interval Widths

We assume $U_0 = 0$. Setting $p = N - 1$ in (2.5), we see that there are transition points at

$$X = b\left(1 - \frac{2j}{N-1}\right), \qquad 1 \leq j \leq N-1$$

Thus no quantization interval can be larger than $2b/(N-1)$. On the other hand, the largest transition point is obtained by setting $(j,p) = (1, N-1)$ in (2.5), since $b - X = 2jb/p$ is minimized by this choice. Therefore the interval near $X = +b$ has width $d_1 = 2b/(N-1)$, that is, it is a quantization interval of maximal width. The same holds for the quantization interval near $X = -b$, so $d_1 = d_C$.

Consider next the intervals close to $X = 0$. Equation (2.5) shows that $(j,p) = (1,2)$ leads to a transition point at $X = 0$ for any sample size $N \geq 3$. The smallest positive transition point is found by maximizing the denominator and minimizing the denominator of (2.5) while keeping it positive. This leads to $(j,p) = ((N-2)/2, N-1)$ and $((N-3)/2, N-1)$ for $N$ even and $N$ odd, respectively, and the interval widths are $b/(N-1)$ and $2b/(N-1)$. By symmetry, the same holds on the other side of $X = 0$. We thus have $d_{C/2} = d_{C/2+1} \geq b/(N-1)$.

Consider now two pairs $(j_1, p_1)$ and $(j_2, p_2)$ satisfying (A.3). The distance between the corresponding transition points is

$$\frac{p_1 - 2j_1 b}{p_1} - \frac{p_2 - 2j_2 b}{p_2} = 2b\frac{p_1 j_2 - p_2 j_1}{p_1 p_2}$$

If $p_1 = p_2$, the smallest absolute distance is $2b/(N-1)$. If $p_1 \neq p_2$, the largest product of $p_1$ and $p_2$ is $(N-1)(N-2)$, and the smallest positive value of the numerator is 1. These extremes are simultaneously achieved by choosing $(j_1, p_1) = (1, N-1)$ and $(j_2, p_2) = (1, N-2)$. Therefore the width of the shortest quantization interval is $d_{\min} = 2b/[(N-1)(N-2)]$, and one interval of this minimal width is adjacent to the interval near $X = +b$. Another interval of the same width is located near $X = -b$, so $d_{C-1} = d_2 = d_{\min}$.

Finally, let us assume that $U_0 > 0$. The largest transition point in (2.5) occurs at $(j,p) = (0, N-1)$. The quantization interval near $X = +b$ thus has width $d_1 = U_0 b/(N-1)$ which is of order $O(N^{-1})$.

## A.2 Derivations for Double-Loop Encoder

Appendices A.2.1 and A.2.2 contain derivations to justify the assertions of sections 2.3.1.1 and 2.3.2.1, respectively.

### A.2.1 Transition Points

Equation (2.15) can be rewritten

$$U_n = U_0 + \sum_{i=1}^{n}(U_i - U_{i-1}) = U_0 + \sum_{i=0}^{n-1} V_i - \sum_{i=0}^{n-1} Q(U_i) \qquad \text{(A.4)}$$

$$V_n = V_0 + \sum_{j=1}^{n}(V_j - V_{j-1}) = V_0 + nX - \sum_{j=1}^{n} Q(U_j) \qquad \text{(A.5)}$$

Summing the left hand side of equation (A.5) we get

$$\sum_{i=0}^{n-1} V_i = nV_0 + \frac{1}{2}n(n-1)X - \sum_{i=1}^{n-1}(n-i)Q(U_i), \quad n \geq 1 \qquad \text{(A.6)}$$

Inserting (A.6) in (A.4) we find

$$U_n = U_0 + nV_0 + \frac{1}{2}n(n-1)X - Q(U_0) - \sum_{i=1}^{n-1}(n-i+1)Q(U_i), \quad n \geq 1 \quad \text{(A.7)}$$

For the single-loop encoder, we now used the fact (2.3). Unfortunately, a similar fact does not hold for the double-loop encoder, which necessitates a less detailed manipulation of (A.7). Its last two terms involving quantizer outputs are more conveniently written

$$Q(U_0) + \sum_{i=1}^{n-1}(n-i+1)Q(U_i) = \frac{1}{2}n(n+1)b - 2jb \qquad \text{(A.8)}$$

where $j$ is an integer in the range $0 \leq j \leq \frac{1}{2}n(n+1)$. The extreme sequences consisting of all $+b$'s or all $-b$'s correspond to $j = 0$ and $j = N-1$, respectively. There are values of $j$ between 0 and $N-1$ for which no $N$-bit codeword exists, but for any codeword there exists a $j$ between 0 and $N-1$. Inserting (A.8) in (A.7), we find

$$U_n = U_0 + nV_0 + \frac{1}{2}n(n-1)X - \frac{1}{2}n(n+1)b + 2jb \qquad \text{(A.9)}$$

The quantizer transition points are the values of $X \in (-b, +b)$ which result in $U_n = 0$ at some time step $n$ between 1 and $N-1$. Unfortunately, the variable $j$ in (A.9) has no simple interpretation as in the single-loop case. Hence we can only solve (A.9) for values of $X$ that *might be* transition points. Counting the number of parameter sets $(j, n)$ for which $U_n = 0$ has a solution $X \in (-b, +b)$ will result in an upper bound on the number of transition points. Setting $U_n$ in equation (A.9) to zero, we obtain

$$X = \frac{\frac{1}{2}n(n+1)b - (2jb + U_0 + nV_0)}{\frac{1}{2}n(n-1)}, \qquad 2 \leq n \leq N-1 \qquad \text{(A.10)}$$

For $n = 0$ and $n = 1$ there are no transition points, since $U_0$ and $U_1$ are independent of $X$. Equation (A.10) is analogous to equation (2.5) for the single-loop encoder.

We now determine an upper bound on the number of codewords. Considering that $X$ must be in the range $(-b, +b)$, it follows from (A.10) that $j$ must satisfy

$$\frac{n - (U_0 + nV_0)/b}{2} < j < \frac{n^2 - (U_0 + nV_0)/b}{2} \tag{A.11}$$

The number of integer values of $j$ satisfying (A.11) is $\leq \frac{1}{2}(n^2 - n)$. Therefore, the number of codewords is upper bounded by

$$C_{\max} \leq 1 + \sum_{n=2}^{N-1} \frac{1}{2}(n^2 - n) = \frac{1}{6}N(N-1)(N-2) + 1, \qquad N \geq 3$$

### A.2.2 Interval Widths

We assume for convenience that the initial states are $U_0 = V_0 = 0$, and show that the largest positive transition point occurs at $X = (N-4)b/(N-2)$. Equations (2.15) together with the initial states imply that

$$Q(U_0) = -b \qquad U_1 = +b \qquad V_1 = X - b$$

$$Q(U_1) = +b \qquad U_2 = X - b < 0 \qquad Q(U_2) = -b$$

This states that the first three output bits are $(-b, +b, +b)$ regardless of the input. At times $n \geq 3$, the quantizer produces only positive bits until $U_n$ becomes negative—the closer $X$ is to $+b$, the longer it takes to generate the first negative bit. We will find the value of $X$ which results in the first negative bit appearing at time $N-1$; this input corresponds to the most positive transition point. Assuming $Q(U_3) = \cdots = Q(U_{N-2}) = +b$, we find from (A.7) that

$$U_{N-1} = \frac{1}{2}(N-1) \cdot [(N-2)X - (N-4)b]$$

Setting $U_{N-1}$ to zero, we conclude that the most positive transition point is at $X = (N-4)b/(N-2)$. The corresponding quantization interval has width $2b/(N-2)$. By symmetry, a similar interval exists from $-b$ to $-(N-4)b/(N-2)$.

## A.3 Proofs of Theorems

### A.3.1 Proof of Theorem 1

From 2.24, the number of negative bits $A$ on an $N$-bit codeword is given by given by

$$A = N - P + \Theta$$

The averaging decoder estimate is therefore

$$\hat{X} = \frac{(N-A)-A}{N}b = -b+(P-\Theta)\Delta$$

so

$$X-\hat{X} = \Theta\Delta - R \qquad \text{(A.12)}$$

On the other hand, we can use (2.20) in (2.21) to show that

$$X = -b+(P-\Theta)\Delta + \frac{U_N-U_0}{N}$$

so

$$X-\hat{X} = \frac{U_N-U_0}{N}$$

Since $U_0$ is uniformly distributed by assumption, and (2.21) can be shown to imply that $U_N$ is also uniformly distributed, we have $E[(X-\hat{X}) \mid X]=0$. By (A.12), this implies

$$0 = E[(\Theta\Delta - R) \mid X] = \Delta \cdot P[\Theta=1 \mid X] - R$$

which shows the theorem.

### A.3.2 Proof of Theorem 2

Using (A.12),

$$\begin{aligned} E\left[(X-\hat{X})^2 \mid X\right] &= \sum_{\theta=0}^{1} P[\Theta=\theta]\cdot(\theta\Delta - R)^2 \\ &= \left(1-\frac{R}{\Delta}\right)(-R)^2 + \frac{R}{\Delta}(\Delta-R)^2 \\ &= R(\Delta-R) \end{aligned}$$

### A.3.3 Proof of Theorem 3

The random variable $R$ is uniformly distributed on $[0,\Delta)$, so $ER = \frac{\Delta}{2}$ and $ER^2 = \frac{\Delta^2}{3}$. Now

$$\begin{aligned} E\left[(X-\hat{X})^2\right] &= E\left[E\left[(X-\hat{X})^2 \mid X\right]\right] \\ &= E[R(\Delta-R)] \\ &= \frac{\Delta}{2}\cdot\Delta - \frac{\Delta^2}{3} \\ &= \frac{\Delta^2}{6} \end{aligned}$$

# Appendix B

# Appendix to Chapter 3

## B.1 Performance Measures

### B.1.1 Ideal Conditions

The number of possible codewords as $X$ sweeps over the dynamic range $D$ is denoted by $C$, and the decoder estimate as a function of the random variable $X$ is denoted by $\hat{X}$. The decoder estimate of the input giving rise to the $i$th codeword is denoted by $\hat{X}_i$; the estimate $\hat{X}_i$ is found as the average of the best interval bounds determined by the zoomer. Finally, the interval corresponding to the $i$th codeword is denoted by $I_i$, and its width by $d_i$.

We assume that the constant input $X$ is uniformly distributed on the dynamic range $D$. This is an analytical convenience, but it is not cardinal to our technique. Any piecewise continuous probability density function for $X$ on $D$ can be incorporated in the analysis below. The performance measures are defined as follows: The MSE is given by

$$\text{MSE} \triangleq E\left[(X - \hat{X})^2\right] \tag{B.1}$$

To obtain a more operational expression for the MSE, we write the contribution from the $i$th interval as

$$\text{MSE}_i = E\left[(X - \hat{X})^2 \mid X \in I_i\right] = \sum_{i=1}^{C} \frac{d_i^2}{12} \tag{B.2}$$

The total MSE is found by taking the weighted sum of these errors,

$$\text{MSE} = \sum_{i=1}^{C} \frac{d_i}{|D|} \cdot \text{MSE}_i = \sum_{i=1}^{C} \frac{d_i^3}{24Kb} \tag{B.3}$$

where $|D|$ is the width of the dynamic range. The average input power is

$$E\left[X^2\right] = \int_{-Kb}^{+Kb} \frac{1}{2Kb} x^2 \, dx = \frac{(Kb)^2}{3} \tag{B.4}$$

Defining $\text{SNR} = 10\log_{10}\left[\frac{E(X^2)}{\text{MSE}}\right]$, we thus have

$$\text{SNR} = 10\log_{10}\frac{8(Kb)^3}{\sum_{i=1}^{C} d_i^3} \tag{B.5}$$

The worst-case error is another important performance measure for non-uniform quantizers such as $\Sigma\Delta$ modulators because it specifies the local performance, as opposed the MSE which is a global or average measure. The worst-case error is defined as

$$\varepsilon \triangleq \max_{X\in D}|X-\hat{X}| = \max_{1\le i\le C}\left\{\frac{d_i}{2}\right\} \tag{B.6}$$

where the last equality follows because the zoomer chooses the quantization interval midpoints as its estimates. The Worst-Case Resolution (WCR) in bits is defined by

$$R = \log_2\frac{|D|}{2\varepsilon} = \log_2\frac{Kb}{\varepsilon} \tag{B.7}$$

### B.1.2 Non-Ideal Conditions

When non-idealities are present, it is necessary to generalize the expressions for MSE and worst-case resolution given in appendix B.1.1 which are derived under the assumption that the original zoomer estimates are the midpoints of the corresponding intervals. We first introduce some additional notation. We use the term "actual" to refer to characteristics of a non-ideal encoder, and we denote the actual input interval corresponding to the $i$th codeword of a non-ideal encoder by $I_i$, and its width by $d_i$. We denote the actual lower and upper bounds on the $i$th interval by $\ell_i$ and $u_i$, respectively. Under ideal circumstances, $L$ and $U$ in the zoomer algorithm correspond exactly to $\ell_i$ and $u_i$, and $\hat{X}_i = (\ell_i + u_i)/2$. Finally, we define $q_i = \hat{X}_i - \ell_i$, $r_i = u_i - \hat{X}_i$ as the signed distances from the original decoder estimate to the edges of the actual interval. If the original zoomer estimate $\hat{X}_i$ is outside of the corresponding actual interval, one of these distances is negative.

The general definitions (B.1) and (B.6) of MSE and worst-case error remain valid, but the MSE contribution from the $i$th quantization interval, previously given by (B.2), is modified to

$$\text{MSE}_i = E\left[(X-\hat{X})^2 \mid X \in I_i\right] = \sum_{i=1}^{C}\frac{q_i^3 + r_i^3}{3(q_i + r_i)}$$

The worst-case error, previously given by (B.6), changes into

$$\varepsilon = \max_{1\le i\le C}\{\max\{q_i, r_i\}\}$$

These generalizations are reflected in an obvious way in the expressions (B.5) and (B.7).

## B.2 Only Codewords Yield Compatible Bounds

We show by contradiction that all non-codewords result in incompatible bounds. Assume that a non-codeword gives rise to zoomer bounds $L' < U'$. Choose any input satisfying $L' < X' < U'$, and determine the corresponding codeword. Since the codeword and the non-codeword are different, we can find the earliest time step $n_0$ at which they differ. The zoomer bounds $\bar{X}'_n$ will be same for the two sequences at all times $n \leq n_0$, so in particular, the best bounds at time $n$, denoted by $L'_n$ and $U'_n$, will be the same up to time $n_0$. At time $n_0$, the new bound $\bar{X}'_{n_0}$ will be a lower bound when decoding one of the sequences, and an upper bound when decoding the other. The new bound is either outside or inside the interval $(L'_{n_0-1}, U'_{n_0-1})$. If it is outside, the new bound cannot be consistent with decoding both sequences. Since the codeword decoding is consistent by definition, the decoding of the non-codeword must be inconsistent. If the new bound is inside the interval limited by the best bounds at time $n_0 - 1$, it will be consistent with these bounds when decoding both sequences, but at time $n_0$, the decoding results in the interval splitting into two disjoint intervals, $(L'_{n_0-1}, \bar{X}'_{n_0})$ and $(\bar{X}'_{n_0}, U'_{n_0-1})$. It is thus impossible that the two sequences are decoded to the same interval. In both cases, the assumptions are violated.

## B.3 Form of Non-Idealities

The discrete-time integrator in Figure 3.1 in practice has a transfer function that can be modeled as

$$g \, \frac{z^{-1}}{1 - \alpha z^{-1}} \tag{B.8}$$

where $g$ is the gain and $\alpha$ is the leak factor; both these are nominally 1.

The gain is usually determined as a capacitor ratio. The leak $\alpha$ in (B.8) is the result of finite operational amplifier (op-amp) gain; if the integrator op-amp gain is $A$, the leak is given by

$$\alpha \;=\; A/(A+1) \approx 1 - 1/A$$

Because we are considering constant inputs or very low-bandwidth applications, large-gain op-amps are probably more easily attainable than in signal acquisition applications.

The zoomer algorithm assumes that the initial integrator states $U_0$, $V_0$ *etc.* are set to zero at the beginning of each conversion cycle. In practice, this can only be achieved with finite precision.

In practice, the 1-bit quantizer $Q$ in (3.1) may not switch between the output values $-b$ and $+b$ at exactly the input value $U = 0$. The 1-bit quantizer is more precisely characterized by a DC offset $q_0$:

$$Q(U) = \begin{cases} -b & \text{if } U \leq q_0 \\ +b & \text{if } U > q_0 \end{cases} \tag{B.9}$$

Real quantizers may also exhibit hysteresis, as in

$$Q_{\text{hyst}}(U_n) = Q[U_n + h_0 Q_{\text{hyst}}(U_{n-1})]$$

where $h_0$ is a measure of the hysteresis effect. Note that $h_0$ introduces additional memory in the system. Simulations for the single-loop encoder indicate that the performance degradation in the presence of hysteresis is close to the degradation seen for non-zero initial states. Compensating the zoomer for known hysteresis is simple. Hysteresis is not further treated in this chapter.

In practice, circuit and external noise must be taken into account when evaluating the performance of decoders. We consider noisy inputs of the form

$$X_i = X + N_i \tag{B.10}$$

where $X$ is a constant, and $\{N_i\}$ is a sequence of independent random variables uniformly distributed on the interval $(-Mb, +Mb)$. $M$ is a constant. The optimal procedure in the presence of this type of noise is difficult to determine analytically. However, a simple way to take noise into account is described in section 3.3.2.5.

For large sample sizes, a slowly varying input signal looks approximately constant, but a more appropriate description of the signal includes a linear or first order term of the form

$$X_i = X_{\text{mid}} + \left(i - \frac{N-1}{2}\right) sb, \quad 0 \leq i \leq N-1$$

where $s$ is the slope, and $X_{\text{mid}}$ is the average input. The signal varies by a total amount of $(N-1)sb$ as $i$ ranges from 0 to $N-1$.

# Appendix C

# Appendix to Chapter 4

## C.1 Energy Concentration

We first consider the energy concentration of the SVD-based method given by (4.6), and its corresponding minimum-energy band-limited extrapolation method (4.5). The energy concentration for the SVD-based method is

$$C_{\mathrm{SVD}}(\mathbf{s}_\Lambda) \triangleq \frac{\|\mathbf{b}_{\mathrm{SVD}}\|_\Lambda^2}{\|\mathbf{b}_{\mathrm{BLX}}\|_{\mathcal{Z}}^2} = \frac{\sum_{n=0}^{r-1}(\mathbf{u}_n^T\mathbf{s}_\Lambda)^2}{\sum_{n=0}^{r-1}(\mathbf{u}_n^T\mathbf{s}_\Lambda)^2/\sigma_n^2}$$

where we have used the facts that the DPSSs $\{\mathbf{U}_n\}$ are orthogonal over $\mathcal{Z}$, and that [96]

$$\|\mathbf{U}_n\|_{\mathcal{Z}}^2 = 1/\sigma_n^2$$

Recall that $1 > \sigma_0^2 > \sigma_1^2 > \cdots > \sigma_{N-1}^2 > 0$. A variational argument shows that $C_{\mathrm{SVD}}(\mathbf{s}_\Lambda)$ is minimized by choosing $\mathbf{s}_\Lambda$ to be proportional to $\mathbf{u}_{r-1}$, so that

$$\min_{\mathbf{s}_\Lambda} C_{\mathrm{SVD}}(\mathbf{s}_\Lambda) = \sigma_{r-1}^2$$

Consider now the general method (4.7) with the same dimension $r$ as in (4.6). We rewrite (4.7) as

$$\begin{aligned}
\mathbf{b}_{\mathrm{general}} &= \sum_{n=0}^{r-1}(\mathbf{a}_n^T\mathbf{s}_\Lambda)\cdot\mathbf{a}_n \\
&= \sum_{n=0}^{r-1}(\mathbf{a}_n^T\mathbf{s}_\Lambda)\cdot\sum_{m=0}^{N-1}(\mathbf{u}_m^T\mathbf{a}_n)\cdot\mathbf{u}_m \\
&= \sum_{m=0}^{N-1}\left(\sum_{n=0}^{r-1}(\mathbf{a}_n^T\mathbf{s}_\Lambda)\cdot(\mathbf{u}_m^T\mathbf{a}_n)\right)\cdot\mathbf{u}_m
\end{aligned}$$

By (4.5), the minimum-energy band-limited extrapolate of $\mathbf{b}_{\text{general}}$ is

$$\mathbf{b}_{\text{BLX,general}} = \sum_{m=0}^{N-1} \left( \sum_{n=0}^{r-1} (\mathbf{a}_n^T \mathbf{s}_\Lambda) \cdot (\mathbf{u}_m^T \mathbf{a}_n) \right) \cdot \mathbf{U}_m$$

The energy concentration (4.7) as a function of $\mathbf{s}_\Lambda$ is defined in (4.8). We will show that

$$\min_{\mathbf{s}_\Lambda} C_{\text{general}}(\mathbf{s}_\Lambda) \leq \sigma_{r-1}^2$$

that is, no band limitation method of the form (4.7) has a larger minimum energy concentration than the SVD method. To show this, we choose $\mathbf{s}_\Lambda$ to be an arbitrary non-zero vector $\mathbf{s}_0$ in the intersection

$$I \triangleq \text{span}(\mathbf{u}_{r-1}, \mathbf{u}_r, \cdots, \mathbf{u}_{N-1}) \bigcap \text{span}(\mathbf{a}_0, \mathbf{a}_1, \cdots, \mathbf{a}_{r-1})$$

Since $I$ is defined as the intersection of an $(N-r+1)$-dimensional and an $r$-dimensional linear subspace of a $N$-dimensional space, $I$ must have at least dimension 1 and thus must contain non-zero elements. For an arbitrary non-zero $\mathbf{s}_0 \in I$, we then have $\mathbf{b}_{\text{general}} = \mathbf{s}_0$ by construction. It also follows by the definition of $I$ that

$$\mathbf{b}_{\text{general}} = \sum_{n=r-1}^{N-1} (\mathbf{u}_n^T \mathbf{s}_0) \cdot \mathbf{u}_n$$

because $\mathbf{u}_n^T \mathbf{s}_0 = 0$ for $0 \leq n < r-1$. The minimum-energy band-limited extrapolate of $\mathbf{b}_{\text{general}}$ is given by

$$\mathbf{b}_{\text{BLX,general}} = \sum_{n=r-1}^{N-1} (\mathbf{u}_n^T \mathbf{s}_0) \cdot \mathbf{U}_n$$

so

$$\begin{aligned} \min_{\mathbf{s}_\Lambda} C_{\text{general}}(\mathbf{s}_\Lambda) &\leq C_{\text{general}}(\mathbf{s}_0) \\ &= \frac{\sum_{n=r-1}^{N-1} (\mathbf{u}_n^T \mathbf{s}_0)^2}{\sum_{n=r-1}^{N-1} (\mathbf{u}_n^T \mathbf{s}_0)^2 / \sigma_n^2} \\ &\leq \frac{\sum_{n=r-1}^{N-1} (\mathbf{u}_n^T \mathbf{s}_0)^2}{\frac{1}{\sigma_{r-1}^2} \sum_{n=r-1}^{N-1} (\mathbf{u}_n^T \mathbf{s}_0)^2} \\ &= \sigma_{r-1}^2 \end{aligned}$$

## C.2 Resolution of the DFT

Consider the DFT of an $N$-point sequence $\mathbf{x} = \{x_0, x_1, \cdots, x_{N-1}\}$ defined by

$$X_k = \sum_{n=0}^{N-1} x_n \exp\left(j\frac{2\pi}{N}kn\right), \quad k \in \Lambda = \{0, 1, \cdots, N-1\}$$

To demonstrate an approximate limit on the SNR of the DFT, we will use an input sequence which is a pure complex exponential at a non-bin baseband frequency,[1]

$$x_n = \exp(j\omega_0 n), \quad n \in \Lambda, \quad 0 < \omega_0 < \frac{\pi}{OSR} \text{ and } \omega_0 \neq \frac{2\pi}{N}k, \quad k \in \Lambda \quad \text{(C.1)}$$

The DFT of the input signal (C.1) is

$$\begin{aligned} X_k &= \sum_{n=0}^{N-1} x_n \exp\left(j\frac{2\pi}{N}kn\right) \\ &= \frac{1 - \exp(jN\omega_0)}{1 - \exp\left(\omega_0 - \frac{2\pi}{N}k\right)}, \quad k \in \Lambda \end{aligned}$$

and the squared magnitudes of the DFT coefficients are

$$|X_k|^2 = \frac{1 - \cos(N\omega_0)}{1 - \cos\left(\omega_0 - \frac{2\pi}{N}k\right)}$$

The fact that the coefficients are non-zero even at high frequencies is referred to as spectral leakage [110]. We assume that OSR $\gg 1$ and $N \gg$ OSR. The signal power equals $E_{\text{signal}} = N^2$. Neglecting details such as whether $N/(2\text{OSR})$ is an integer, the error power equals

$$\begin{aligned} E_{\text{error}} &= \sum_{k=N/(2\text{OSR})}^{N-N/(2\text{OSR})} |X_k|^2 \\ &\approx 2\int_{N/(2\text{OSR})}^{N/2} \frac{1 - \cos(N\omega_0)}{1 - \cos\left(\omega_0 - \frac{2\pi}{N}x\right)} dx \\ &= \frac{N}{\pi}\left[1 - \cos(N\omega_0)\right] \cdot \left[-\cot\left(\frac{\pi}{N}x - \frac{\omega_0}{2}\right)\right]_{N/(2\text{OSR})}^{N/2} \\ &\approx \frac{2N \cdot \text{OSR}}{\pi^2} \cdot \left(1 - \cos\frac{N\omega_0}{2}\right) \end{aligned}$$

where we have assumed $\omega \ll \pi/\text{OSR}$ so that the signal frequency is safely within baseband. By choosing $\omega_0$ to be some small odd multiple of $\pi/N$, we

[1] The derivations are facilitated by using a complex exponential as the input, rather than its real part.

find that the SNR can be made as small as

$$\mathrm{SNR}_{\mathrm{min}} \approx 10\log_{10}\left(\frac{N\pi^2}{4\mathrm{OSR}}\right)$$

which is the desired result.

## C.3 Effect of Windowing

Consider an input sequence at a baseband bin frequency,

$$x_n = \exp(j\omega_0 n), \qquad n \in \Lambda, \qquad 0 < \omega_0 < \frac{\pi}{OSR} \text{ and } \omega_0 = \frac{2\pi}{N}k_0$$

where $k_0 \in \{0, 1, \cdots, N/\mathrm{OSR}\}$. The DFT coefficients of this sequence are

$$X_k = \begin{cases} N & \text{for } k = k_0 \\ 0 & \text{otherwise} \end{cases}$$

Applying the Hanning window [110], the windowed DFT coefficients are

$$X'_k = \begin{cases} \frac{1}{2}N & \text{for } k = k_0 \\ -\frac{1}{4}N & \text{for } k = k_0 \pm 1 \\ 0 & \text{otherwise} \end{cases}$$

The error power between $\{X_k\}$ and $\{X'_k\}$ equals $\frac{3}{8}N^2$, so the SNR equals $10\log_{10}\frac{8}{3} = 4.3$ dB independent of OSR and sample size. The SNR can be improved slightly by introducing a scaling factor $\beta$ and using $\beta\{X'_k\}$ as the estimate of $\{X_k\}$. In this case, the optimal choice of $\beta$ is 4/3, which results in an SNR of $10\log_{10}3 = 4.8$ dB. Similar calculations can be made for other windows. The conclusion is that windowing introduces large signal frequency errors.

# Appendix D

# Appendix to Chapter 5

## D.1 Quadratic Programming Approximation

We first rewrite (5.1) by partitioning all vectors and matrices into blocks of $L$ samples, where $L$ divides $N$. Specifically, we set $P = N/L$ and

$$\begin{aligned}
\mathbf{x} &= \{\mathbf{x}_1, \cdots, \mathbf{x}_P\} \\
\mathbf{y} &= \{\mathbf{y}_1, \cdots, \mathbf{y}_P\} \\
\mathbf{z} &= \{\mathbf{z}_1, \cdots, \mathbf{z}_P\} \triangleq \mathbf{Zs} \\
\mathbf{Q} &= \text{diag}(\mathbf{Q}_1, \cdots, \mathbf{Q}_P\} \\
\mathbf{H} &= \begin{bmatrix} \mathbf{H}_1 & 0 & \cdots & 0 & 0 \\ \mathbf{H}_2 & \mathbf{H}_1 & \cdots & 0 & 0 \\ \vdots & \vdots & & \vdots & \vdots \\ \mathbf{H}_{P-1} & \mathbf{H}_{P-2} & \cdots & \mathbf{H}_1 & 0 \\ \mathbf{H}_P & \mathbf{H}_{P-1} & \cdots & \mathbf{H}_2 & \mathbf{H}_1 \end{bmatrix}
\end{aligned}$$

Then (5.1) can be written

$$\mathbf{Q}_n \sum_{i=1}^{n} \mathbf{H}_{n-i+1}\mathbf{x}_i \le \mathbf{Q}_n \left[ \sum_{i=1}^{n} \mathbf{H}_{n-i+1}\mathbf{y}_i - \mathbf{z}_n \right], \quad n = 1, 2, \cdots, P \tag{D.1}$$

and the objective is to find the sequence $\hat{\mathbf{x}}$ that satisfies (D.1) and minimizes

$$d^2 = \sum_{n=1}^{P} \|\hat{\mathbf{x}}_n - \mathbf{x}_n\|_2^2$$

We can solve this $N$-dimensional QP problem approximately by solving the following $N/L$ $L$-dimensional QP problems in sequence: Minimize $\|\hat{\mathbf{x}}_n - \mathbf{x}_n\|_2^2$

subject to

$$\mathbf{Q}_n\mathbf{H}_1\mathbf{x}_n \leq \mathbf{Q}_n \left[\sum_{i=1}^{n-1} \mathbf{H}_{n-i+1}(\mathbf{y}_i - \hat{\mathbf{x}}_i) + \mathbf{H}_1\mathbf{y}_n - \mathbf{z}_n\right], \quad n = 1, 2, \cdots, P$$

To avoid large errors at the beginnings of $L$-dimensional QP sub-blocks, we can overlap the QP sub-blocks so that each sub-block continues into the immediately following sub-block. Specifically, we denote the length of the $n$th QP sub-block by $\lambda(n)$, where

$$\lambda(n) \triangleq \begin{cases} L & \text{for } n \neq N/L \\ \ell & \text{for } n = N/L \end{cases}$$

and we define the index set over which the $n$th QP problem is solved by

$$\Lambda(n) \triangleq \{(n-1)\ell, \cdots, (n-1)\ell + \lambda(n) - 1\}$$

We require that the length $\ell$ divides $N$, and we solve $N/\ell$ QP problems of which all but one have dimension $L$, and the last one has dimension $\ell$. The last one is shorter than the others because we only have $N$ and not $N + L - \ell$ samples available. We use the index set $\Lambda(n)$ as a subscript of sequences to denote the index limitation to $\Lambda(n)$; for instance, $\mathbf{x}_{\Lambda(n)} = \{x_{(n-1)\ell}, \cdots, x_{(n-1)\ell+\lambda(n)-1}\}$. We also use $\Lambda(n)$ as a subscript for matrices to denote the index limitation in both dimensions to $\Lambda(n)$; for instance,

$$\mathbf{Q}_{\Lambda(n)} = \text{diag}(y_{\Lambda(n)}) = \{y_{(n-1)\ell}, \cdots, y_{(n-1)\ell+\lambda(n)-1}\}$$

From the solution of each QP sub-problem, we only keep the $\ell$ first samples, that is, we set $\hat{\mathbf{x}}_n$ equal to the first $\ell$ samples of $\mathbf{x}_{\Lambda(n)}$. We thus solve the following $N/\ell$ QP problems in sequence: Minimize $\|\hat{\mathbf{x}}_{\Lambda(n)} - \mathbf{x}_{\Lambda(n)}\|_2^2$ subject to

$$\mathbf{Q}_{\Lambda(n)}\mathbf{H}_{\Lambda(n)}\mathbf{x}_{\Lambda(n)} \leq \mathbf{Q}_{\Lambda(n)} \left[\sum_{i=1}^{n-1} \mathbf{H}_{n-i+1}(\mathbf{y}_i - \hat{\mathbf{x}}_i) + \mathbf{H}_{\Lambda(n)}\mathbf{y}_{\Lambda(n)} - \mathbf{z}_{\Lambda(n)}\right]$$

where $n = 1, 2, \cdots, N/\ell$.

# Appendix E

# Appendix to Chapter 6

## E.1 Stability of Error Diffusion Encoders

In this appendix we show that the quantization error in an error diffusion encoder is bounded to $[-L_{\text{mid}}, +L_{\text{mid}}]$, provided that

- The input mask is a Kronecker $\delta$-function.
- The absolute sum of the output mask coefficients is less than or equal to 1.
- All input samples are in the range $[0, L_{\text{high}}]$.

Using the same technique, we then show that if an input sample $x(n_1, n_2)$ equals zero, the corresponding output sample $y(n_1, n_2)$ also equals zero.

We show the claims by induction. We first note that if $(n_1, n_2) \in R_{\text{BC}}$, then $e(n_1, n_2) = 0$ which belongs to the interval $[L_{\text{mid}} - L_{\text{high}}, L_{\text{mid}}] = [-L_{\text{mid}}, +L_{\text{mid}}]$. We next establish the induction step. Consider the evaluation of a particular quantizer input sample $u(n_1, n_2)$. The largest possible value of $u(n_1, n_2)$ occurs if the corresponding input sample $x(n_1, n_2)$ is as large as possible, and if the contributions of the quantization errors through $G(z_1, x_2)$ are as large as possible. Thus, in an obvious notation,

$$
\begin{aligned}
u(n_1, n_2) &= x(n_1, n_2) + \sum_{m_1, m_2} \text{output mask coeff.} \cdot e(m_1, m_2) \\
&\leq \max x(n_1, n_2) + \max_{m_1, m_2} |e(m_1, m_2)| \cdot \sum_{m_1, m_2} |\text{output mask coeff.}| \\
&= L_{\text{high}} + L_{\text{mid}} \qquad \text{(E.1)}
\end{aligned}
$$

where we have used the upper bound on $x(n_1, n_2)$, the upper and lower bounds on the quantization error and the upper bound on the absolute sum of the output mask coefficients. Similarly, we find that

$$\begin{aligned} u(n_1, n_2) &= x(n_1, n_2) + \sum_{m_1, m_2} \text{output mask coeff.} \cdot e(m_1, m_2) \\ &\geq \min x(n_1, n_2) - \max_{m_1, m_2} |e(m_1, m_2)| \cdot \sum_{m_1, m_2} |\text{output mask coeff.}| \\ &= -L_{\text{mid}} \end{aligned} \tag{E.2}$$

Combining (E.1) and (E.2), we get

$$u(n_1, n_2) \in [-L_{\text{mid}}, L_{\text{high}} + L_{\text{mid}}] \tag{E.3}$$

Now for all quantizer inputs in this range,

$$e(n_1, n_2) = Q(u) - u \in [-L_{\text{mid}}, +L_{\text{mid}}] \tag{E.4}$$

which establishes the induction step and shows that (E.4) is valid for all pairs $(n_1, n_2)$ under the stated assumptions.

Consider next the case of a zero input sample, $x(n_1, n_2) = 0$, as occurs for any pair $(n_1, n_2) \notin R_x$. The upper bound (E.1) on $u(n_1, n_2)$ can then be strengthened to

$$u(n_1, n_2) \leq L_{\text{mid}} \tag{E.5}$$

As the quantizer output is zero for any quantizer input less than or equal to $L_{\text{mid}}$, equation (E.5) establishes the claim that $y(n_1, n_2) = 0$. In the absence of this result, the halftone output bits $y(n_1, n_2)$ outside of $R_x$ would carry information necessary to derive bounds on the input image from (6.3), and thus it would not be possible to perform the space-domain projection using only $\{y(n_1, n_2)\}$ over $R_x$.

## E.2 Two-Dimensional SVD Band Limitation

In this appendix we describe the generalization to two dimensions of the SVD-based band limitation method presented in chapter 4 for one-dimensional signals. As in that chapter, we denote the one-dimensional truncated Discrete Prolate Spheroidal Sequences (DPSSs) corresponding to a particular OverSampling Ratio (OSR) and sample size $N$ by $\{\mathbf{u}_0, \cdots, \mathbf{u}_{N-1}\}$. The SVD transform $\{X(m_1, m_2)\}$ of an image $\{x(n_1, n_2)\}$ is then given by

$$X(m_1, m_2) = \sum_{n_1=0}^{N-1} \sum_{n_2=0}^{N-1} x(n_1, n_2) u_{m_1}(n_1) u_{m_2}(n_2) \tag{E.6}$$

with the inversion formula

$$x(n_1, n_2) = \sum_{m_1=0}^{N-1} \sum_{n_2=0}^{N-1} X(m_1, m_2) u_{m_1}(n_1) u_{m_2}(n_2) \tag{E.7}$$

The image $\{x(n_1, n_2)\}$ can be band-limited by restricting the summation limits in (E.7). In the special case of a separable band limitation, that is, if the summation limits on $m_1$ and $m_2$ are independent, the image $\{x(n_1, n_2)\}$ can be band-limited successively in the two dimensions as follows:

$$x^{(1)}(n_1, n_2) = \sum_{i=0}^{r-1} \left( \sum_{m_1=0}^{N-1} x(m_1, n_2) u_i(m_1) \right) u_i(n_1) \tag{E.8}$$

$$x^{(2)}(n_1, n_2) = \sum_{j=0}^{r-1} \left( \sum_{m_2=0}^{N-1} x^{(1)}(n_1, m_2) u_j(m_2) \right) u_j(n_2) \tag{E.9}$$

This is the form in which we will use the SVD band limitation. The method is characterized by the OSR, the sample size $N$ and the number of singular vectors $r$. Unlike for most one-dimensional signals, there is no natural bandwidth for a given image, so choosing the OSR must be done empirically. We find that an OSR of 2.5 and a number $r = 210$ of singular vectors gives adequate results for an image size of $N \times N = 512 \times 512$.

# Appendix F

# Appendix to Chapter 7

## F.1 Upper Bounds on Double-Loop Limit Cycles

In this appendix we derive the upper bounds stated without proof in section 7.3.2. An outline of the proof is as follows: We divide the state space of pairs $(U_n, V_n)$ into two half planes according to the value of $Q(U_n)$. Any trajectory, that is, any sequence of state variable pairs $\{(U_n, V_n)\}$, can be decomposed into a number of trajectory segments over which the quantizer outputs are identical. A segment is referred to as positive or negative depending on whether the quantizer outputs all equal $+1$ or $-1$ on that segment, that is, on whether the segment lies wholly in the positive or negative half plane. In each half plane, the difference equations describing the encoder can be solved in closed form. Each segment defines a number, called the potential, which is preserved for all state variable pairs on that segment. The potential is referred to as the positive or negative potential depending on the segment type; the potential itself can be defined so that it is always a non-negative number. We derive an analytical expression for the minimum and maximum potential which can occur in one half plane as a function of the potential in the other. Using this information, we derive bounds on the positive potential as a function of the positive potential on the previous positive segment, and similarly, we derive bounds on the negative potential as a function of the previous negative potential.

### F.1.1 Potential Bounds

We assume without loss of generality that the normalized constant input $\xi$ satisfies $0 \leq \xi < 1$. We also assume $\gamma > 1$, that is, the internal stabilizing feedback is not too weak. The standard double-loop encoder corresponds to $\gamma = 2$. The difference equations describing the normalized double-loop encoder are

$$\begin{aligned} U_n &= U_{n-1} + gV_{n-1} - \frac{g\gamma\phi}{2} Q(U_{n-1}) \\ V_n &= V_{n-1} + \phi\xi - \phi Q(U_n) \end{aligned} \tag{F.1}$$

Consider a segment on which a number of quantizer outputs are constant, $Q(U_0) = Q(U_1) = \cdots = Q(U_n) = a$, where $a = \pm 1$. Then the last equation in (F.1) yields

$$V_n = V_0 - n\phi(a - \xi) \Rightarrow n \frac{V_0 - V_n}{\phi(a - \xi)}$$

The first equation in (F.1) yields

$$\begin{aligned} U_n &= U_0 - n\frac{g\gamma\phi}{2} a + g\sum_{i=0}^{n-1} V_i \\ &= U_0 - n\frac{g\gamma\phi}{2} a + g\left(nV_0 - \frac{1}{2}n(n-1)\phi(a-\xi)\right) \\ &= U_0 + g\frac{V_0 - V_n}{\phi(a-\xi)}\left[-\frac{1}{2}\gamma\phi a + V_0 - \frac{1}{2}\phi(a-\xi)\left(\frac{V_0 - V_n}{\phi(a-\xi)} - 1\right)\right] \end{aligned}$$

This can be rewritten

$$\phi(a-\xi)U_n + \frac{g}{2}\left[V_n - \frac{\phi}{2}(\xi + a(\gamma - 1))\right]^2 = \phi(a-\xi)U_0 + \frac{g}{2}\left[V_0 - \frac{\phi}{2}(\xi + a(\gamma - 1))\right]^2$$

Therefore the following non-negative quantity, referred to as the potential, is preserved on a segment passing through the point $(U_n, V_n)$ in state space:

$$P_a = \phi(a - \xi)U_n + \frac{g}{2}\left[V_n - \frac{\phi}{2}(\xi + a(\gamma - 1))\right]^2 \tag{F.2}$$

We define the quantity

$$Q_a = \sqrt{\frac{2P_a}{g}}$$

When there is no possibility of confusion, we also refer to $Q_a$ as the potential. We now consider the transition from a segment characterized by the constant quantizer output $a$ to the other possible quantizer output $c = -a$. If $Q(U_n) = a$ and $Q(U_{n+1}) = c$, we have from (F.1) that

$$\begin{aligned} U_{n+1} &= U_n + gV_n - \frac{g\gamma\phi}{2} a \\ V_{n+1} &= V_n - \phi(c - \xi) \end{aligned} \tag{F.3}$$

The potential $P_c$ on the segment characterized by the constant quantizer output $c$ is thus

$$\begin{aligned} P_c &= \phi(a-\xi)U_{n+1} + \frac{g}{2}\left[V_{n+1} - \frac{\phi}{2}(\xi + a(\gamma-1))\right]^2 \\ &= \phi(a-\xi)\left(U_n + gV_n - \frac{g\gamma\phi}{2}a\right) \\ &+ \frac{g}{2}\left[V_n - \phi(c-\xi) - \frac{\phi}{2}(\xi + a(\gamma-1))\right]^2 \end{aligned} \tag{F.4}$$

We can isolate $U_n$ in (F.2) and substitute it in (F.4). Rearranging the result, we find that the potential transition from quantizer output $a$ to $c$ obeys the law

$$\frac{(a-\xi)P_c - (c-\xi)P_a}{g} = \frac{a-c}{2}\left[\left(V_n - \frac{\gamma\phi\xi}{2}\right)^2 - \frac{\phi^2}{2}(\gamma+1)^2(a-\xi)(c-\xi)\right] \tag{F.5}$$

Consider the transition from $a = +1$ to $c = -1$. The constraint $a = Q(U_n) = +1$ implies that $U_n > 0$, and (F.2) therefore implies that $V_n$ must lie in the interval

$$V_n \in (M_1 - Q_1, M_1 + Q_1) \tag{F.6}$$

where $M_1 = (\xi+\gamma-1)\phi/2$. The constraint $c = Q(U_{n+1}) = -1$ implies $U_{n+1} \leq 0$, which by the first equation in (F.3) implies a bound on $U_n + gV_n$. The state variable $U_n$ can be isolated in (F.3), so the bound on $U_n + gV_n$ can be expressed in terms of $V_n$ and $P_1$ or equivalently $Q_1$. This bound is

$$V_n \notin (M_2 - Q_1, M_2 + Q_1) \tag{F.7}$$

where $M_2 = (-\xi + \gamma + 1)\phi/2$. For a given $P_1$ or $Q_1$, we will derive upper and lower bounds on the achievable $P_{-1}$ on the negative segment following the positive one. We must therefore consider all values of $V_n$ that satisfy both (F.6) and (F.7), and find the value which results in the minimum and maximum values of $(V_n - \gamma\phi\xi/2)^2$ in (F.5). As $M_2 - M_1 = (1-\xi)\phi > 0$ and the intervals in (F.6) and (F.7) have the same length, there is no value satisfying (F.6) for which $V_n \geq M_2 + Q_1$. Therefore (F.7) reduces to the bound $V_n \leq M_2 - Q_1$. This upper bound is weaker than the upper bound $V_n < M_1 + Q_1$ in (F.6) if and only if $0 \leq Q_1 \leq (M_2 - M_1)/2 = (1-\xi)\phi/2$. The lower bound $V_n > M_1 - Q_1$ in (F.6) is always below both upper bounds $M_1 + Q_1$ and $M_2 - Q_1$. Thus

$$\begin{aligned} M_1 - Q_1 < V_n < M_1 + Q_1 &\quad \text{for} \quad 0 \leq Q_1 \leq (1-\xi)\phi/2 \\ M_1 - Q_1 < V_n \leq M_2 - Q_1 &\quad \text{for} \quad (1-\xi)\phi/2 < Q_1 \end{aligned}$$

and so

$$\begin{aligned} N_1 - Q_1 < V_n - \frac{\gamma\phi\xi}{2} < N_1 + Q_1 &\quad \text{for} \quad 0 \leq Q_1 \leq (1-\xi)\phi/2 \\ N_1 - Q_1 < V_n - \frac{\gamma\phi\xi}{2} \leq N_2 - Q_1 &\quad \text{for} \quad (1-\xi)\phi/2 < Q_1 \end{aligned}$$

where $N_1 = M_1 - \gamma\phi\xi/2 = (\gamma-1)(1-\xi)\phi/2 > 0$, $N_2 = M_2 - \gamma\phi\xi/2 = (\gamma+1)(1-\xi)\phi/2 > N_1$. When both the lower and upper bounds on $V_n - \gamma\phi\xi/2$ are positive, the square of these bounds are the lower and upper bounds on $(V_n - \gamma\phi\xi/2)^2$, respectively. When both the lower and upper bounds on $V_n - \gamma\phi\xi/2$ are negative, the square of these bounds are the upper and lower bounds on $(V_n - \gamma\phi\xi/2)^2$, respectively; note the reverse order. When the lower bound on $V_n - \gamma\phi\xi/2$ is negative and the upper bound is positive, as is the case for $N_1 < Q_1 < N_2$, the square of the bound with the largest absolute value is the upper bound on $(V_n - \gamma\phi\xi/2)^2$, and the square of the other bound on $V_n - \gamma\phi\xi/2$ is the lower bound on $(V_n - \gamma\phi\xi/2)^2$. To summarize,

$$\begin{aligned}
(N_1 - Q_1)^2 < (V_n - \gamma\phi\xi/2)^2 &< (N_1 + Q_1)^2 \quad \text{for} \quad 0 \le Q_1 \le (1-\xi)\phi/2 \\
(N_1 - Q_1)^2 < (V_n - \gamma\phi\xi/2)^2 &\le (N_2 - Q_1)^2 \quad \text{for} \quad (1-\xi)\phi/2 < Q_1 \le N_1 \\
0 \le (V_n - \gamma\phi\xi/2)^2 &\le (N_2 - Q_1)^2 \quad \text{for} \quad N_1 < Q_1 \le (N_1+N_2)/2 \\
0 \le (V_n - \gamma\phi\xi/2)^2 &< (N_1 - Q_1)^2 \quad \text{for} \quad (N_1+N_2)/2 < Q_1 \le N_2 \\
(N_2 - Q_1) \le (V_n - \gamma\phi\xi/2)^2 &< (N_1 - Q_1)^2 \quad \text{for} \quad N_2 < Q_1
\end{aligned} \tag{F.8}$$

Inserting each of the lower bounds on $(V_n - \gamma\phi\xi/2)^2$ in (F.5), we arrive at the following lower bounds on $Q_{-1}$ as a function of $Q_1$:

$$\begin{aligned}
Q_{-1} &> \sqrt{[Q_1 - \phi(\gamma-1)]^2 + 2\phi^2\gamma(1+\xi)} \quad \text{for} \quad 0 \le Q_1 \le N_1 \\
Q_{-1} &\ge \sqrt{-\frac{1+\xi}{1-\xi}Q_1 + \frac{\phi^2}{2}(1+\xi)(\gamma+1)^2} \quad \text{for} \quad N_1 < Q_1 \le N_2 \\
Q_{-1} &\ge |Q_1 - \phi(\gamma+1)| \quad \text{for} \quad N_2 < Q_1
\end{aligned}$$

The lower bound on $Q_{-1}$ is a decreasing function of $Q_1$ for $0 \le Q_1 \le \phi(\gamma+1)$. For our purposes, it is sufficient to use the weaker bound

$$\begin{aligned}
Q_{-1} &\ge N_{-2} \quad \text{for} \quad 0 \le Q_1 \le N_2 \\
Q_{-1} &\ge \phi(\gamma+1) - Q_1 \quad \text{for} \quad N_2 < Q_1 \le \phi(\gamma+1) \\
Q_{-1} &\ge 0 \quad \text{for} \quad \phi(\gamma+1) < Q_1
\end{aligned} \tag{F.9}$$

where we have defined $N_{-2} = (\gamma+1)(1+\xi)\phi/2$, analogously to the definition of $N_2$. Note that $N_2 + N_{-2} = \phi(\gamma+1)$. For future use, we also define $N_{-1} = (\gamma-1)(1+\xi)\phi/2 < N_{-2}$. Inserting each of the upper bounds on $(V_n - \gamma\phi\xi/2)^2$ from (F.8) in (F.5), we arrive at the following upper bounds on $Q_{-1}$ as a function of $Q_1$:

$$\begin{aligned}
Q_{-1} &< \sqrt{[Q_1 + \phi(\gamma-1)]^2 + 2\phi^2\gamma(1+\xi)} \quad \text{for} \quad 0 \le Q_1 \le \frac{(1-\xi)\phi}{2} \\
Q_{-1} &< \phi(\gamma+1) - Q_1 \quad \text{for} \quad \frac{(1-\xi)\phi}{2} < Q_1 \le \frac{N_1+N_2}{2} \\
Q_{-1} &< \sqrt{[Q_1 - \phi(\gamma-1)]^2 + 2\phi^2\gamma(1+\xi)} \quad \text{for} \quad \frac{N_1+N_2}{2} < Q_1
\end{aligned}$$

For our purposes, it is sufficient to use the weaker bound

$$\begin{aligned} Q_{-1} &< \phi(\gamma+1) - Q_1 \quad \text{for} \quad 0 \leq Q_1 \leq (N_1+N_2)/2 \\ Q_{-1} &< \sqrt{[Q_1 - \phi(\gamma-1)]^2 + 2\phi^2\gamma(1+\xi)} \quad \text{for} \quad (N_1+N_2)/2 < Q_1 \end{aligned} \tag{F.10}$$

The second bound on $Q_{-1}$ in (F.10) is always above $\phi(\gamma+1) - Q_1$ in its interval of validity. For the potential transition from $a = -1$ to $c = +1$, we can similarly derive upper and lower bounds on the positive potential as a function of the negative potential on the previous negative segment. The results analogous to (F.9) and (F.10) are

$$\begin{aligned} Q_1 &\geq N_2 \quad \text{for} \quad 0 \leq Q_{-1} \leq N_{-2} \\ Q_1 &\geq \phi(\gamma+1) - Q_{-1} \quad \text{for} \quad N_{-2} < Q_{-1} \leq \phi(\gamma+1) \\ Q_1 &\geq 0 \quad \text{for} \quad \phi(\gamma+1) < Q_{-1} \end{aligned} \tag{F.11}$$

and

$$\begin{aligned} Q_1 &< \phi(\gamma+1) - Q_{-1} \quad \text{for} \quad 0 \leq Q_{-1} \leq \frac{N_{-1}+N_{-2}}{2} \\ Q_1 &< \sqrt{[Q_{-1} - \phi(\gamma-1)]^2 + 2\phi^2\gamma(1-\xi)} \quad \text{for} \quad \frac{N_{-1}+N_{-2}}{2} < Q_{-1} \end{aligned} \tag{F.12}$$

Note that the second bound on $Q_1$ in (F.12) is always above $\phi(\gamma+1) - Q_{-1}$ in its interval of validity. We now consider a trajectory as it goes from a positive segment over a negative segment back to a positive segment. We call the potentials on the two positive segments $Q_1^{\text{old}}$ and $Q_1^{\text{new}}$, respectively, and the potential on the intervening negative segment $Q_{-1}^{\text{mid}}$. We can find upper and lower bounds on $Q_1^{\text{new}}$ as a function of $Q_1^{\text{old}}$ by using the bounds (F.9), (F.10), (F.11) and (F.12). The lower bound on $Q_1$ in (F.11) is non-increasing, so the lower bound on $Q_1^{\text{new}}$ corresponds to choosing $Q_{-1}^{\text{mid}}$ as large as possible as a function of $Q_1^{\text{old}}$ in (F.10). For $0 \leq Q_1^{\text{old}} \leq (N_1+N_2)/2$, the upper bound $\phi(\gamma+1) - Q_1^{\text{old}}$ on $Q_{-1}^{\text{mid}}$ is between $N_{-2}$ and $\phi(\gamma+1)$, so $Q_1^{\text{new}}$ is lower bounded by $Q_1^{\text{old}}$. For $Q_1^{\text{old}} > (N_1+N_2)/2$, we must use the other upper bound on $Q_{-1}^{\text{mid}}$ in (F.10). This upper bound can be below $N_{-2}$ if and only if $\xi \geq -1 + 8\gamma/(1+\gamma)^2$; note that $-1 \leq -1 + 8\gamma/(1+\gamma)^2 < 1$ for $\gamma > 1$. Similarly, the upper bound on $Q_{-1}^{\text{mid}}$ is above $\phi(\gamma+1)$ for $Q_1 \geq \phi(\gamma-1) + \phi\sqrt{\gamma^2 - 2\gamma\xi + 1}$ which always exists for $0 \leq \xi < 1$. To summarize, we find that

$$\begin{aligned} Q_1^{\text{new}} &> Q_1^{\text{old}} \quad \text{for} \quad 0 \leq Q_1^{\text{old}} \leq (N_1+N_2)/2 \\ Q_1^{\text{new}} &> \min\{N_2, \phi(\gamma+1) - K_1\} \quad \text{for} \quad (N_1+N_2)/2 \leq Q_1^{\text{old}} \leq N_3 \\ Q_1^{\text{new}} &\geq 0 \quad \text{for} \quad N_3 < Q_1^{\text{old}} \end{aligned} \tag{F.13}$$

where we have defined $N_3 = \phi(\gamma-1) + \phi\sqrt{\gamma^2 - 2\gamma\xi + 1}$ and

$$K_1 = \sqrt{[Q_1^{\text{old}} - \phi(\gamma-1)]^2 + 2\phi^2\gamma(1+\xi)}$$

The upper bound on $Q_1^{\text{new}}$ as a function of $Q_1^{\text{old}}$ is found in a similar way. The upper bound (F.12) on $Q_1^{\text{new}}$ as a function of $Q_{-1}^{\text{mid}}$ is decreasing for $0 \leq Q_{-1}^{\text{mid}} \leq \max\{\gamma(1+\xi)\phi/2, (\gamma-1)\phi\}$, and increasing for $\max\{\gamma(1+\xi)\phi/2, (\gamma-1)\phi\} < Q_{-1}^{\text{mid}}$, so for a given interval of possible values of $Q_{-1}^{\text{mid}}$, the largest value of $Q_1^{\text{new}}$ always occurs at one of the endpoints of the interval. We first consider the ways in which we can get $Q_1^{\text{new}} \geq Q_1^{\text{old}}$ for $Q_1^{\text{old}} > N_2$. For $Q_1^{\text{old}} > \phi(\gamma+1)$, choosing the lower bound on $Q_{-1}^{\text{mid}}$ in (F.9) will not work, because the largest possible value of $Q_1^{\text{new}}$ for $Q_{-1}^{\text{mid}} = 0$ is $\phi(\gamma+1)$. For $(\gamma-\gamma\xi+2)\phi/2 < Q_1^{\text{mid}} \leq \phi(\gamma+1)$, the lower bound $\phi(\gamma+1)-Q_1^{\text{old}}$ on $Q_{-1}^{\text{mid}}$ is below $(N_{-1}+N_{-2})/2$, so the resulting largest $Q_1^{\text{new}}$ is less than $Q_1^{\text{old}}$. For $N_2 < Q_1^{\text{old}} \leq (\gamma-\gamma\xi+2)\phi/2$, however, the lower bound on $Q_{-1}^{\text{mid}}$ is above $(N_{-1}+N_{-2})/2$, and so by (F.12), it is possible to get a value of $Q_1^{\text{new}}$ above $Q_1^{\text{old}}$. We can also attempt to get $Q_1^{\text{new}} \geq Q_1^{\text{old}}$ by choosing the upper bound on $Q_{-1}^{\text{mid}}$ in (F.9), and use that value of $Q_{-1}^{\text{mid}}$ in (F.12) to find the largest value of $Q_1^{\text{new}}$ achievable in this way. It can be shown that in this way, we can achieve $Q_1^{\text{new}} \geq Q_1^{\text{old}}$ if and only if

$$Q_1^{\text{old}} \leq \frac{\phi}{2}\left((\gamma-1)(1-\xi)+\frac{2\gamma}{\gamma-1}\right) \triangleq B_1$$

We then consider the interval $(N_1+N_2)/2 \leq Q_1^{\text{old}} \leq \max\{B_1, B_2\}$ where we have defined $B_2 = (\gamma-\gamma\xi+2)\phi/2$. The lower bound on the corresponding interval for $Q_{-1}^{\text{mid}}$ is achieved at $Q_1^{\text{old}} = \max\{B_1, B_2\}$, because the lower bound on $Q_{-1}^{\text{mid}}$ is a non-increasing function of $Q_1^{\text{old}}$. The upper bound on the corresponding interval for $Q_{-1}^{\text{mid}}$ is achieved at either $Q_1^{\text{mid}} = (N_1+N_2)/2$ or at $Q_1^{\text{mid}} = \max\{B_1, B_2\}$. In the former case, inserting $Q_{-1}^{\text{mid}} = (\gamma+\gamma\xi+2)\phi/2 \geq (N_{-1}+N_{-2})/2$ in (F.12) yields the upper bound $Q_1^{\text{new}} \leq \sqrt{16+\gamma^2(1-\xi)^2} \cdot \phi/2 \triangleq B_3$ for $(N_1+N_2)/2 \leq Q_1^{\text{old}} \leq \max\{B_1, B_2\}$, and in the latter case, we know from above that $Q_{-1}^{\text{mid}} \leq \max\{B_1, B_2\}$,

Finally, consider the interval $0 \leq Q_1^{\text{old}} < (N_1+N_2)/2$. For a given $Q_1^{\text{old}}$ in this interval, the possible interval for $Q_{-1}^{\text{mid}}$ according to (F.9) and (F.10) is a superset of the possible interval for $Q_{-1}^{\text{mid}}$ at $Q_1^{\text{old}} = (N_1+N_2)/2$. Therefore, the largest achievable value of $Q_1^{\text{new}}$ for a given $Q_1^{\text{old}}$ is the maximum of $C_1 \triangleq \max\{B_1, B_2, B_3\}$ and the value we get by choosing the upper bound for $Q_{-1}^{\text{mid}}$ in (F.10) and the corresponding upper bound for $Q_1^{\text{new}}$ in (F.12). To summarize,

$$\begin{aligned} Q_1^{\text{new}} &\leq \max\left\{C_1, \sqrt{\left(Q_1^{\text{old}}-2\phi\right)^2+2\phi^2\gamma(1-\xi)}\right\} && \text{for} \quad 0 \leq Q_1^{\text{old}} \leq \frac{\gamma(1-\xi)\phi}{2} \\ Q_1^{\text{new}} &\leq C_1 && \text{for} \quad \frac{\gamma(1-\xi)\phi}{2} < Q_1^{\text{old}} \leq C_1 \\ Q_1^{\text{new}} &< Q_1^{\text{old}} && \text{for} \quad C_1 < Q_1^{\text{old}} \end{aligned} \qquad \text{(F.14)}$$

For the potential transition from a quantizer output of $-1$ over $+1$ back to $-1$, we can similarly derive upper and lower bounds on the potential on a negative segment, $Q_{-1}^{\text{new}}$, as a function of the potential on the previous negative segment,

$Q_{-1}^{\text{old}}$. The results analogous to (F.13) and (F.14) are

$$\begin{aligned} Q_{-1}^{\text{new}} &> Q_{-1}^{\text{old}} \quad \text{for} \quad 0 \le Q_{-1}^{\text{old}} \le (N_{-1}+N_{-2})/2 \\ Q_{-1}^{\text{new}} &> \min\{N_{-2}, \phi(\gamma+1) - K_{-1}\} \quad \text{for} \quad (N_{-1}+N_{-2})/2 \le Q_{-1}^{\text{old}} \le N_{-3} \\ Q_{-1}^{\text{new}} &\ge 0 \quad \text{for} \quad N_{-3} < Q_{-1}^{\text{old}} \end{aligned} \tag{F.15}$$

where we have defined $N_{-3} = \phi(\gamma-1) + \phi\sqrt{\gamma^2 + 2\gamma\xi + 1}$ and

$$K_{-1} = \sqrt{[Q_{-1}^{\text{old}} - \phi(\gamma-1)]^2 + 2\phi^2\gamma(1-\xi)}$$

and

$$\begin{aligned} Q_{-1}^{\text{new}} &\le \max\left\{C_{-1}, \sqrt{\left(Q_{-1}^{\text{old}} - 2\phi\right)^2 + 2\phi^2\gamma(1+\xi)}\right\} \quad \text{for} \quad 0 \le Q_{-1}^{\text{old}} \le \frac{\gamma(1+\xi)\phi}{2} \\ Q_{-1}^{\text{new}} &\le C_{-1} \quad \text{for} \quad \frac{\gamma(1+\xi)\phi}{2} < Q_{-1}^{\text{old}} \le C_{-1} \\ Q_{-1}^{\text{new}} &< Q_{-1}^{\text{old}} \quad \text{for} \quad C_{-1} < Q_{-1}^{\text{old}} \end{aligned}$$

where we have defined

$$B_{-1} = \frac{\phi}{2}\left((\gamma-1)(1+\xi) + \frac{2\gamma}{\gamma-1}\right)$$

$$B_{-2} = \frac{\phi}{2}(\gamma + \gamma\xi + 2), \quad B_{-3} = \frac{\phi}{2}\sqrt{16 + \gamma^2(1-\xi)^2}$$

and $C_{-1} = \max\{B_{-1}, B_{-2}, B_{-3}\}$. We now show that on any limit cycle, the positive potential is bounded by $Q_1 \le C_1$ and the negative potential is bounded by $Q_{-1} \le C_{-1}$. Consider first the positive potential, and suppose that a positive potential $Q_1 = A_1 > C_1$ is the largest positive potential occurring on a limit cycle. The limit cycle must also contain a positive potential $A_2$ for which the upper bound on $Q_1^{\text{new}}$ in (F.14) equals $A_1$ when $Q_1^{\text{old}} = A_2$. Equation (F.14) shows that $A_2$ cannot be between $\gamma(1-\xi)\phi/2$ and $C_1$, because $Q_1^{\text{new}} \le C_1$ in this interval of $Q_1^{\text{old}}$. Nor can $A_2$ be above $C_1$, because $Q_1^{\text{new}} < Q_1^{\text{old}}$ when $Q_1^{\text{old}} > C_1$, so $A_2$ would have to be above $A_1$, and we have assumed that $A_1$ is the largest positive potential occurring on a limit cycle. Therefore $A_2$ must be between 0 and $\gamma(1-\xi)\phi/2$.

To get $A_1 > C_1$, the negative potential $Q_{-1}^{\text{mid}}$ generating $Q_1^{\text{new}} = A_1$ must be above $B_{-1}$. We define $D_{-1}$ to be the smallest negative potential $Q_{-1}^{\text{mid}}$ exceeding $B_{-1}$ such that the next positive potential $Q_1^{\text{new}}$ can reach $A_1$. Since we must have $Q_1^{\text{old}} \le \gamma(1-\xi)\phi/2$, the largest value of $Q_1^{\text{old}}$ for which $Q_1^{\text{new}} = A_1$ is achievable, must satisfy $Q_1^{\text{old}} \le \phi(\gamma+1) - D_{-1}$, so in particular, $A_2 \le \phi(\gamma+1) - D_{-1}$. We will now show that such values of $Q_1^{\text{old}}$ could in turn only be generated by values of $Q_1$ that exceed $A_1$, thus violating the assumption that $A_1$ is the largest occurring positive potential.

To see this, consider (F.15). To generate a positive potential $Q_1^{\text{new}} \le \gamma(1-\xi)\phi/2$, the previous negative potential must satisfy $Q_{-1}^{\text{mid}} > \phi(\gamma+1) - Q_1^{\text{new}} \ge$

$(\gamma+\gamma\xi+2)\phi/2$. We are trying to generate a positive potential $Q_1^{\text{new}} \leq \phi(\gamma+1)-D_{-1}$, so the previous negative potential must satisfy $Q_{-1}^{\text{mid}} > D_{-1}$. If such a large negative potential is generated by a positive potential exceeding $\gamma(1-\xi)\phi/2$, the negative potential must also exceed $B_1$; we define $D_1$ to be the smallest positive potential $Q_1^{\text{old}}$ for which the next negative potential $Q_{-1}^{\text{mid}}$ equals $D_{-1}$. Now $D_1 > B_1$, and $Q_1^{\text{old}} = D_1$ maps into $Q_{-1}^{\text{mid}} = D_{-1}$ using the upper bound in (F.10), and $Q_{-1}^{\text{mid}}$ maps into $Q_1^{\text{new}} = A_1$ using the upper bound in (F.12). Therefore $D_1 > A_1$, proving that if the negative potential $D_{-1}$ is generated by a positive potential above $\gamma(1-\xi)\phi/2$, the same positive potential must also exceed $A_1$. This violates the assumption defining $A_1$. Alternatively, a negative potential exceeding $\phi(\gamma+1) - Q_1^{\text{new}}$ may be generated for a positive potential $Q_1^{\text{old}}$ below $\gamma(1-\xi)\phi/2$; however, we must then also have $Q_1^{\text{old}} < Q_1^{\text{new}}$. Using the same argument on $Q_1^{\text{old}}$ and going backwards in time by another positive segment, the previous positive potential must either exceed $A_1$ or be below $Q_1^{\text{old}}$. On the other hand, $A_1$ is assumed to be a positive potential on a limit cycle, so the positive potential $A_1$ must recur with a periodicity that corresponds to a finite number of positive segments. Therefore, going backwards in time by a sufficiently large number of positive segments must result in the positive potential $A_1 > \gamma(1-\xi)\phi/2$ by assumption. This establishes a contradiction, showing that $A_1 > C_1$ cannot be a positive potential on a limit cycle, and that $C_1 = \max\{B_1, B_2, B_3\}$ is an upper bound on positive potentials on limit cycles. A similar argument shows that $C_{-1} = \max\{B_{-1}, B_{-2}, B_{-3}\}$ is an upper bound on negative potentials on limit cycles.

The potential bounds can be written in more detail as follows:

$$\frac{2Q_1}{\phi} \leq \begin{cases} (\gamma-1)(1-\xi) + \dfrac{2\gamma}{\gamma-1} & \text{for } 1 < \gamma \leq 1 + \dfrac{2}{1+\xi} \\ \sqrt{16+\gamma^2(1-\xi)^2} & \text{for } 1 + \dfrac{2}{1+\xi} < \gamma \leq \dfrac{3}{1-\xi} \\ \gamma(1-\xi)+2 & \text{for } \dfrac{3}{1-\xi} < \gamma \end{cases} \tag{F.16}$$

$$\frac{2Q_{-1}}{\phi} \leq \begin{cases} (\gamma-1)(1+\xi) + \dfrac{2\gamma}{\gamma-1} & \text{for } 1 < \gamma \leq 1 + \dfrac{2}{1+\xi} \\ \gamma(1+\xi)+2 & \text{for } 1 + \dfrac{2}{1+\xi} < \gamma \end{cases} \tag{F.17}$$

### F.1.2 State Variable Bounds

In this section we use the potential bounds (F.16) and (F.17) to derive bounds on the largest absolute values of the state variables $U_n$ and $V_n$ on limit cycles for the double-loop encoder.

The definition of the potential (F.2) shows that for a given potential $P_a$, the largest absolute value of the state variable $U_n$ occurs when $V_n = (\xi+a(\gamma-1))\phi/2$ and equals $P_a/(\phi|a-\xi|)$. We must consider both the negative and the positive potential in order to arrive at an upper bound on the absolute value of $U_n$. We find the following result:

$$\frac{8(1-\xi)|U_n|}{\phi g} \leq \begin{cases} \left[(\gamma-1)(1-\xi)+\frac{2\gamma}{\gamma-1}\right]^2 & \text{for } 1<\gamma\leq 1+\frac{2}{1+\xi} \\ 16+\gamma^2(1-\xi)^2 & \text{for } 1+\frac{2}{1+\xi}<\gamma\leq f_2(\xi) \\ [\gamma(1+\xi)+2]^2 & \text{for } f_2(\xi)<\gamma \end{cases} \tag{F.18}$$

where

$$f_2(\xi) = \frac{-1+\sqrt{1+2\xi\left(\frac{4}{1-\xi}-\frac{1}{1+\xi}\right)}}{\xi}$$

An upper bound on the absolute value of $V_n$ on limit cycles can be found as follows. The state equation (F.1) for $V_n$ shows that on negative segments $V_n$ increases, while on positive segments $V_n$ decreases. Furthermore, $V_n$ decreases when making a transition from a negative to a positive segment, and increases when making a transition from a positive to a negative segment. Therefore the largest positive value of $V_n$ is bounded by the largest positive value of $V_n$ on a negative segment. The definition of the potential (F.2) shows that this value equals $(\xi-\gamma+1)\phi/2+Q_{-1}$. Similarly, the negative value of $V_n$ with the largest absolute value occurs on a positive segment, and this absolute value equals $-(\xi+\gamma-1)\phi/2+Q_1$. Choosing the largest of these two absolute values for $V_n$ yields

$$\frac{2|V_n|}{\phi} \leq \begin{cases} \gamma\left(\xi+\frac{2}{\gamma-1}\right) & \text{for } 1<\gamma\leq 1+\frac{2}{1+\xi} \\ (\gamma+1)\xi+3 & \text{for } 1+\frac{2}{1+\xi}<\gamma \end{cases} \tag{F.19}$$

The derived bounds are only upper bounds for two reasons, both related to the discrete nature of the encoder: First, the trajectories in state space may not go through the parabola extremes specified by the potentials. Second, it is in general not possible for a trajectory to attain the maximum negative potential $P_{-1}$ for a given positive potential $P_1$, and subsequently to obtain the maximum $P_1$ for that $P_{-1}$.

# F.2 Design of Double-Loop Encoder

## F.2.1 Approximate SNR Measure

We will show that the product $gG$ is an approximate measure of the SNR performance of the double-loop encoder. Consider a linearized model of the double-loop encoder in which the quantizer is viewed as an independent noise source.[1] The transfer function between the $z$-transform of the input and output

[1] For the purposes of the present discussion, we need not make the assumption of white noise, but an assumption of independence is implicit. As is well known, these assumptions are in many respects inadequate [41].

sequences, $X(z)$ and $Y(z)$, is then

$$H_X(z) = \frac{Y(z)}{X(z)} = \frac{gGz^{-1}}{1+(gb+gGB-2)z^{-1}+(1-gb)z^{-2}}; \qquad H_X(1) = \frac{1}{B} \tag{F.20}$$

and the transfer function between the $z$-transforms of the noise and output sequences is

$$H_E(z) = \frac{Y(z)}{E(z)} = \frac{(1-z^{-1})^2}{1+(gb+gGB-2)z^{-1}+(1-gb)z^{-2}} \tag{F.21}$$

If the scaling factors are chosen such that the poles of the transfer functions are safely out of baseband, the baseband transfer functions are approximately given by

$$H_X(z) \approx \frac{z^{-1}}{B} \tag{F.22}$$

$$H_E(z) \approx \frac{(1-z^{-1})^2}{gGB} \tag{F.23}$$

From (F.22), the noise shaping function of the encoder is seen to be preserved if the approximation leading to (F.22) is valid. From (F.23), the product $gG$ adjusts the amount of baseband noise suppression, and hence we will use this product as an approximate measure of SNR to be maximized.

## F.2.2 Design Trade-Off

To perform the design optimization presented in section 7.3.2.2, we will consider $\xi_{\max}$ and $\gamma$ to be the two independent variables. Our optimization must take into account the validity regions of the different bounds in (F.18) and (F.19), but we first perform some general manipulations. Let us denote the upper bound on $8|U|_n/(\phi g)$ in (F.18) by $h_1(\xi,\gamma)$, and the upper bound on $2|V|_n/\phi$ in (F.19) by $h_2(\xi,\gamma)$. The constraint $V_{\max} = L$ implies that

$$\phi = \frac{2L}{h_2(\xi_{\max},\gamma)}, \quad G = \frac{\phi}{B} = \frac{2(1+\xi_{\max})}{h_2(\xi_{\max},\gamma)}$$

where we have used the constraint $B = L/(1+\xi_{\max})$. The constraint $U_{\max} = L$ implies that

$$g = \frac{8L}{\phi h_1(\xi_{\max},\gamma)} = \frac{4h_2(\xi_{\max},\gamma)}{h_1(\xi_{\max},\gamma)}$$

Therefore the product $gG$ equals

$$gG = \frac{8(1+\xi_{\max})}{h_1(\xi_{\max},\gamma)}$$

Maximizing the product $gG$ for a fixed value of $\xi_{\max}$ with respect to $\gamma$ is equivalent to minimizing $h_1(\xi_{\max},\gamma)$. Equation (F.18) shows that for $\gamma >$

$1 + 2/(1 + \xi_{\max})$, $h_1(\xi_{\max}, \gamma)$ is an increasing function of $\gamma$. Therefore it is advantageous to choose $\gamma$ as small as possible, that is, $\gamma = 1 + 2/(1 + \xi_{\max})$, given that $\gamma$ cannot be below that value. But we still need to consider the range $1 < \gamma \leq 1 + 2/(1 + \xi_{\max})$. We can show that $h_1(\xi_{\max}, \gamma)$ has a minimum at $\gamma = 1 + \sqrt{2/(1 - \xi_{\max})}$, which is below $1 + 2/(1 + \xi_{\max})$ for $0 \leq \xi_{\max} < \sqrt{5} - 2 \approx 0.23$. Therefore the optimum choice of $\gamma$ is

$$\gamma = \begin{cases} 1 + \sqrt{\dfrac{2}{1 - \xi_{\max}}} & \text{for } 0 \leq \xi_{\max} \leq \sqrt{5} - 2 \\ 1 + \dfrac{2}{1 + \xi_{\max}} & \text{for } \sqrt{5} - 2 < \xi_{\max} < 1 \end{cases} \tag{F.24}$$

where $\xi_{\max}$ is understood to be the largest constant normalized input for which we design. We want to maximize the largest unnormalized constant input $X_{\max} = B\xi_{\max}$, that is,

$$X_{\max} = L \cdot \frac{\xi_{\max}}{1 + \xi_{\max}} \tag{F.25}$$

which is equivalent to maximizing $\xi_{\max}$. However, we also want to optimize the performance product $gG$ which depends on $\xi_{\max}$ directly as well as through $\gamma$ and (F.24). It can be shown that setting $\gamma = 1 + \sqrt{2/(1 - \xi_{\max})}$ results in the product

$$gG = \frac{2(1 - \xi_{\max}^2)}{[1 + \sqrt{2(1 - \xi_{\max})}]^2} \tag{F.26}$$

whose only extremum between 0 and 1 is a maximum at $\xi_{\max} \approx 0.3611 > \sqrt{5} - 2$, so for $0 \leq \gamma \leq 1 + \sqrt{2/(1 - \xi_{\max})}$, $gG$ is an increasing function of $\xi_{\max}$, and within the range $0 \leq \xi_{\max} \leq \sqrt{5} - 2$, the upper limit on $\xi_{\max}$ is optimal. For $\xi_{\max} > \sqrt{5} - 2$, we find that

$$gG = \frac{8(1 + \xi_{\max})^3(1 - \xi_{\max})}{[4 + (1 + \xi_{\max})^2]^2} \tag{F.27}$$

whose only extremum between 0 and 1 is a rather broad maximum at $\xi_{\max} = \sqrt{28} - 5 \approx 0.2915$. Figure 7.8 shows the product $gG$ a function of $\xi_{\max}$, taking into account the ranges of validity of (F.26) and (F.27). The figure also shows $X_{\max}/L$ from (F.25) which measures the designed dynamic range. As a trade-off between the conflicting requirements of maximizing the dynamic range and maximizing the approximate performance measure, we choose $\xi_{\max} = 0.5$ in section 7.3.2.2, which leads to

$$\gamma = 1 + \frac{2}{1 + \xi_{\max}} = 2.33, \quad \phi = \frac{2L}{(\gamma + 1)\xi_{\max} + 3} = 0.429L$$

The corresponding physical scaling factors are

$$B = \frac{L}{1 + \xi_{\max}} = 0.667L, \quad g = \frac{L}{\phi} \cdot \frac{8(1 - \xi_{\max})}{16 + \gamma^2(1 - \xi_{\max})^2} = 0.538,$$

$$G = \frac{\phi}{B} = 0.643, \quad b = \frac{\gamma\phi}{2} = 0.500L \tag{F.28}$$

which results in a maximum permissible constant input of $X_{\max} = B\xi_{\max} = 0.333L$, and a performance product of $gG = 0.346$. The main result of the appendix is thus that (F.28) dictates the choice of the scaling factors for a given clipping level $L$.

## F.3 Describing Function Method

Consider the set-up of section 7.4.2.1. It can be shown that the DC component and first harmonic of $Q(U(t))$ equal

$$Q(U(t)) = \frac{2}{\pi}\arcsin\frac{C}{A} - \frac{4}{\pi}\sqrt{1-\left(\frac{C}{A}\right)^2}\sin\omega_0 t + \cdots \tag{F.29}$$

If the frequency $\omega_0$ is chosen such that $\angle H(e^{j\omega_0}) = -180^{\circ}$, and if furthermore all higher harmonics of $\omega_0$ are neglected, we find that the assumption $U(t) = C - A\sin(\omega_0 t)$ is consistent if and only if

$$\begin{aligned} C &= k_{\mathrm{DC}}\left(X - \frac{2}{\pi}\arcsin\frac{C}{A}\right) \\ A &= k_{\mathrm{GM}}\cdot\frac{4}{\pi}\sqrt{1-\left(\frac{C}{A}\right)^2} \end{aligned} \tag{F.30}$$

where we have introduced the DC gain $k_{\mathrm{DC}} = H(1)$ and the gain margin $k_{\mathrm{GM}} = |H(e^{j\omega_0})|$. It is convenient to introduce the new variable

$$D = \sqrt{1-\left(\frac{C}{A}\right)^2}, \quad 0 < D \leq 1$$

implying

$$\frac{C}{A} = \sqrt{1-D^2}, \quad C^2 = A^2(1-D^2)$$

Using this and the fact that $\arcsin t = \arccos\sqrt{1-t^2}$, equations (F.30) become

$$A = \frac{4}{\pi}k_{\mathrm{GM}}D \tag{F.31}$$

$$D\sqrt{1-D^2} = g\left(X - \frac{2}{\pi}\arccos D\right), \quad g = \frac{\pi}{4}\frac{k_{\mathrm{DC}}}{k_{\mathrm{GM}}} \tag{F.32}$$

The latter equation can be solved numerically for $D$, but insight can be gained from a graphical representation. Figure F.1 shows the left and right hand sides of (F.32), labeled LHS and RHS, as functions of $D$; the specific value used for $g$ corresponds to the fourth order interpolative encoder of [32] with its pole moduli scaled by 0.98, while the constant input is chosen to be $X = 0.7$. The constant $C$ is proportional to the LHS with proportionality constant $4k_{\mathrm{GM}}/\pi$.

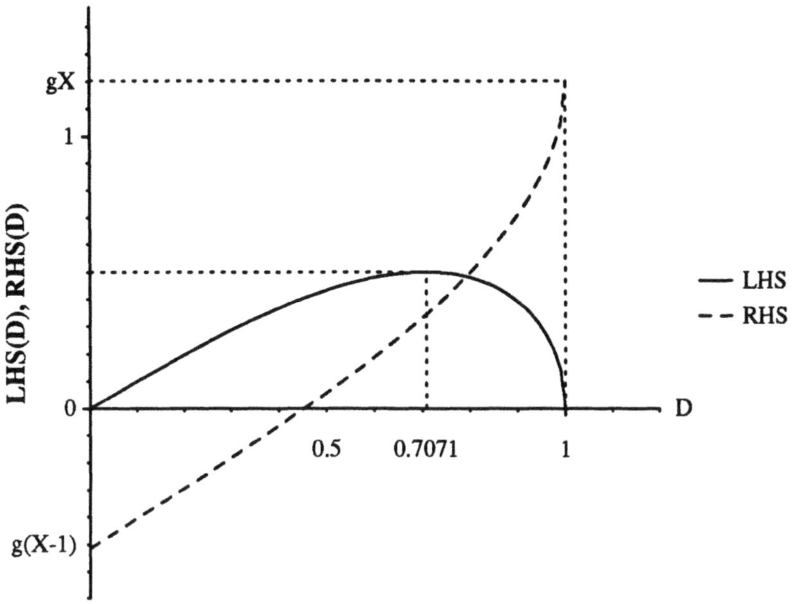


Figure F.1: *Figure showing the left and right hand sides of (F.32).*

We can show that the LHS reaches its maximum value $\frac{1}{2}$ at $D = \frac{\sqrt{2}}{2}$, and that for any $0 \le X < 1$, $g > 0$, the RHS and LHS have a unique intersection point between 0 and 1. The latter fact follows from the observation that the RHS is only positive for $\cos(\pi X/2) < D \le 1$, and in this interval the derivative of the RHS with respect to $D$ always exceeds the derivative of the LHS.

From the figure and equations (F.31), (F.32) we observe that for $X = 0$, the constants are given by $D = 1$, $C = 0$ and $A = 4k_{\mathrm{GM}}/\pi \approx 1.27k_{\mathrm{GM}}$. As $X$ increases, $C$ increases while $D$ and $A$ decrease; as $D$ passes $\frac{\sqrt{2}}{2}$, the LHS and $C$ begin to decrease along with $A$. In terms of limit cycle amplitudes, this has the following interpretation: There is a value of $X$ between 0 and 1 which maximizes the amplitude. This result is confirmed by table 7.2. Changing the open-loop transfer function such that $k_{\mathrm{GM}}$ decreases will increase $g$, which decreases $D$ and $A$; the effect on $C$ depends on the particular value of $D$.

We next find the largest limit cycle amplitude that can occur for a given open-loop transfer function and a constant input $0 \le X < 1$. We observe that $A$ and $C$ are proportional to the abscissa and ordinate of the LHS curve, respectively, and that in both cases, the proportionality constant is $4k_{\mathrm{GM}}/\pi$. The maximum sum of $D$ and $D\sqrt{1-D^2}$ subject to $0 \le D \le 1$ can easily be shown to occur at $D = \sqrt{3}/2$, and the resulting maximum value of $U_n$ is

$$U_{\max} = \max_X(A + C) = \frac{3\sqrt{3}}{\pi} k_{\mathrm{GM}} \approx 1.65 k_{\mathrm{GM}}$$

Finally, we consider the case of multi-bit quantization. Specifically, let a two-bit quantizer be defined by

$$Q_2(U) = \begin{cases} +1 & \text{for } \frac{2}{3} < U \\ +\frac{1}{3} & \text{for } 0 < U \leq \frac{2}{3} \\ -\frac{1}{3} & \text{for } -\frac{2}{3} < U \leq 0 \\ -1 & \text{for } U \leq -\frac{2}{3} \end{cases}$$

We can show that

$$Q_2(C - A\sin(\omega_0 t)) = \frac{2}{3\pi}\left[\arcsin\left(\frac{C - \frac{2}{3}}{A}\right) + \arcsin\frac{C}{A} + \arcsin\left(\frac{C + \frac{2}{3}}{A}\right)\right]$$
$$-\frac{4}{3\pi}\left[\sqrt{1 - \left(\frac{C - \frac{2}{3}}{A}\right)^2} + \sqrt{1 - \left(\frac{C}{A}\right)^2} + \sqrt{1 - \left(\frac{C + \frac{2}{3}}{A}\right)^2}\right]\sin(\omega_0 t) + \cdots \tag{F.33}$$

To first order in $1/A$ and $1/C$, we can show that (F.33) reduces to

$$Q_2(C - A\sin(\omega_0 t)) \approx \frac{2}{\pi}\arcsin\frac{C}{A} - \frac{4}{\pi}\sqrt{1 - \left(\frac{C}{A}\right)^2}\sin\omega_0 t + \cdots$$

which is identical to (F.29). This implies that if the solution to (F.30) for the one-bit quantizer has large values of $A$ and $C$, then approximately the same solution is valid for the two-bit quantizer.

In the limit as the number of bits in the quantizer goes to infinity, the quantizer approaches the saturation characteristic

$$Q_\infty(U) = \begin{cases} +1 & \text{for } 1 < U \\ U & \text{for } -1 \leq U \leq 1 \\ -1 & \text{for } U < -1 \end{cases}$$

We can show that

$$Q_\infty(C - A\sin(\omega_0 t)) = \frac{A}{\pi}\left[\frac{C+1}{A}\arcsin\left(\frac{C+1}{A}\right) - \frac{C-1}{A}\arcsin\left(\frac{C-1}{A}\right)\right.$$
$$\left. + \sqrt{1 - \left(\frac{C+1}{A}\right)^2} - \sqrt{1 - \left(\frac{C-1}{A}\right)^2}\right]$$
$$-\frac{A}{\pi}\left[\arcsin\left(\frac{C+1}{A}\right) - \arcsin\left(\frac{C-1}{A}\right)\right.$$

$$+\frac{C+1}{A}\sqrt{1-\left(\frac{C+1}{A}\right)^2}-\frac{C-1}{A}\sqrt{1-\left(\frac{C-1}{A}\right)^2}\Bigg]\sin(\omega_0 t)+\cdots \tag{F.34}$$

To first order in $1/A$ and $1/C$, we can show that (F.34) reduces to (F.29). This implies that if the solution to (F.30) for the one-bit quantizer has large values of $A$ and $C$, then approximately the same solution is valid for a quantizer with infinitely many bits.

The conclusions for the two-bit and infinite-bit quantizers show that if an encoder with a one-bit quantizer has large-amplitude sinusoidal limit cycles, then the encoder will have the same stability problems even with multi-bit quantizers.

# F.4 Proofs of Statements in Section 7.5.2.2

## F.4.1 The General Case $s > 1$

We will show that using the outlined technique with $s > 1$ positive segments produces an upper bound on limit cycle amplitudes. This can be seen by considering a limit cycle with amplitude $U_{\text{peak0}}$, and dividing it into positive segments. Say that the peak value $U_{\text{peak0}}$ occurs every $t$ positive segments. If we start at time zero with a peak value of $U_{\text{peak0}}$ and go through any multiple of $t$ positive segments, the peak value $U'_{\text{peak0}}$ must again be $U_{\text{peak0}}$. This is true in particular for traversing $st$ positive segments. Consider on the other hand a plot of the form in figure 7.15, only modified by waiting $s$ positive segments rather than only one, before registering $U'_{\text{peak}}$ vs. $U_{\text{peak}}$. If $U_{\text{peak0}}$ is greater than $U_{\max}$, then all subsequent peak values after multiples of $s$ positive segments must be below $U_{\text{peak0}}$. This holds in particular after $st$ positive segments. Putting together the two statements on the peak value after $st$ positive segments, it follows that $U_{\max}$ in the modified figure is an upper bound on limit cycle amplitudes as asserted.

## F.4.2 Bounds with $s > 1$ Are Tighter

In general, if one value of $U_{\max}$ satisfies requirements R1 and R2, then all values $U > U_{\max}$ also satisfy the requirements. Let $U_{\max}(s)$ denote the infimum of all the possible values of $U_{\max}$ in the two requirements, for a given value of $s$. We want to show that if $s > 1$, $U_{\max}(s) \leq U_{\max}(1)$. A more general statement would be that if $s_2 > s_1$, $U_{\max}(s_2) \leq U_{\max}(s_1)$. This, however, does not appear to be true.

Let us denote the peak values in figure 7.15 by $U_{\text{peak}}(s)$ and $U'_{\text{peak}}(s)$ for a given value of $s$. Let us further introduce the function $f_s$ for the curve in the figure, $U'_{\text{peak}}(s) = f_s(U_{\text{peak}})$. The maximum value $U_{\max}(s)$ is uniquely determined by $f_s$. Consider first any $U > U_{\max}(1)$; clearly, $f_1(U) < U$. If $f_1(U) > U_{\max}(1)$, any trajectory with initial positive peak value will have all subsequent peak values below $f_1(U)$. In particular, the trajectory that corresponds to the point $(U, f_s(U))$ in the plot of $f_s$, must have $f_s(U) < U$. If instead $f_1(U) \leq U_{\max}(1)$, then any trajectory with initial positive peak value $U$ will have all subsequent positive peak values $\leq U_{\max}(1)$. In both cases, $f_s(U) < U$, so requirement R1 for $f_s$ is satisfied with $U_{\max} = U_{\max}(1)$. Consider next any $0 \leq U \leq U_{\max}$. No trajectory with this initial peak value can ever exceed $U_{\max}(1)$. In particular, the trajectory corresponding to the point $(U, f_s(U))$ must have $f_s(U) \leq U_{\max}(1)$, so requirement R2 for $f_s$ is again satisfied with $U_{\max} = U_{\max}(1)$. Since $U_{\max}(s)$ is defined as the infimum of all possible values of $U_{\max}$, we have shown that $U_{\max}(s) \leq U_{\max}(1)$ for $s > 1$.

### F.4.3 Solution to Simplified System Equation

If $H(z)$ has a pole of order $\ell$ at DC, a particular solution to (7.24) is of the form $U_{n,\text{part}} = k_0 n^\ell$ where we can solve for $k_0$ by inserting $U_{n,\text{part}}$ in (7.24).[2] If $H(z)$ has no poles at DC, the particular solution reduces to a constant. The solution to the homogeneous equation obtained by setting $K = 0$ is determined by the poles of $H(z)$. Assume for simplicity that $N$ is even, that all poles are simple with non-zero imaginary part, and denote them by $\{p_m, \bar{p}_m = r_m e^{\pm j\omega_m} : 1 \leq m \leq N/2\}$. The homogeneous solution can then be written

$$U_{n,\text{hom}} = \sum_{m=1}^{N/2} r_m^n \left\{k_m \cos(\omega_m n) + \ell_m \sin(\omega_m n)\right\}$$

where $\{k_m, \ell_m : 1 \leq m \leq N/2\}$ are $N$ arbitrary constants. The complete solution to (7.24) is then

$$U_n = U_{n,\text{part}} + U_{n,\text{hom}}$$

We can find the trajectories with property P1 by enforcing the linear constraint $U_0 = U_{\text{peak}}$ on the arbitrary constants. To find a necessary condition for property P2 to approximately hold, we consider the continuous-time function

$$U(t) = k_0 + \sum_{m=1}^{N/2} r_m^n \left\{k_m \cos(\omega_m t) + \ell_m \sin(\omega_m t)\right\}$$

We can obtain another linear constraint on the arbitrary constants by setting the derivative of $U(t)$ at $t = 0$, $U'(0)$ to zero.[3] To check whether property P2 holds for a trajectory with arbitrary constants satisfying both linear constraints, we must then check whether the trajectory actually attains its segment maximum at $n = 0$, rather than having a minimum, a local extremum or an inflection point. This check is done by running the difference equation backwards and forwards in time from $n = 0$ until $U_n \leq 0$.

[2] In fact, $k_0 = (X-1)H(1) < 0$ for positive segments, $k_0 = (X+1)H(1) > 0$ for negative segments.

[3] Enforcing $U'(0) = 0$ is an approximation since the discrete-time sequence $\{U_n\}$ may not peak exactly at the points where $U'(t) = 0$, even if $t$ is an integer.

# Bibliography


[1] F. Goodenough, "20-bit Delta-Sigma ADCs vie for integrator jobs," *Electronic Design*, pp. 93–96F, April 25 1991.

[2] C. H. Giancarlo and C. G. Sodini, "A slope adaptive Delta modulator for VLSI signal processing systems," *IEEE Transactions on Circuits and Systems*, vol. CAS–33, pp. 51–58, Jan. 1986.

[3] D. A. Kerth and D. S. Piasecki, "An oversampling converter for strain gauge transducers," in *International Solid-State Circuits Conference*, pp. 42–43, Feb. 1992.

[4] J. C. Candy, B. A. Wooley, and O. J. Benjamin, "A voiceband codec with digital filtering," *IEEE Transactions on Communications*, vol. COM–29, pp. 815–830, June 1981.

[5] L. Longo and M. A. Copeland, "A 13 bit ISDN oversampled ADC using two-stage third-order noise shaping," in *IEEE Custom Integrated Circuits Conference*, pp. 21.2.1–21.2.4, 1988.

[6] N.-S. Lin, B. Fraenkel, G. Jacobs, L. Yiu, D. Senderowicz, and R. Ulmer, "Sigma-Delta A/D and D/A for high-speed voiceband modems," in *International Conference on Acoustics, Speech and Signal Processing*, pp. 2472–2475, 1989.

[7] R. W. Adams, "Design and implementation of an audio 18-bit analog-to-digital converter using oversampling techniques," *Journal of the Audio Engineering Society*, vol. 34, pp. 153–166, Mar. 1986.

[8] D. R. Welland, B. P. DelSignore, E. J. Swanson, T. Tanaka, K. Hamashita, S. Hara, and K. Takasuka, "A stereo 16-bit Delta-Sigma A/D converter for digital audio," *Journal of the Audio Engineering Society*, vol. 37, pp. 476–486, June 1989.

[9] A. W. Swager, "Technique bolsters dc-to-audio converters," *EDN*, pp. 77–86, September 2 1991.

[10] B. P. Brandt, D. E. Wingard, and B. A. Wooley, "Second-order Sigma-Delta modulation for digital-audio signal acquisition," *IEEE Journal on Solid-State Circuits*, vol. 26, pp. 618–627, Apr. 1991.

[11] S. H. Lewis, "Video-rate analog-to-digital conversion using pipelined architectures," UCB/ERL memo M87/90, University of California at Berkeley, 18 November 1987.

[12] J. van Valburg and R. J. van de Plassche, "An 8b 650MHz folding ADC," in *International Solid-State Circuits Conference*, pp. 30–31, Feb. 1992.

[13] C. C. Cutler, "Transmission systems employing quantization." U. S. Patent No. 2,927,962, March 8 1960.

[14] H. Inose and Y. Yasuda, "A unity feedback coding method by negative feedback," *Proceedings of the IEEE*, vol. 51, pp. 1524–1535, Nov. 1963.

[15] Analog Devices, *Data-Acquisition Handbook*, vol. 1. Analog Devices: Norwood, MA, 1982.

[16] P. H. Garrett, *Analog I/O Design*. Reston Publishing Company: Reston, Virginia, 1981.

[17] A. Gersho, "Principles of quantization," *IEEE Transactions on Circuits and Systems*, vol. CAS–25, pp. 427–436, July 1978.

[18] R. M. Gray, "Quantization noise spectra," *IEEE Transactions on Information Theory*, vol. IT–36, pp. 1220–1244, Nov. 1990.

[19] R. M. Gray, *Source Coding Theory*. Kluwer Academic Press, 1990. Chapter 6.

[20] T. F. Darling and M. O. J. Hawksford, "Oversampled data conversion techniques," *Journal of the Audio Engineering Society*, vol. 38, pp. 924–943, Dec. 1990.

[21] M. W. Hauser, "Principles of oversampling A/D converters," *Journal of the Audio Engineering Society*, vol. 39, pp. 3–26, January/February 1991.

[22] P. Horowitz and W. Hill, *The Art of Electronics*. Cambridge University Press: Cambridge, MA, 1980.

[23] N. S. Jayant and P. Noll, *Digital Coding of Waveforms*. Prentice-Hall: Englewood Cliffs, NJ, 1984.

[24] A. V. Oppenheim and R. W. Schafer, *Discrete-Time Signal Processing*. Prentice-Hall: Englewood Cliffs, NJ, 1989.

[25] D. H. Sheingold, ed., *Analog-Digital Conversion Handbook*. Prentice-Hall: Englewood Cliffs, NJ, 1986.

[26] B. Widrow, "A study of rough amplitude quantization by means of Nyquist sampling theory," *IRE Transactions on Circuit Theory*, vol. CT–3, pp. 266–276, Dec. 1956.

[27] A. Zakhor, *Reconstruction of Multidimensional Signals from Multiple Level Threshold Crossings.* PhD thesis, Massachusetts Institute of Technology, Oct. 1987.

[28] A. Zakhor and A. V. Oppenheim, "Sampling schemes for reconstruction of multidimensional signals from multiple level threshold crossings," in *International Conference on Acoustics, Speech and Signal Processing*, pp. 721–724, Apr. 1988. New York, NY.

[29] A. Zakhor and A. V. Oppenheim, "Sampling and reconstruction schemes for multidimensional signals," *Proceedings of the IEEE*, vol. 78, pp. 31–55, Jan. 1990.

[30] S. Haykin, *Modern Filters.* Macmillan Publishing Company: New York, 1989.

[31] J. C. Candy, "A use of double integration in Sigma-Delta modulation," *IEEE Transactions on Communications*, vol. COM–33, pp. 249–258, Mar. 1985.

[32] K. C. Chao, S. Nadeem, W. L. Lee, and C. G. Sodini, "A higher order topology for interpolative modulators for oversampling A/D converters," *IEEE Transactions on Circuits and Systems*, vol. 37, pp. 309–318, Mar. 1990.

[33] R. M. Oliver, J. R. Pierce, and C. E. Shannon, "The philosophy of PCM," *Proceedings of the IRE*, vol. 36, pp. 1324–1331, Nov. 1948.

[34] J. C. Candy, "A use of limit cycle oscillations to obtain robust analog-to-digital converters," *IEEE Transactions on Communications*, vol. COM–22, pp. 298–305, Mar. 1974.

[35] J. C. Candy, W. H. Ninke, and B. A.Wooley, "A per-channel A/D converter having 15-segment $\mu - 255$ companding," *IEEE Transactions on Communications*, vol. COM–24, pp. 33–42, 1 1976.

[36] J. C. Candy, Y. C. Ching, and D. S. Alexander, "Using triangular weighted interpolation to get 13 bit PCM from a Sigma-Delta modulator," *IEEE Transactions on Communications*, vol. COM–24, pp. 1268–1275, Nov. 1976.

[37] J. C. Candy and O. J. Benjamin, "The structure of quantization noise from Sigma-Delta modulation," *IEEE Transactions on Communications*, vol. COM–29, pp. 1316–1323, Sept. 1981.

[38] J. C. Candy, "Decimation for Sigma Delta modulation," *IEEE Transactions on Communications*, vol. COM–34, pp. 72–76, Jan. 1986.

[39] J. C. Candy and A.-N. Huynh, "Double interpolation for digital-to-analog conversion," *IEEE Transactions on Communications*, vol. COM–34, pp. 77–81, Jan. 1986.

[40] M. W. Hauser, "On the terminology "Delta Sigma" and "Sigma Delta"," tech. rep., Cornell University, Apr. 1990.

[41] R. M. Gray, "Oversampled Sigma-Delta modulation," *IEEE Transactions on Communications*, vol. COM–35, pp. 481–488, May 1987.

[42] R. M. Gray, "Spectral analysis of quantization noise in a single-loop Sigma-Delta modulator with dc input," *IEEE Transactions on Communications*, vol. COM–37, pp. 588–599, June 1989.

[43] W. Chou, P. W. Wong, and R. M. Gray, "Multistage Sigma-Delta modulation," *IEEE Transactions on Information Theory*, vol. 35, pp. 784–796, July 1989.

[44] R. M. Gray, W. Chou, and P. W. Wong, "Quantization noise in single-loop Sigma-Delta modulation with sinusoidal inputs," *IEEE Transactions on Communications*, vol. COM–37, pp. 956–968, Sept. 1989.

[45] W. Chou and R. M. Gray, "Dithering and its effects on Sigma-Delta and multi-stage Sigma-Delta modulation," tech. rep., Information Systems Laboratory, Dept. of Electrical Eng., Stanford Univ., Nov. 1989.

[46] W. Chou, T. H. Meng, and R. M. Gray, "Time domain analysis of Sigma Delta modulation," in *International Conference on Acoustics, Speech and Signal Processing*, pp. 1751–1754, 1990.

[47] P. W. Wong and R. M. Gray, "Two-stage Sigma-Delta modulation," *IEEE Transactions on Acoustics, Speech and Signal Processing*, vol. 38, pp. 1937–1952, Nov. 1990.

[48] S. J. Park and R. M. Gray, "Sigma Delta modulation with leaky integration and constant input." Preprint, May 8 1991.

[49] P. W. Wong, *Oversampled Sigma-Delta Modulation: Analysis and Applications.* PhD thesis, Stanford University, June 1989.

[50] P. F. Ferguson, Jr., A. Ganesan, and R. W. Adams, "One-bit higher order Sigma-Delta A/D converters," in *International Symposium on Circuits and Systems*, pp. 890–893, 1990.

[51] R. W. Adams, P. F. Ferguson, Jr., A. Ganesan, S. Vincelette, A. Volpe, and R. Lipert, "Theory and practical implementation of a fifth-order Sigma-Delta A/D converter," *Journal of the Audio Engineering Society*, vol. 39, pp. 515–528, July/August 1991.

[52] W. R. Bennett, "Spectra of quantized signals," *Bell Systems Technical Journal*, vol. 27, pp. 446–472, July 1948.

[53] Chrystal Semiconductor, "CSZ 5316 data sheet," Aug. 1987.

[54] Chrystal Semiconductor, "CDB 5316 data sheet," Sept. 1987.

[55] P. Ferguson, Jr., A. Ganesan, R. Adams, S. Vincelette, R. Libert, A. Volpe, D. Andreas, A. Charpentier, and J. Dattorro, "An 18b 20kHz dual $\Sigma\Delta$ A/D converter," in *International Solid-State Circuits Conference*, pp. 68–69, 1991.

[56] J. J. Paulos, G. T. Brauns, M. B. Steer, and S. H. Ardalan, "Improved signal-to-noise ratio using tri-level Delta-Sigma modulation," in *International Symposium on Circuits and Systems*, pp. 463–466, May 1987.

[57] J. G. Kenney and L. R. Carley, "Design of multibit noise-shaping data converters," research report no. CMUCAD–90–35, Carnegie Mellon University, Aug. 1990.

[58] O. Feely and L. O. Chua, "Multilevel and non-ideal quantization in $\Sigma-\Delta$ modulation," UCB/ERL memo M91/54, University of California at Berkeley, June 1991.

[59] P. J. A. Naus and E. C. Dijkmans, "Multibit oversampled $\Sigma\Delta$ A/D converters as front end for CD players," *IEEE Journal on Solid-State Circuits*, vol. 26, pp. 905–909, July 1991.

[60] Y. Sakina, "Multi-bit analog-to-digital converters with nonlinearity correction using dynamic barrel shifting," Master's thesis, University of California at Berkeley, June 1990.

[61] T. Cataltepe, A. R. Kramer, L. E. Larson, G. C. Temes, and R. H. Walden, "Digitally corrected multi-bit $\Sigma\Delta$ data converters," in *International Symposium on Circuits and Systems*, pp. 647–650, 1989.

[62] Z. X. Zhang and G. C. Temes, "Multibit oversampled $\Sigma-\Delta$ A/D converter with nonuniform quantisation," *Electronics Letters*, vol. 27, pp. 528–529, 14th March 1991.

[63] A. Hairapetian, G. C. Temes, and Z. X. Zhang, "Multibit Sigma-Delta modulator with reduced sensitivity to DAC nonlinearity," *Electronics Letters*, vol. 27, pp. 990–991, 23rd May 1991.

[64] J. Yu, M. B. Handler, and R. E. Hawken, "Adaptive quantization for one-bit Sigma-Delta modulation." To appear in IEE Proceedings, Part G, 1991.

[65] F. O. Eynde, G. M. Yin, and W. Sansen, "A CMOS fourth-order 14b 500k-sample/s Sigma-Delta ADC converter," in *International Solid-State Circuits Conference*, pp. 62–63, 1991.

[66] E. F. Stikvoort, "Some remarks on the stability and performance of the noise shaper or Sigma-Delta modulator," *IEEE Transactions on Communications*, vol. 36, pp. 1157–1162, Oct. 1988.

[67] S. H. Ardalan and J. J. Paulos, "An analysis of nonlinear behavior in Delta-Sigma modulators," *IEEE Transactions on Circuits and Systems*, vol. CAS–34, pp. 593–603, June 1987.

[68] R. Schreier and M. Snelgrove, "Bandpass Sigma-Delta modulation," *Electronics Letters*, vol. 25, pp. 1560–1561, 9th November 1989.

[69] R. Schreier and M. Snelgrove, "Decimation for bandpass Sigma-Delta analog-to-digital conversion," in *International Symposium on Circuits and Systems*, pp. 1801–1804, 1990.

[70] R. Schreier, *Noise-Shaped Coding.* PhD thesis, University of Toronto, 1991.

[71] H.-J. Dressler, "Bandpass A/D-conversion–experimental results," *Electronics Letters*, vol. 26, pp. 1652–1653, 27 September 1990.

[72] H.-J. Dressler, "Interpolative bandpass-A/D conversion," *Signal Processing*, vol. 22, pp. 139–151, 1991.

[73] L. R. Carley and J. G. Kenney, "A 16-bit 4'th order noise-shaping D/A converter," in *Custem Integrated Circuits Conference*, May 1988.

[74] V. Eerola, T. Ritoniemi, and H. Tenhunen, "Second-order sampling and oversampled A/D- and D/A–converters in digital data transmission," in *International Symposium on Circuits and Systems*, pp. 1521–1524, June 1991.

[75] Y. Shoji and T. Suzuki, "Improved double integration Delta-Sigma modulations for A to D and D to A conversion," in *International Symposium on Circuits and Systems*, pp. 451–454, 1987.

[76] Y. Matsuya, K. Uchimura, A. Iwata, T. Kobayashi, M. Ishikawa, and T. Yoshitome, "A 16-bit oversampling A-to-D conversion technology using triple-integration noise shaping," *IEEE Journal on Solid-State Circuits*, vol. SC–22, pp. 921–928, Dec. 1987.

[77] K. Uchimura, T. Hayashi, T. Kimura, and A. Iwata, "Oversampling A-to-D and D-to-A converters with multistage noise shaping modulators," *IEEE Transactions on Acoustics, Speech and Signal Processing*, vol. 36, pp. 1899–1905, Dec. 1988.

[78] B. F. Logan, Jr., "Information in the zero-crossings of bandpass signals," *Bell Systems Technical Journal*, vol. 56, pp. 487–510, Apr. 1977.

[79] D. Anastassiou, "Error diffusion coding for A/D conversion," *IEEE Transactions on Circuits and Systems*, vol. 36, pp. 1175–1186, Sept. 1989.

[80] I. Daubechies, "The wavelet transform, time-frequency localization and signal analysis," *IEEE Transactions on Information Theory*, vol. 36, pp. 961–1005, Sept. 1990.

[81] N. J. Munch, "Noise reduction in tight Weyl-Heisenberg frames," *IEEE Transactions on Information Theory*, vol. 38, pp. 608–616, Mar. 1992.

[82] D. C. Youla and H. Webb, "Image restoration by the method of convex projections: Part I—theory," *IEEE Transactions on Medical Imaging*, vol. MI-1, pp. 81–94, Oct. 1982.

[83] M. I. Sezan and H. Stark, "Image restoration by the method of convex projections: Part II—applications and numerical results," *IEEE Transactions on Medical Imaging*, vol. MI–1, pp. 95–101, Oct. 1982.

[84] C. A. Desoer and M. Vidyasagar, *Feedback Systems: Input-Output Properties.* Academic Press: New York, 1975.

[85] A. I. Mees, *Dynamics of Feedback Systems.* John Wiley & Sons: New York, 1981.

[86] S. Hein and A. Zakhor, "Lower bounds on the MSE of the single and double loop Sigma Delta modulators," in *International Symposium on Circuits and Systems*, pp. 1851–1855, May 1990. New Orleans, LA.

[87] S. Hein, K. Ibraham, and A. Zakhor, "New properties of Sigma Delta modulators with DC inputs." *IEEE Transactions on Communications*, vol. 40, pp. 1375–1387, August 1992.

[88] A. N. Netravali, "Optimum digital filters for interpolative A/D converters," *Bell Systems Technical Journal*, vol. 56, pp. 1629–1641, Nov. 1977.

[89] G. W. Hardy and E. M. Wright, *An Introduction to the Theory of Numbers.* Oxford University Press: New York, 1979.

[90] R. M. Gray and A. H. Gray, Jr., "Asymptotically optimum quantizers," *IEEE Transactions on Information Theory*, vol. 23, pp. 143–144, Jan. 1977.

[91] S. Hein and A. Zakhor, "Optimal decoding for data acquisition applications of Sigma Delta modulators." Accepted in publication in *IEEE Transactions on Signal Processing*, Feb. 1993.

[92] S. Hein and A. Zakhor, "Optimal decoding for data acquisition applications of Sigma Delta modulators," in *Asilomar Conference on Signals, Systems and Computers*, pp. 520–527, Nov. 1990. Pacific Grove, CA.

[93] S. Hein and A. Zakhor, "Optimal decoding for data acquisition applications of Sigma Delta modulators." U. S. Patent Application No. 07/694,294, Apr. 1991.

[94] S. Hein and A. Zakhor, "Theoretical and numerical aspects of an SVD-based method for band-limiting finite extent sequences." Submitted to *IEEE Transactions on Signal Processing*, Dec. 1991.

[95] J. W. Cooley and J. W. Tukey, "An algorithm for the machine computation of complex Fourier series," *Mathematics of Computation*, vol. 19, pp. 297–301, Apr. 1965.

[96] D. Slepian, "Prolate spheroidal wave functions, Fourier analysis, and uncertainly—v: The discrete case," *Bell Systems Technical Journal*, vol. 57, pp. 1371–1430, May/June 1978.

[97] A. T. Walden, "Accurate approximation of a 0th order discrete prolate spheroidal sequence for filtering and data tapering," *Signal Processing*, vol. 18, pp. 341–348, 1989.

[98] E. J. Diethorn and D. C. Munson, Jr., "A linear time-varying system framework for noniterative discrete-time band-limited signal extrapolation," *IEEE Transactions on Signal Processing*, vol. 39, pp. 55–68, Jan. 91.

[99] D. W. Tufts and J. T. Francis, "Designing digital low-pass filters—comparison of some methods and criteria," *IEEE Transactions on Audio and Electroacoustics*, vol. AU–18, pp. 487–494, Dec. 1970.

[100] A. Papoulis, "A new algorithm in spectral analysis and band-limited extrapolation," *IEEE Transactions on Circuits and Systems*, vol. CAS–22, pp. 735–742, Sept. 1975.

[101] T. S. Huang and P. M. Narenda, "Image restoration by singular value decomposition," *Applied Optics*, vol. 14, pp. 2213–2216, Sept. 1975.

[102] Y. S. Shim and Z. H. Cho, "SVD pseudoinversion image reconstruction," *IEEE Transactions on Acoustics, Speech and Signal Processing*, vol. ASSP–29, pp. 904–909, Aug. 1981.

[103] A. K. Jain and S. Ranganath, "Extrapolation algorithms for discrete signals with application in spectral estimation," *IEEE Transactions on Acoustics, Speech and Signal Processing*, vol. ASSP–29, pp. 830–845, Aug. 1981.

[104] G. H. Golub and C. F. V. Loan, *Matrix Computations.* Johns Hopkins University Press: MD, 1989.

[105] T. F. Chan and P. C. Hansen, "Computing truncated singular value decomposition least squares solutions by rank revealing QR-factorizations," *SIAM Journal Sci. Stat. Comput.*, pp. 519–530, May 1990.

[106] B. N. Parlett, *The Symmetric Eigenvalue Problem.* Prentice-Hall: Englewood Cliffs, NJ, 1980.

[107] B. N. Parlett, Aug. 1991. Personal communication.

[108] S. Hein and A. Zakhor, "Reconstruction of oversampled band-limited signals from Sigma Delta encoded binary sequences," in *International Conference on Acoustics, Speech and Signal Processing*, Mar. 1992. San Francisco, CA.

[109] B. J. Sullivan and B. Liu, "On the use of singular value decomposition and decimation in discrete-time band-limited signal extrapolation," *IEEE Transactions on Acoustics, Speech and Signal Processing*, vol. ASSP–32, pp. 1201–1212, Dec. 1984.

[110] F. J. Harris, "On the use of windows for harmonic analysis with the Discrete Fourier Transform," *Proceedings of the IEEE*, vol. 66, pp. 51–83, Jan. 1978.

[111] T. W. Parks and J. H. McClellan, "Chebyshev approximation for nonrecursive digital filters with linear phase," *IEEE Transactions on Circuit Theory*, vol. CT–19, pp. 189–194, Mar. 1972.

[112] W. L. Lee, "A novel higher order interpolative modulator topology for high resolution oversampling A/D converters," Master's thesis, Massachusetts Institute of Technology, June 1987.

[113] S. Hein and A. Zakhor, "Reconstruction of oversampled band-limited signals from Sigma Delta encoded binary sequences." Submitted to *IEEE Transactions on Signal Processing*, Dec. 1991.

[114] S. Hein and A. Zakhor, "Reconstruction of oversampled band-limited signals from Sigma Delta encoded binary sequences," in *Asilomar Conference on Signals, Systems and Computers*, pp. 866–872, Nov. 1991. Pacific Grove, CA.

[115] S. Hein and A. Zakhor, "Iterative reconstruction of Sigma Delta encoded signals," in *International Conference on Industrial and Applied Mathematics*, p. 88, July 1991. Washington, D. C.

[116] P. R. Gray, D. Pederson, and A. Zakhor, "National science foundation proposal." University of California at Berkeley, Oct. 1988.

[117] N. T. Thao and M. Vetterli, "Oversampled A/D conversion using alternate projections," in *Twenty-Fifth Annual Conference on Information Sciences and Systems*, pp. 241–248, 1991.

[118] N. T. Thao and M. Vetterli, "Convex coders and oversampled A/D conversion: Theory and algorithms," CU/CTR/TR 289-91-81, Columbia University, December 16 1991. Submitted to *IEEE Transactions on Signal Processing.*

[119] N. T. Thao and M. Vetterli, "Optimal MSE signal reconstruction in oversampled A/D conversion, part I: Deterministic analysis of A/D conversion and convexity, part II: Algorithms and numerical experiments." Submitted to *IEEE Transactions on Signal Processing*, March 13 1992.

[120] N. T. Thao and M. Vetterli, "Optimal MSE signal reconstruction in oversampled A/D conversion using convexity," in *International Conference on Acoustics, Speech and Signal Processing*, Mar. 1992.

[121] D. G. Luenberger, *Linear and Nonlinear Programming.* Addison-Wesley Publishing Company: Reading, MA, second ed., 1984.

[122] B. E. Boser, *Design and Implementation of Oversampled Analog-to-Digital Converters.* PhD thesis, Stanford University, Oct. 1988.

[123] B. E. Boser and B. A. Wooley, "Quantization error spectrum of $\Sigma\Delta$ modulators," in *International Symposium on Circuits and Systems*, 1988.

[124] S. R. Norsworthy, "Effective dithering of Sigma-Delta modulators," in *International Symposium on Circuits and Systems*, pp. 1304–1307, May 1992.

[125] S. Hein and A. Zakhor, "Halftone to continuous-tone conversion of error-diffusion coded images." Submitted to *IEEE Transactions on Image Processing*, June 1992.

[126] Y.-M. Chao, *An Investigation into the Coding of Halftone Pictures.* PhD thesis, Massachusetts Institute of Technology, May 1982.

[127] J. C. Stoffel and J. F. Moreland, "A survey of electronic techniques for pictorial image reproduction," *IEEE Transactions on Communications*, vol. COM–29, pp. 1898–1925, Dec. 1981.

[128] R. A. Ulichney, *Digital Halftoning.* MIT Press: MA, 1987.

[129] M. Analoui and J. Allebach, "New results on reconstruction of continuous-tone from halftone," in *International Conference on Acoustics, Speech and Signal Processing*, pp. III: 313–316, Mar. 1992.

[130] R. W. Floyd and L. Steinberg, "An adaptive algorithm for spatial greyscale," in *SID International Symp. Dig. Tech. Papers*, pp. 36–37, 1975.

[131] Z. Fan, "Retrieval of images from digital halftones," in *International Symposium on Circuits and Systems*, pp. 2477–2480, May 1992.

[132] J. S. Lim, *Two-Dimensional Signal and Image Processing.* Prentice-Hall: Englewood Cliffs, NJ, 1990.

[133] D. C. Youla, "Mathematical theory of image restoration by the method of convex projections," in *Image Recovery: Theory and Application* (H. Stark, ed.), ch. 2, pp. 29–78, Academic Press, Inc.: New York, 1987.

[134] M. R. Civanlar and H. J. Trussell, "Digital image restoration using fuzzy sets," *IEEE Transactions on Acoustics, Speech and Signal Processing,* vol. 34, pp. 919–936, 1986.

[135] M. I. Sezan and H. J. Trussell, "Prototype image constraints for set-theoretic image restoration," *IEEE Transactions on Signal Processing,* vol. 39, pp. 2275–2285, Oct. 1991.

[136] P. L. Combettes and H. J. Trussell, "The use of noise properties in set theoretic estimation," *IEEE Transactions on Signal Processing*, vol. 39, pp. 1630–1641, July 1991.

[137] P. P. Vaidyanathan, "Multirate digital filters, filter banks, polyphase networks, and applications: A tutorial," *Proceedings of the IEEE*, vol. 78, pp. 56–93, Jan. 1990.

[138] D. Taubman and A. Zakhor, "A multi-start algorithm for signal adaptive subband systems," in *International Conference on Acoustics, Speech and Signal Processing*, Mar. 1992.

[139] R. Rosenholtz and A. Zakhor, "Iterative procedures for reduction of blocking effects in transform image coding," *IEEE Transactions on Circuits and Systems for Video Technology*, vol. 2, pp. 91–95, Mar. 1992.

[140] S. Hein and A. Zakhor, "On the stability of interpolative Sigma Delta modulators," in *International Symposium on Circuits and Systems*, pp. 1621–1624, June 1991. Singapore.

[141] S. Hein and A. Zakhor, "Stability and scaling of double loop Sigma Delta modulators," in *International Symposium on Circuits and Systems*, May 1992. San Diego, CA.

[142] S. Hein and A. Zakhor, "On the stability of Sigma Delta modulators." Submitted to *IEEE Transactions on Signal Processing*, Aug. 1991.

[143] B. P. Brandt and B. A. Wooley, "A CMOS oversampling A/D converter with 12b resolution at conversion rates above 1 MHz," in *International Solid-State Circuits Conference*, pp. 64–65, 1991.

[144] G. T. Brauns, R. J. Bishop, M. B. Steer, and J. J. Paulos, "Table-based modeling of Delta-Sigma modulators using ZSIM," *IEEE Transactions on Computer-Aided Design of ICs and Systems*, vol. 9, pp. 142–150, Feb. 1990.

[145] V. Friedman, "The structure of limit cycles in Sigma Delta modulation," *IEEE Transactions on Communications*, vol. COM–36, pp. 972–979, Aug. 1988.

[146] A. Gersho, "Stochastic stability of delta modulation," *Bell Systems Technical Journal*, vol. 51, pp. 821–841, Apr. 1972.

[147] P. T. Nielsen, "On the stability of a double integration delta modulator," *IEEE Transactions on Communications*, vol. COM–19, pp. 364–366, June 1971.

[148] R. Steele, *Delta Modulation Systems.* John Wiley & Sons: New York, 1975. Chapter 2.

[149] N. He, F. Kuhlmann, and A. Buzo, "Double-loop Sigma-Delta modulation with dc input," *IEEE Transactions on Communications*, vol. COM–38, pp. 487–495, Apr. 1990.

[150] D. B. Ribner, R. D. Baertsch, S. L. Garverick, D. T. McGrath, J. E. Krisciunas, and T. Fuji, "16b third-order Sigma Delta modulator with reduced sensitivity to nonidealities," in *International Solid-State Circuits Conference*, pp. 66–667, 1991.

[151] E. I. Jury, *Theory and Applications of the z-Transform Method.* Robert E. Krieger Publishing Company, 1964.

[152] M. H. H. Höfelt, "On the stability of a 1-bit-quantized feedback system," in *International Conference on Acoustics, Speech and Signal Processing*, pp. 844–848, 1979.

[153] T. S. Parker and L. O. Chua, "Chaos: A tutorial for engineers," *Proceedings of the IEEE*, vol. 75, pp. 982–1008, Aug. 1987.

[154] M. E. Pai, "Oscillations in nonlinear sampled-data systems," *IRE Transactions on Automatic Control*, pp. 350–355, Jan. 1963.

[155] A. R. Bergen and R. L. Franks, "Justification of the describing function method," *SIAM Journal on Control*, vol. 9, pp. 568–589, Nov. 1971.

[156] P. W. Wong, "FIR digital filter with Sigma-Delta modulated input and impulse response," in *International Conference on Acoustics, Speech and Signal Processing*, pp. 1755–1758, 1990.

[157] D. A. Johns and D. M. Lewis, "IIR filtering on Sigma-Delta modulated signals," *Electronics Letters*, vol. 27, pp. 307–308, 14th February 1991.

[158] M. Freedman and D. G. Zrilić, "Nonlinear arithmetic operations on the Delta Sigma pulse stream," *Signal Processing*, vol. 21, pp. 25–35, 1990.

[159] M. R. Bateman and B. Liu, "An approach to programmable CTD filters using coefficients 0, +1 and −1," *IEEE Transactions on Circuits and Systems*, vol. CAS–27, pp. 451–456, June 1980.

[160] N. Benvenuto, L. E. Franks, and F. S. Hill, Jr., "Realization of finite impulse response filters using coefficients +1, 0 and −1," *IEEE Transactions on Communications*, vol. COM–33, pp. 1117–1125, Oct. 1985.

[161] N. Benvenuto, L. E. Franks, and F. S. Hill, Jr., "Dynamic programming methods for designing FIR filters using coefficients −1, 0 and +1," *IEEE Transactions on Acoustics, Speech and Signal Processing*, vol. ASSP–34, pp. 785–792, Aug. 1986.

[162] O. Feely and L. O. Chua, "Nonlinear dynamics of a class of analog-to-digital converters," UCB/ERL memo M91/30, University of California at Berkeley, Apr. 1991.

[163] O. Feely and L. O. Chua, "The effect of integrator leak in $\Sigma - \Delta$ modulation," *IEEE Transactions on Circuits and Systems*, vol. 38, pp. 1293–1305, Nov. 1991.

[164] M. O. J. Hawksford, "Chaos, oversampling and noise shaping in digital-to-analog conversion," *Journal of the Audio Engineering Society*, vol. 37, pp. 980–1001, Dec. 1989. See also the comments by R. W. Adams and the reply by the author, pp. 766-768, October 1990.

[165] J. C. Kieffer, "Analysis of dc input response for a class of one-bit feedback encoders," *IEEE Transactions on Communications*, vol. 38, pp. 337–340, Mar. 1990.

[166] J.-J. E. Slotine, J. K. Hedrick, and E. A. Mlsawa, "On sliding observers for nonlinear systems," in *American Control Conference*, 1986. Seattle, WA.

[167] L. Schuchman, "Dither signals and their effect on quantization noise," *IEEE Transactions on Communications*, vol. COM–12, pp. 162–165, Dec. 1964.

[168] O. C. Feely, *An Analytical Study of the Nonlinear Dynamics of Sigma-Delta Analog-to-Digital Conversion.* PhD thesis, University of California at Berkeley, May 1992.

# Index

GPSR Compliance
The European Union's (EU) General Product Safety Regulation (GPSR) is a set of rules that requires consumer products to be safe and our obligations to ensure this.

If you have any concerns about our products, you can contact us on

ProductSafety@springernature.com

In case Publisher is established outside the EU, the EU authorized representative is:

Springer Nature Customer Service Center GmbH
Europaplatz 3
69115 Heidelberg, Germany

www.ingramcontent.com/pod-product-compliance
Ingram Content Group UK Ltd.
Pitfield, Milton Keynes, MK11 3LW, UK
UKHW020802290926
15470UKWH00003B/905
* 9 7 8 0 7 9 2 3 9 3 0 9 2 *